SUSPENSION ACOUSTICS

This book examines, from a fundamental point of view, the response of single particles in fluids and uses the results of such a detailed examination to consider suspension motions as a whole – paying particular attention to acoustic motions (i.e., to the propagation of sound waves). Such propagation is examined from different perspectives in a unified manner that applies to several particle-fluid combinations. Among the possible applications of the theory presented, this book discusses the characterization of suspensions by acoustic means and the agglomeration of particles with sound waves.

Samuel Temkin is Professor Emeritus, Department of Mechanical and Aerospace Engineering, Rutgers University.

T0212846

Suspension Acoustics

An Introduction to the Physics of Suspensions

SAMUEL TEMKIN

Rutgers University

CAMBRIDGE
UNIVERSITY PRESS

CAMBRIDGE UNIVERSITY PRESS
Cambridge, New York, Melbourne, Madrid, Cape Town, Singapore,
São Paulo, Delhi, Dubai, Tokyo, Mexico City

Cambridge University Press
32 Avenue of the Americas, New York, NY 10013-2473, USA

www.cambridge.org
Information on this title: www.cambridge.org/9780521174473

© Cambridge University Press 2005

This publication is in copyright. Subject to statutory exception
and to the provisions of relevant collective licensing agreements,
no reproduction of any part may take place without the written
permission of Cambridge University Press.

First published 2005
First paperback edition 2011

A catalog record for this publication is available from the British Library

Library of Congress Cataloging in Publication data

Temkin, Samuel, 1936–
Suspension acoustics : an introduction to the physics of suspensions / Samuel Temkin.
   p.   cm.
Includes bibliographical references and indexes.
ISBN 0-521-84757-5
1. Suspensions (Chemistry)   2. Molecular acoustics.   I. Title.
QD549.T38   2005
541'.34 – dc22   2004020377

ISBN 978-0-521-84757-5 Hardback
ISBN 978-0-521-17447-3 Paperback

Cambridge University Press has no responsibility for the persistence or
accuracy of URLs for external or third-party internet websites referred to in
this publication, and does not guarantee that any content on such websites is,
or will remain, accurate or appropriate.

Figures 4.5.1, 4.9.1, 4.10.2, 4.10.3, 4.10.4, 4.10.5, 4.10.6, 4.10.7, 4.12.4, 4.12.5,
4.12.8, 5.4.2, and 6.6.4 are reproduced by permission of Cambridge University
Press.

Figures 1.1.1, 1.1.2, 1.1.3, 4.4.1, 9.2.1, 9.4.8, 9.4.9, and 9.4.10 are reproduced by
permission of the Acoustical Society of America.

Figure 4.5.2 is reproduced by permission of the Royal Society of London.

Figures 8.9.6, 8.9.7, and 8.9.8 are reproduced by permission of Marcel Dekker.

Contents

List of Figures and Tables

Figures

Tables

Preface

It is customary for authors to explain their reasons for writing a book. Mine are twofold. The first is to provide, under one cover, a detailed discussion of the propagation of sound waves in suspensions and of the foundations upon which that discussion is based. The second is due to the historical development of the field and takes longer to explain.

Nearly 100 years ago, there appeared in the research literature an article about the extinction of sound waves by small particles. Since then, very significant advances have been made in the understanding of sound wave propagation in suspensions of various kinds. These advances have occurred as a result of specific needs at a particular time. For example, the first theory for sound wave propagation in bubbly liquids was developed during World War II, as a part of an effort to understand the pressure waves produced by underwater explosions. Similarly, propagation in aerosols received increased attention during the moon-landing program in the 1960s, as a result of certain instability problems encountered in rocket propulsion. Currently, because of industrial and environmental needs, there is a renewed interest in the acoustics of emulsions and hydrosols. One unintended result of this divided interest is that the theories for sound propagation in each type of suspension developed nearly independently from one another. At some level, this has been advantageous because individual suspension types usually require simpler mathematical models for their description. However, the division has resulted in our discarding some physical details that could be important in some contexts.

This book is an attempt to treat suspensions in a unified manner and takes the view that all suspensions have, at a certain level, much in common. Of course, given the extraordinarily wide range of particle materials and sizes that exist in suspensions, it is not possible to include all types under the same umbrella. Because of this, I exclude from the discussion certain topics, as mentioned later. What is left is an account of basic suspension mechanics, thermodynamics, heat, and sound. The last of these – sound – is based on the first three and provides the main focus of the book. As it turns out, sound waves provide a relatively simple stage, wherein a discussion of many of the basic aspects of suspension physics can take place without having to limit the conversation too much, while at the same time being able to treat important issues in a broader manner than is customary.

The book has been written for persons who have a need, or desire, to understand the basics and who are not satisfied with using a formula for this or that without understanding its origins and limitations. Such people are students doing graduate work in the physical sciences and individuals in research laboratories who are new to suspensions, as well as more experienced research workers in related fields. Such fields include *Acoustics*, *Atmospheric Science* (clouds, marine aerosols), *Earth Sciences* (volcanic emissions), *Chemical Engineering* (industrial particulates), *Environmental Science* (dusts, smokes, and their control), *Mechanical Engineering* (propulsion, transportation of particles), *Metrology* (particle size measurements), and *Oceanography* (bubbles in the ocean). Parts of the book could be used as a text for a graduate-level course dealing with small particles in fluids. For that purpose, a small number of exercises have been included.

The book is divided into three conceptual parts. The first, Chapters 1 and 2, deals with basic ideas and equations. Chapters 3–6 deal with single particles, and Chapters 7 and beyond deal with dilute suspensions – where the material developed in Chapters 3–6 is put to use. A bibliography is placed after Chapter 10. Though fairly large, the bibliography is not inclusive. Given that the book deals with spheres in fluids, such a list would probably be longer than the text itself. Rather, the bibliography consists of books and research articles I felt had a close connection to the material covered in the book, or that could direct the reader to important topics not considered in it. I am sure, however, that I missed many important works.

Because of the focus of the book and because suspension motions are generally unsteady, the book emphasizes unsteady motions. For the same reasons, the discussion is mostly limited to linear motions. This may appear as a serious limitation, but in reality suspension motions are, on the whole, linear; furthermore, the linear world allows for the inclusion of effects that are not otherwise possible. Finally, the suspensions treated are dilute. This can be a drawback in some situations; but, given that dilute suspensions are the most common, the exclusion of nondilute suspensions is probably not too harmful at this level.

On the other hand, I exclude from the discussion several important topics. Included here are colloidal suspensions, as well as chemical and electromagnetic effects. Colloidal suspensions are, in fact, a very important class of suspensions, but their discussion here is not warranted, as they have been treated in an excellent manner in the books by Hunter (1986) and by Russel, Saville, and Schowalter (1989). Electromagnetic theory, as applied to suspensions, is important because it includes optical scattering by small particles, but would have required considerable additional space. Furthermore, the excellent book by van de Hulst on that topic continues to be available.

The notation used in the book deserves some explanation. Because suspensions involve two or more materials, the notation used in most works tends to be difficult to follow. Mine is no exception. However, the notation I have adopted can, with a little patience, become both clear and simple to use. Essentially, I denote suspension quantities without suffixes, and affix the symbols f and p to them to denote the corresponding fluid and particle quantities. For example, the density of a suspension is simply denoted by ρ, whereas the densities of the particle and fluid materials are denoted by ρ_p and ρ_f, respectively. Things get a bit more involved when a physical quantity already contains a suffix. Here, the same prescription is followed. For example, the isentropic sound speed in a suspension is denoted by c_s, whereas the corresponding speeds in the fluid and particle materials are denoted by c_{sf} and c_{sp}, respectively. Difficulties may occur when a quantity uses either the particle or the fluid suffixes (e.g., the specific heat at constant pressure). Rather than introduce a new symbol, I stay with the customary one, and denote that quantity for the suspension as c_p. In such cases, the corresponding particle and fluid quantities would be denoted by a second Latin suffix – thus, c_{pp} and c_{pf}. Of course, given the large number of quantities that appear in the text, some symbols are used for more than one purpose. To facilitate the rapid identification of a given symbol, I have added a symbol index at the end of the book.

In writing this book, I have been influenced by the works – general and specialized – of many others. Among the general works I have profited from, I should mention volumes 1, 5, and 6 of Landau and Lifshitz's outstanding *Course of Theoretical Physics*; *The Physics of Vibration* by A. B. Pippard; *An Introduction to Fluid Mechanics* by G. K. Batchelor; *Cloud Physics* by B. J. Mason; *The Thermodynamics of Fluid Systems* by L. C. Wood, and *The Mechanics of Aerosols* by N. A. Fuchs.

The specialized literature has also, of course, been very influential; but, it is hard to acknowledge each important paper that has had an effect on my thinking. However, certain bodies of work stand up in this regard. These include the works of P. S. Epstein, for attenuation in suspensions and emulsions; E. H. Kennard, for the propagation of sound in bubbly liquids; F. E. Marble and G. Rudinger, for dusty gases; A. Prosperetti for bubble motions; and L. van Wijngaarden for bubbly liquids.

Although I have profited, considerably, from these works, the approach presented here regarding those specialized topics is mine. Also, the book includes a fair amount of research by myself alone or with my students. This is particularly true of Chapter 9, where a detailed treatment of sound propagation in suspensions is presented using some of the methods I have developed for that purpose. The reason for doing this is not to overemphasize my work, but to show how different perspectives offer additional insight. The book also contains new material not yet published (e.g., the use, in Section 9.4, of sound emission theory to study sound propagation in suspensions).

In closing, I would like to thank Cambridge University Press, the Acoustical Society of America, the Royal Society of London, and Marcel Dekker for their permission to reproduce some figures that originally appeared in other works.

S. TEMKIN

1

Preliminaries

1.1 Introduction

This book is concerned with the physical behavior of a collection of small particles in fluids and with the motions they can sustain, particularly sound waves. By a small *particle*, we mean a material body of small dimensions, and made of a gas (bubble), a liquid (droplet), or a solid. Roughly speaking, the particle size range we will be interested in is between one and a few hundred microns. The combination of many such particles and a fluid is referred to as a *suspension*.

Examples of suspensions of particles in fluids abound, whether we are aware of them or not. Thus, the majestic beauty of a cumulus cloud is due to both its size and to its large number of droplets; the murmurs of a brook are due to their small bubbles; and the colors of paints and inks, both natural and man-made, are due to small particles. Fogs, smokes, dusts, salad dressings, milk, blood, and many other familiar examples are also suspensions. This list makes it evident that suspensions are important in many fields, both basic and applied. For example, the settling of small particles is important in both cloud physics and in the manufacture of paints. Particle settling, or sedimentation, is but one out of many motions sustained by suspensions. This book is concerned with a subset of those motions, namely sound waves. Although apparently limited in scope, this subset includes heat transfer, mechanics, thermodynamics, and, of course, acoustics.

Before delving into these matters in more detail, we ask the reader to perform the following simple experiment. Take two equal glasses of water, filled to about the same level. Tapping either glass with a metallic object – say a teaspoon – will usually produce a high-pitch sound. We now sprinkle the water in one of the two glasses with a small amount of Alka Seltzer, noticing that many bubbles are produced. At this moment, tap the two glasses again, noting the drastic differences in the pitch of the sounds emitted. Eventually, the water becomes clear again, at which time pitch differences nearly disappear, indicating that the differences were produced by the bubbles.

Although a detailed explanation of the observations requires additional knowledge, a qualitative one may be given in the following manner. First, note that, if sound is

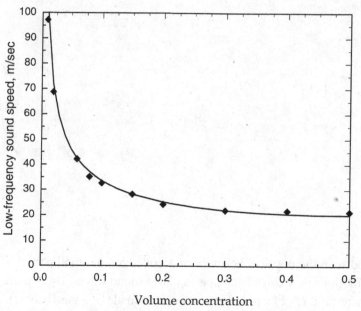

Figure 1.1.1. Low-frequency sound speed in a bubbly liquid. Diamonds represent the experimental data of Karplus (1961), while the line is the theoretical prediction, equation (7.4.3), (Temkin, 1992).

emitted after the glass is tapped, it is because the system – glass, water, and bubbles – is vibrating. The frequency of these vibrations depends on the compressibility of the whole system, but the observed change in the frequency of the vibrations is due to changes produced in the water by the bubbles. Now, the compressibility of a medium depends on both its density and its sound speed. If the volume occupied by the bubbles is small – as is the case if the amount of powder placed in the liquid is small – the density of the bubbly water is essentially equal to that of the water alone. Hence, the changes in the emitted frequency are due to variations in the speed of sound in the water. Given the drastic changes in pitch of the note emitted, the sound speed in the bubbly liquid must be very different from that in the water without bubbles.

The above explanation for the observed changes is, of course, only qualitative. A quantitative explanation requires an understanding of the motions of small particles – bubbles in the above experiment – in fluids. This understanding will be developed in the chapters to come. But, to put the problem in focus, we show here some graphs related to the propagation of sound waves in bubbly liquids. First, we note that sound propagation in suspensions is generally dispersive, meaning that the phase velocity of the waves depends strongly on the frequency of the waves. It is therefore helpful to separate speed results at low frequencies, where dispersion is absent, from those at higher frequencies, where dispersion plays a more significant role.

Figure 1.1.1 shows theoretical and experimental results for the low-frequency sound speed in water containing air bubbles, as a function of the bubble volume concentration. Experimental data were obtained by Karplus (1961). The theoretical

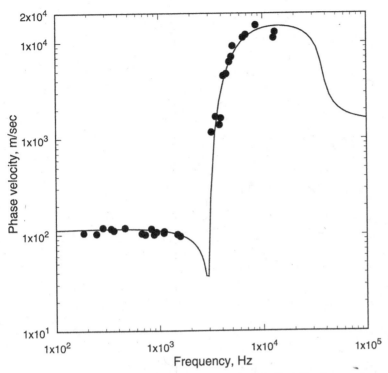

Figure 1.1.2. Phase velocity of plane waves in a bubbly liquid having a volume concentration equal to 0.01 and a bubble radius of 1,100 μm. Circles represent the experimental results of Cheyne et al. (1995), while the line is the theoretical prediction, equation (9.4.15). (Temkin, 2002).

prediction is presented in Chapter 7. Both theory and experiments refer to bubbly liquids at normal temperature and pressure, and show the drastic changes produced by the bubbles. The smallest concentration data point appearing in the graph is about 1%. Yet, the corresponding speed of sound is only 100 m/sec, or 1/15th of the value in water without bubbles! The reason for the low speed is the large compressibility of the bubbly liquid. But this compressibility is influenced by both mechanical and thermal effects. At low frequencies, these are relatively easy to understand and quantify because it is then unnecessary to consider relative motions and temperatures between fluid and bubbles. But at higher frequencies, the relative motions are more significant and make the sound speed, or rather, the phase velocity of the waves, frequency dependent. This is true in all suspensions; but, in the case of bubbly liquids, the frequency dependence is very pronounced, owing to the resonant nature of the bubbles' response to the sound. Figure 1.1.2 shows the phase velocity for plane sound waves in water containing 1% air bubbles by volume. The figure shows several important features. First, as mentioned previously, when the frequency is small, the sound speed is quite low, compared with the sound speed in pure water. Second, as the frequency increases, we first note a decrease, but at some point, which as shown later coincides with the resonance frequency for radial pulsations of the bubbles, the phase velocity increases

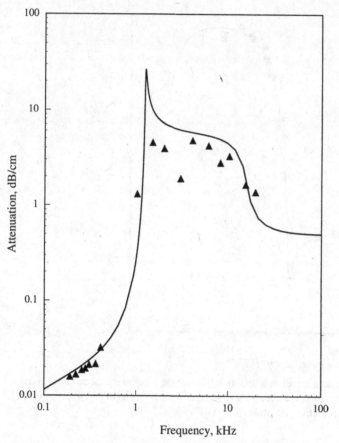

Figure 1.1.3. Attenuation in a bubbly liquid containing a 0.01 volume fraction of bubbles having a radius equal to 0.264 cm. Solid triangles represent the experimental results of Silberman (1957). Solid line is the theoretical description, equation (9.4.15). (Temkin, 2002).

rapidly, becoming soon afterward rather large. Finally, at very large frequencies, the phase velocity approaches the sound speed in pure water, just as if no bubbles were present.

This remarkable behavior of the phase velocity is accompanied by an equally remarkable behavior of the attenuation experienced by the waves, that is, of the rapidity with which their amplitude is reduced as they travel. This is shown in Fig. 1.1.3 for the same bubble volume concentration of 1%, but for a different bubble diameter. Here we see that the attenuation is rather large at certain frequencies and that it has a sharp peak at a well-defined frequency. This correspond to the resonance frequency mentioned previously.

What causes such remarkable behavior? Which mechanisms are responsible for the large attenuation observed? How do we quantify them? What are the effects of sound waves on the small particles? These questions are important in many contexts, acoustic and nonacoustic. The acoustic context, emphasized here, is important because sound waves are important as a diagnostic tool, as an agent for both the modification and the production of small particles, and as a means for the study of certain fundamental quantities that affect suspension motions. To answer these questions, we require an understanding of the physical properties of suspensions, of the dynamics of small particles responding to external changes, of the interactions between a particle and the fluid outside it, of the behavior of a cloud of particles in fluids, and, in some cases, of the interactions between the particles themselves. These issues are discussed in this book.

The book is broadly divided into three parts. The first, containing this and the next chapter, deals with definitions and other basic material required in the remaining chapters. The second, Chapters 3–6, deals with single, spherical particles in fluids, their motions, their temperature changes, and with the mechanisms driving them – namely the fluid force and the heat transfer rate. Because the focus of the book is acoustic motions, considerable attention is given to small-amplitude oscillatory motions. But, as pointed out before, these are also important in nonacoustic situations.

The single-particle material is put to use in the remaining chapters of the book. After a discussion of equilibrium properties in Chapter 7, we begin our study of suspensions in motion. The discussion is limited to dilute suspensions. These are not the only type of suspension, but are the most common.

1.2 Types of Suspensions

Several classifications are used in the literature to describe suspensions. They are based on the composition of the particles and fluid around them, as well as on the chemical and physical effects that affect their behavior. In this work, suspensions are divided into four types that refer to physically distinct cases. These are:

Aerosols. These consist of solid or liquid particles in gases (e.g., atmospheric clouds, dusts, and smokes).

Bubbly liquids. As the name implies, these are liquids with bubbles in them (e.g., effervescent liquids).

Emulsions. These consist of immiscible liquid droplets in a host liquid (e.g., milk).

Hydrosols. These consist of solid particles in a host liquid (e.g., paint pigments).

As this short list shows, suspensions are generally composed of particles in one state of matter, immersed in a fluid in a different state. But exceptions occur, such as with emulsions or when the particles contain more than one phase (e.g., small soap bubbles in air). The range of particle sizes found in suspensions is very wide, including both submicron and millimeter particle sizes. However, in specific suspensions, the range is more narrow, as shown in Table 1.2.1. These sizes may be put in perspective by comparison with other length scales of interest. Thus, the mean free path in gases, which is a measure of the distance traveled by a gas molecule between collisions and which presents a limiting scale to the applicability of continuum theories in gases, is of the order of $0.1\ \mu$m for air at 1 atm and 15°C. Light waves in the visible region have

Table 1.2.1. *Approximate particle size ranges for some suspensions*

Type	Suspension	Diameter range (μm)
Aerosols	Clouds	1–100
	Dusts	0.1–100
	Smoke	0.1–10
Emulsions	Milk	1–10
Hydrosols	Paints	0.1–10

wavelengths ranging from about 0.4 μm (violet) to about 0.8 μm (red). The diameter of a red blood cell is about 10 μm, and the thickness of a human hair is of the order of 200 μm.

Occasionally, we shall refer to suspensions as *dispersions*, because the suspended particles are often far apart from one another and may be thus considered dispersed in the fluid. Further, when the number of particles is very large, suspensions can be regarded as multiphase media, in which case the particles are referred to as the particulate, or the *dispersed*, phase. The corresponding name for the fluid phase would then be the *continuous* phase.

We have described the nature of the particles in very vague terms, namely solid, liquid, and gaseous. More common names for the last two are, of course, droplets and bubbles, respectively. But whatever name is used, it is clear that the physical properties of the particle material are, for the description of suspensions, as important as those of the surrounding fluid. For example, whether the particles are compressible or not can have very drastic effects on the response of a suspension, as shown by the simple experiment described previously.

The origins of the particles found in suspensions include vapor condensation, combustion, breakup of liquid surfaces, air-entrainment in liquids, and a host of man-made processes, such as grinding solid surfaces. These processes are important on their own, and indeed some of them have received much attention. Thus, atmospheric clouds owe their existence largely to vapor condensation processes that form small droplets. Because of the obvious importance of this, vapor condensation has been studied in considerable detail. Similarly, concern with the environment has resulted in much interest on combustion-generated particles. The production of droplets by the breakup of liquid surfaces is of considerable importance in the generation of sprays and marine aerosols, in the formation of small droplets out of larger ones in atmospheric clouds, as well as in some industrial processes, such as ink-jet printing. Finally, the production of powders, such as paint pigments, and some medical products are of considerable industrial importance.

In this work, we shall be more interested in the behavior of the particles than in their origin. Furthermore, we shall be mainly concerned with physical effects. That is, we will not study those effects that are primarily chemical in nature, nor shall we be concerned with mass transfer between particles and the exterior fluid. To be sure, these effects may be important in certain cases, for example in clouds of very small droplets, where condensation and evaporation are a dominant effect, or in suspensions of very small particles in liquids, called *colloidal suspensions*, where chemical and electrochemical effects play a very significant role. For a discussion of this important type of suspension, the reader is referred to books on the subject, some of which are mentioned in the bibliography.

1.3 Particle Concentrations

Although our main interest in suspensions is their dynamical response to applied forces and temperature fields, it is useful to first introduce some measures that serve to quantify their particulate content. Among all possible types of suspensions, we will consider only those that have a large number of particles per unit volume and that are homogeneous in the sense that any small volume element will have the same distribution of particles regardless of the location of the element within the suspension. Such suspensions may be described in terms of average properties that can be either measured or calculated.

At the most basic level, the quantities we are concerned with are purely geometrical (e.g., the shape of the particles, their dimensions, and their volumes). Generally speaking, the shape of the particles is not spherical. However, as a result of surface tension effects, small droplets and bubbles do have a spherical shape when no forces are applied to them. Similarly, small solid particles that result from the solidification of high-temperature droplets, are also spherical. But particles that result from grinding are not, nor are combustion-generated solid particles, which are aggregates of smaller solid particles. In those cases, a representative lateral dimension can usually be introduced that may serve as an equivalent diameter.

Now, to quantify the amount of particulate material in a given suspension, we may specify the material, or materials, they are made of and the *number concentration*, that is, the number of particles in a unit volume of the suspension. Most of the examples listed in Table 1.2.1 can have widely ranging concentrations, but, typically, the number of particles is rather large. For example, 1 cm^3 of cigarette smoke can have between 10^4 and 10^5 particles, whereas the same volume in an atmospheric cloud could have as many as 1,000 droplets.

In general, we will assume that the number concentration is sufficiently large so as to warrant the use of statistical methods to describe certain physical quantities in terms of their averages over a representative suspension volume. In a homogeneous suspension at rest, we can always increase the size of the sample by increasing the size of the representative volume; but, with dynamic situations in mind, we will be interested in volume sizes that are small, compared with the length scales over which variations are significant. Broadly speaking, then, the volume elements we will consider will be assumed to have representative lateral sides that are small, compared with other scale lengths of interest, but which at the same time contain a large number of particles.

Particle-Size Distributions

Given that the number of particles is large, we may obtain local averages of several relevant quantities on the basis of the distribution of particle sizes present in the volume. If the particles were identical, the number of particles per unit volume, N_0, would be represented by a vertical line in a graph showing the number as a function of size. Such suspensions are said to be *monodisperse*. Actual suspensions are normally *polydisperse*, in the sense that they contain particles of several sizes. In many cases, the number of sizes present is so large that their variation is best modeled in terms of continuous size distributions. To do this, we consider a distribution function, $N_D(D)$ such that $N_D(D)dD$ is the number of particles in unit volume of suspension, having sizes in the range D and $D + dD$. It should be noted that $N_D(D)$ is not the number of particles in the volume element. That number is given by integration over the sizes present. To include all possible sizes, we integrate between zero and infinity, keeping in mind that no distribution has particles at either limit of the range of integration. Thus,

$$N_0 = \int_0^\infty N_D dD \qquad (1.3.1)$$

The function $N_D(D)$ contains the desired information on the distribution of sizes. However, it is more convenient to work with fractional numbers. Thus, the fractional number of particles of size D in a strip of width dD, which includes the size D, is $n_D dD = N_D dD/N_0$. The quantity n_D is sometimes called a particle size distribution function. We may also interpret $n_D dD$ as the probability of finding a particle in the size range D and $D + dD$. We note that n_D satisfies the normalization condition $\int_0^\infty n_D dD = 1$.

In some applications, it is advantageous to work with the probability of finding a particle of a volume v_p in the range of volumes v, $v_p + dv_p$. We denote this quantity as n_v, so that $n_v dv_p$ gives their number in that range. As with n_D, n_v satisfies the normalization condition $\int_0^\infty n_v dv_p = 1$. These size distribution functions are normally determined by fitting experimental data. In some cases, simple mathematical expressions have been found that describe particular cases. For the purposes of this book, however, we will assume that the size distribution function corresponding to any particular suspension is known. A knowledge of n_D, or of n_v, is sufficient to obtain certain average quantities. For example, the arithmetic mean size, $\overline{D}$, in a distribution is given by $\overline{D} = \int_0^\infty n_D(D)D dD$. This is the simplest of all mean sizes; but, depending on the application, other mean values may be more significant. Some of these are used in Chapter 10.

Volume Concentration

At the lowest level, a suspension can be characterized by the volumes that the particles and the fluid occupy in the whole suspension, or more conveniently, in a volume element in it. Let us consider the volume of the particles in a volume δV of the suspension. This is given by $\delta V_p = N_0 \bar{v}_p \delta V$, where $\bar{v}_p$ is the arithmetic-average particle volume. If the particles are spherical, we would simply have, $\bar{v}_p = \pi \overline{D^3}/6$.

A more convenient way to specify the volume occupied by the particles is in terms of the fraction of the suspension volume that is occupied by the particles. Thus, if δV is an element of volume in the suspension and δV_p denotes that part of δV occupied by the particles, then the particle *volume concentration* is defined as

$$\phi_v = \delta V_p/\delta V \tag{1.3.2}$$

so that $\phi_v = N_0 \bar{v}_p$. As stated previously, δV must be sufficiently large so as to contain many particles, and yet be sufficiently small for the volume to be macroscopically regarded as a point in the suspension. This is important for suspensions that are not uniform throughout, as is usually the case in dynamic conditions. In such cases, the concentration will vary from place to place, or with time at a given location, in which case the volume fraction will depend on position within the suspension. In later portions of this work, we will introduce means that can be used to compute the variations in space and in time of such locally defined quantities.

Returning now to (1.3.2), it is clear that, if the particles and the fluid share no space – as is the case when the particles are porous – the suspension volume is the sum of the volumes occupied by the particles and by the fluid. It then follows that the volume concentration of the fluid is given by $1 - \phi_v$. In what follows, we will assume that δV_p is measured in such a way that no part of it is occupied by the fluid.

A numerical example may be useful to visualize the magnitudes of these quantities. Suppose that, in an aerosol, each cubic centimeter contains 10^6 particles, all of $1 \mu m$ in diameter. If the aerosol fills a finite region, it will appear to the eye as being very dense, as a result of the strong absorption of light it produces. And yet, the corresponding particle volume concentration would be equal to about 10^{-6}. Suspensions having small volume concentrations are called *dilute*.

Mass Concentration

A suspension can also be characterized in terms of its mass concentration, that is of the mass of particles relative to the mass of fluid in a given element. Let the mass of particles and fluid in δV be δM_p and δM_f, respectively, and let $\delta M = \delta M_f + \delta M_p$ be the total mass in the element. Then, the particle mass concentration is defined by

$$\phi_m = \delta M_p/\delta M \tag{1.3.3}$$

The corresponding value for the fluid is $1 - \phi_m$. As with the volume concentration, ϕ_m also varies between 0 and 1; but, in most suspensions, its value is small.

Mass and Volume Loadings

Other measures of the amounts of particles in a suspension are provided by the relative masses and volumes of particles and fluid in a given volume element. The first, called the *mass loading*, η_m, is defined as the ratio of particle to fluid mass in a given volume, i.e.,

$$\eta_m = \delta M_p / \delta M_f \tag{1.3.4}$$

We may express this in terms of measurable quantities. For example, if the particles in the suspension are made of the same material, each has a well-defined density, ρ_p. Hence, the mass of all particles in δV is $\delta M_p = \rho_p \delta V_p$. Similarly, if density of the fluid is ρ_f, we have $\delta M_f = \rho_f \delta V_f$. Thus,

$$\eta_m = \rho_p \phi_v / \rho_f (1 - \phi_v) \tag{1.3.5}$$

For dilute suspensions, this may be written as $\rho_p \phi_v / \rho_f$, showing that the mass loading is not necessarily small, even for dilute suspensions, when the density of the particle material is much larger than that of the fluid around it, as is the case in aerosols.

Similarly, the volume loading is defined by

$$\eta_v = \delta V_p / \delta V_f \tag{1.3.6}$$

In dilute suspensions, this is essentially equal to ϕ_v. As a result, the volume loading does not play a significant role in suspension dynamics.

Suspension Density

We may also obtain the average density of a suspension. This is defined as usual by dividing the mass by the volume in a given volume element of the suspension. Thus, in volume element δV of the suspension, there is an amount of mass δM, then

$$\rho = \delta M / \delta V \tag{1.3.7}$$

This is sometimes referred to as the *effective* density of a suspension. Since δM is the sum of fluid, δM_f, and particle, δM_p, masses, we may express ρ in terms of both volume and mass concentrations. Thus, as with the mass loading, we note that when all of the particles are made of the same material, the mass of all particles in δV is $\delta M_p = \rho_p \delta V_p$. The corresponding mass of fluid is $\delta M_f = \rho_f \delta V (1 - \phi_v)$. Hence, the density of the suspension is

$$\rho = \rho_f (1 - \phi_v) + \rho_p \phi_v \tag{1.3.8}$$

Similarly, the *specific volume* of the suspension, $\delta V/\delta M$ can be expressed as

$$\frac{1}{\rho} = \frac{1 - \phi_m}{\rho_f} + \frac{\phi_m}{\rho_p} \qquad (1.3.9)$$

These two equations may be combined to give yet another expression for ρ, namely

$$\rho = \rho_f(1 - \phi_v)/(1 - \phi_m) \qquad (1.3.10)$$

These expressions show that the density varies from ρ_f to ρ_p as the volume concentration is increased from zero to unity. However, it should be remembered that, by assumption, the suspension is homogeneous throughout and that the element contains many particles. These conditions are unlikely to be met at large volume concentrations, as it may be seen by considering a bubbly liquid. At low concentrations, the bubbles may be uniformly dispersed in the suspension; but, at large concentrations, it is likely that they will coalesce forming, in some parts of the suspension, large pockets of gas, leaving others without bubbles. The aforementioned formula will still give the density of the gas-liquid mixture, but the mixture will not be a suspension in the sense used here.

As these expressions show, the density and the mass loading depend, for a given volume concentration on the material density ratio, ρ_f/ρ_p. This ratio appears repeatedly in this book. It is therefore convenient to define another quantity to easily refer to it. Thus, we put

$$\delta = \rho_f/\rho_p \qquad (1.3.11)$$

noticing that it is very small for bubbly liquids, very large for aerosols, and of order 1 for emulsions and hydrosols.

Returning to (1.3.8), we note that, when the volume concentration is very small, the density of a bubbly liquid will be very nearly equal to that of the liquid – but that the density of an aerosol having the same small concentration can be higher than that of the gas in the aerosol because of the large density of liquid and solid particles. This is shown in Figs. 1.3.1 and 1.3.2. The first of these shows the variations of ρ/ρ_f with the volume concentration for an aerosol composed of water droplets in air. The figure also displays the variations of the corresponding mass concentration and loading. It is observed that the mass loading can become large even for small values of the volume concentration. This has important implications in the dynamic behavior of aerosols. Figure 1.3.2 shows the variations of ρ/ρ_f for three liquid suspensions: a bubbly liquid, an emulsion, and a hydrosol. As we see, the suspension density for these differs little from that of the liquid host for concentrations as large as 0.01.

The concentrations, loadings, and density introduced previously are related to one another, as shown by the given expressions for the density. These relationships are sometimes needed in our derivations. It is therefore useful to have them readily

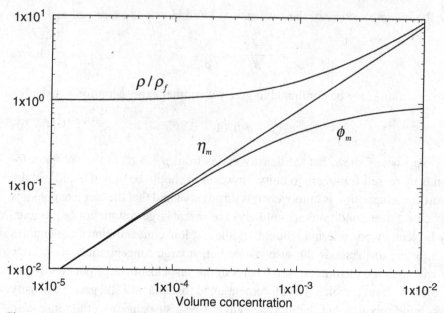

Figure 1.3.1. Density, mass concentration, and mass loading for an aerosol composed of water droplets in air.

available when needed. For that purpose, we list in Table 1.3.1 some of the most useful relations. All entries in the table refer to suspensions where the particles are made of the same material. Extensions to more than one type of material can be obtained by means of the given definitions. For example, the density of a suspension containing n types

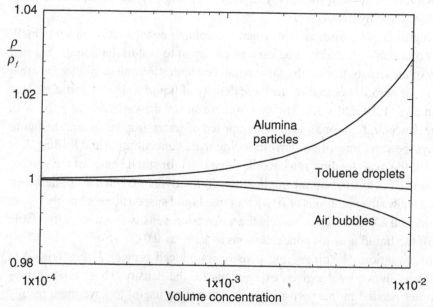

Figure 1.3.2. Density of three liquid suspensions.

Table 1.3.1. *Relationships between suspension density, concentration, and mass loading*

	ϕ_v	ϕ_m	η_m
ρ	$(1-\phi_v)\rho_f + \phi_v\rho_p$	$\dfrac{1}{\rho} = \dfrac{1-\phi_m}{\rho_f} + \dfrac{\phi_m}{\rho_p}$	$\rho_f\dfrac{1+\eta_m}{1+\eta_m\delta}$
ϕ_v		$\dfrac{\phi_m\delta}{1+\phi_m(\delta-1)}$	$\dfrac{\eta_m\delta}{1+\eta_m\delta}$
ϕ_m	$\dfrac{\phi_v/\delta}{1-\phi_v+\phi_v/\delta}$		$\dfrac{\eta_m}{1+\eta_m}$
η_m	$\dfrac{\phi_v/\delta}{1-\phi_v}$	$\dfrac{\phi_m}{1-\phi_m}$	

of particles is $\rho = (1-\phi_v)\rho_f + \sum_{i=1}^{n} \phi_v^{(i)}\rho_p^{(i)}$, where $\phi_v^{(i)}$ is the volume concentration for the *i*th type, and $\rho_p^{(i)}$ is the density of the corresponding material.

1.4 Interparticle Separation

It is evident that in very dilute suspensions, the particles are, on average, widely separated. This means, for example, that particle collisions and coalescence – that is, the joining of two or more particles – seldom occurs. As the concentration increases, however, such particle interactive effects begin to take place. But even if collisions do not occur, decreasing the distance between particles may produce other types of interactions that affect the overall dynamic behavior of a suspension. These interactions owe their origin to the motions induced by the particles in the surrounding fluid. These motions usually decay very slowly with distance from one particle and may, therefore, modify the fields in the vicinity of other particles. Such modifications naturally complicate the analytical determination of the forces acting on any given particle.

While we are not yet in a position to determine how far do these induced motions extend into the surrounding fluid, we can estimate the distance between neighboring particles and make a *guess* as to the extent of the interactions that may exist in a given situation. For this purpose, we consider a suspension having a number N_0 of spherical particles per unit volume, each having the average diameter $\overline{D}$. Suppose the volume element consists of a cube of unit side and divide it into N_0 cubic cells, placing a single particle at the center of each. It then follows that the lateral dimensions of each cell can be used as a measure of the distance, λ_p, between the centers of two *neighboring* spheres. This interparticle distance is therefore of the order of $N_0^{-1/3}$. Also, since the particles are spherical, the particle volume concentration can be written as $\phi_v = N_0\pi\overline{D}^3/6$. Therefore, the interparticle distance may be expressed as

$$\lambda_p/\overline{D} = (\pi/6\phi_v)^{1/3} \tag{1.4.1}$$

A numerical example may be useful to visualize these distances. Thus, considering again an aerosol containing 10^6 particles/cm^3, each having a diameter of 1μm, we see that the volume concentration is about 5×10^{-7} and that the interparticle distance would be about 100 diameters. Further, $\lambda_p/\overline{D}$ increases relatively slowly as the concentration increases. Thus, increasing ϕ_v by a factor of 10, decreases $\lambda_p/\overline{D}$ only by a factor of about 2. Although we do not yet know how far the motion of one particle extends into the surrounding fluid, it may be anticipated that, at such large separations, the interactions between particles produced by such motions are small and can be neglected. The matter is revisited in Chapter 4, after we consider the fluid force on an isolated particle.

The assumption that the single particle results apply in suspensions greatly simplifies their description because it means that the basic mechanisms that determine the behavior of small particles in fluids – namely the force exerted by the fluid on a given particle and the heat transfer between particle and fluid – can be obtained by ignoring the presence of other particles. This is done in Chapters 3–6. But first, we give in the next chapter a listing of the basic equations that are needed to study the motions of isolated particles.

2

Conservation Equations

2.1 Introduction

We begin our study by considering suspensions so dilute that each particle behaves independently of others. In this situation, the dynamic behavior of a particle is determined by its physical properties and by the pressure, velocity, and temperature fields in the fluid surrounding it. When no particles are present, those fields may be regarded as known. But particles introduce generally unknown disturbances that modify them and that obey the same equations as the main, or background, field. These are the equations of fluid mechanics; they will be needed throughout the book and are presented herein without derivation.

Equations of Motion

The equations of fluid mechanics are based on conservation laws of mass, momentum, and energy – supplemented by an equation of state and by a relation between the stress and the rate of strain. To express these laws mathematically, we will usually use the Eulerian, or field, description, in which the quantities describing the motion of the fluid are specified at a fixed point in space. For example, $\rho_f(\mathbf{x},t)$, $\mathbf{u}_f(\mathbf{x},t)$, $p_f(\mathbf{x},t)$, and $T_f(\mathbf{x},t)$ denote the fluid's density, velocity, pressure, and temperature, respectively, at time t, at a point whose position vector – relative to the origin of a system of coordinates – is $\mathbf{x}$. When necessary, the Lagrangian description will also be used. Here, the conservation equations are written for an element of fluid composed of the same molecules, whose volume, momentum, and energy; but, not its mass, can change as the element moves. Such an element is usually called a "fluid particle"; but, to avoid confusion with the particles considered here, we shall refer to it as a *fluid element*.

It will be important to distinguish the rates of change of physical quantities appearing in both descriptions. In the Lagrangian description, the rate of change of a physical property of a material element – for example, the volume v – is normally denoted by dv/dt, or by a dot over the symbol representing the variable (i.e., $\dot{v}$). In the Eulerian description, the time rate of change is expressed in terms of the time rate of change of that quantity at a fixed point, plus the change due to the motion of the fluid. To avoid confusion with the Lagrangian derivative, the total rate of change is customarily denoted by as D/Dt, where

$$\frac{D}{Dt} = \frac{\partial}{\partial t} + \mathbf{u}_f \cdot \nabla \tag{2.1.1}$$

The distinction is particularly important for fluids with small particles because both descriptions are sometimes used simultaneously.

The motions will be referred to a Cartesian system of coordinates, and the indicial notation will be sometimes used to simplify the writing. Thus, if the components of some vector $\mathbf{q}$ are q_1, q_2, and q_3 with respect to such a system, a fact that will be represented by $\mathbf{q} = \{q_1, q_2, q_3\}$, then the components of $\mathbf{q}$ will be denoted by q_i, where the index i runs from 1 to 3. The indicial notation will also be used to identify the components of second and higher order Cartesian tensors. To avoid confusion with other suffixes, we will only use Latin suffixes ranging from i to n to denote such quantities.

Because the same basic equations also apply to the fluids in some particles (e.g., droplets and bubbles), we will denote the *external* fluid variables, as well as its properties, with the suffix f. Particle variables will be denoted with the suffix p (e.g., T_p for the particle temperature). Generally speaking, all quantities thus denoted will depend on position and time. However, unless specifically stated to the contrary, all fluid properties, except the density, will be taken as constant. These include the thermal conductivity k_f, shear viscosity μ_f, thermal expansion β_f, as well as the specific heats. This does represent a limitation of the theories developed here; but, as shown later, the changes of pressure and temperature in many suspensions motions are usually small. Hence, the constant-property assumption is a reasonable point of departure. We will also assume that the fluids are Newtonian and that they have zero expansive, or bulk, viscosity. For such fluids, then, the Eulerian equations of motion are:

Continuity

$$\frac{\partial \rho_f}{\partial t} + \nabla \cdot (\rho_f \mathbf{u}_f) = 0 \tag{2.1.2}$$

Navier-Stokes

$$\rho_f \left(\frac{\partial \mathbf{u}_f}{\partial t} + \mathbf{u}_f \cdot \nabla \mathbf{u}_f \right) = \rho_f \mathbf{f}_{Bf} - \nabla p_f + \mu_f \left[\nabla^2 \mathbf{u}_f + \tfrac{1}{3}\nabla(\nabla \cdot \mathbf{u}_f) \right] \tag{2.1.3}$$

where $\mathbf{f}_{Bf}$ is the external *body force* per unit mass acting on the fluid.

Internal Energy

$$\rho_f \frac{De_f}{Dt} = -p_f \nabla \cdot \mathbf{u_f} + \rho_f \Phi_f + k_f \nabla^2 T_f \qquad (2.1.4)$$

where Φ_f is the *viscous dissipation function*, defined by

$$\Phi_f = \frac{2\mu_f}{\rho_f} \left(e_{ij} e_{ij} - \tfrac{1}{3} \Delta^2 \right)_f \qquad (2.1.5)$$

Here, $\Delta_f = \nabla \cdot \mathbf{u}_f$ is the *rate of expansion*, and e_{ij} is the i, j component of the *rate-of-strain tensor*, defined by

$$e_{ij} = \frac{1}{2} \left(\frac{\partial u_i}{\partial x_j} + \frac{\partial u_j}{\partial x_i} \right) \qquad (2.1.6)$$

The left-hand side of (2.1.4) can, with the aid of thermodynamics, be written as

$$\rho_f c_{pf} \frac{DT_f}{Dt} - \beta_f T_f \frac{Dp_f}{Dt} = \rho_f \Phi_f + k_f \nabla^2 T_f \qquad (2.1.7)$$

The right-hand side of this equation represents the rate at which heat is added to the fluid element by viscous dissipation and heat conduction, respectively. It is noted, from (2.1.5) and (2.1.6), that the viscous dissipation function, Φ_f, is proportional to the squares of the spatial derivatives of the velocity components. When these are small, as is usually the case in acoustic motions, the term may be disregarded – a fact that considerably simplifies the analysis.

State. Already assumed in writing, (2.1.7) is the existence of a relation between the entropy of the element and its pressure and temperature, i.e.,

$$s_f = s_f(p_f, T_f) \qquad (2.1.8)$$

where s_f is the element's entropy per unit mass. A similar equation may be written for the internal energy. More commonly, the thermodynamic state of a material element is described by a relation between pressure, density, and temperature called the equation of state; for example,

$$p_f = p_f(\rho_f, T_f) \qquad (2.1.9)$$

Implicit in these equations is the assumption that the fluid element is always in thermodynamic equilibrium. Among other implications, this means that the thermodynamic pressure p_f appearing in (2.1.9) has been taken to be equal to the dynamic pressure appearing in the Navier-Stokes equation. As is known from fluid mechanics, this assumption breaks down when extremely rapid variations occur, but these relate to situations outside the scope of this work.

Boundary Conditions

The system of equations is complemented by boundary conditions on one or more of the field variables. In the next section, we consider the most common ones.

Mechanical. We adopt the so-called *no-slip* condition. This requires fluids to adhere to the external boundaries they touch (e.g., the surface of a particle immersed in it). Thus, if $\mathbf{U}_s$ denotes the velocity of the boundary at some point $\mathbf{x}_s$ on it, the velocity boundary condition is

$$\mathbf{u}_f(\mathbf{x}_s, t) = \mathbf{U}_s(\mathbf{x}_s, t) \tag{2.1.10}$$

When the effects of viscosity are ignored, we can only require that the normal component of the velocity of the fluid be equal to that of the boundary, so that $\mathbf{u}_f \cdot \mathbf{n} = \mathbf{U}_s \cdot \mathbf{n}$, where $\mathbf{n}$ is the *outward* unit normal vector to the boundary. In suspension dynamics, however, the effects of viscosity can seldom be neglected.

In addition to the kinematic condition (2.1.10), we need some conditions on the forces acting at an interface between two media, usually two different fluids. To introduce these forces, we first recall that if the interface is curved, surface tension effects come into play. For static conditions, these effects produce a pressure difference across it, whose magnitude is given by the *Young-Laplace* equation, e.g.,

$$p^{(1)} - p^{(2)} = \sigma \left(\frac{1}{R_1} + \frac{1}{R_2} \right) \tag{2.1.11}$$

Here, $p^{(1)}$ and $p^{(2)}$ and the pressures in mediums 1 and 2, R_1 and R_2, are the radii of curvature of the interface, and σ is the coefficient of surface tension for that interface. By definition, σ is the force per unit length acting along a line drawn on the interface, on a direction normal to the line and tangential to the surface. Thus, a curved interface is in a state of tension, and this translates into a pressure difference across the interface. It should be noted that the sum inside the parentheses in the previous equation – called the *curvature* of the surface – is a signed quantity. This means that the pressure difference can be either positive or negative. As a mnemonic device, it is helpful to remember that the pressure is higher on that side of the interface containing the center of curvature. It should also be added that the interface is actually a thin region having molecular dimensions. For our purposes, it is sufficient to regard the transition zone as having zero thickness. Seen in this manner, the interface is an imaginary surface separating two different media. That is, the interface is massless.

When the interface is in motion, other forces appear. To express the conditions at the interface, it is useful to first ignore surface tension effects. The zero-thickness interface divides fluid (1) from fluid (2), and has unit normals $\mathbf{n}^{(1)}$ and $\mathbf{n}^{(2)}$ ($= -\mathbf{n}^{(1)}$) that point away from the corresponding sides. At each point on the surface, the local forces are, respectively, $\mathbf{\Sigma}^{(1)}$ and $\mathbf{\Sigma}^{(2)}$, where $\mathbf{\Sigma}^{(1)}$ is the force that the fluid on side (2) exerts on the fluid on side (1). The sign convention on these forces should be noted. In it, a positive $\mathbf{\Sigma}$ indicates a tension.

Because the orientation of the unit normal vector **n** can vary from point to point on the interface, it is convenient to express the local force in terms of the stress tensor by means of

$$\Sigma_i = \sigma_{ij} n_j \qquad (2.1.12)$$

where σ_{ij} is the i, j component of the stress tensor, and n_j is the jth component of the unit normal vector **n**. The physical interpretation of a component, σ_{mn} say, of the stress tensor is that it represents the force per unit area acting along the direction of the mth axis (first index) of an orthogonal coordinate system, on a surface that is perpendicular to the nth axis (second index) of that system. As shown in fluid mechanics texts, mechanical stability requires that the stress tensor be symmetric. Thus, $\sigma_{ij} = \sigma_{ji}$. The average value of the diagonal components of σ_{ij} is taken as the mechanical definition of pressure. Thus, $p = -\sigma_{kk}/3$, where the repeated index implies summation over that index and where the negative sign is due to the sign convention on the surface force. In view of these definitions, the components of the stress tensor may be written as

$$\sigma_{ij} = -p\delta_{ij} + d_{ij} \qquad (2.1.13)$$

where the second-order tensor δ_{ij} is known as the *Kronecker delta*. This is unity if $i = j$ and zero otherwise. The first part of σ_{ij}, namely $-p\delta_{ij}$ represents an isotropic force per unit area, whereas the second, d_{ij}, called the *deviatoric* part, depends on direction. It specifies the off-diagonal terms of the stress tensor and vanishes upon contraction. That is, $d_{ii} = 0$. For a Newtonian fluid, d_{ij} is given by

$$d_{ij} = 2\mu_f\left(e_{ij} - \tfrac{1}{3}\Delta\delta_{ij}\right), \qquad (2.1.14)$$

As a result of its definition, d_{ij} is identically zero during uniform compressions or expansions. In view of this, we anticipate that d_{ij} vanishes for radial motions of small bubbles and droplets, meaning that their internal motion is controlled by the pressure alone.

We now consider the conditions of the local force. These derive from Newton's third law and require that all components of $\Sigma^{(1)}$ and $\Sigma^{(2)}$ be equal and opposite. We write the conditions on the normal components explicitly. First, we use the indicial notation to express $\Sigma \cdot \mathbf{n}$ as $\Sigma_k n_k$, with $\Sigma_k = \sigma_{kj} n_j$. We then use (2.1.13) and the properties of the Kronecker delta and get

$$\Sigma_k n_k = -p + d_{ij} n_{ij} \qquad (2.1.15)$$

Hence,

$$p^{(1)} - p^{(2)} = d_{ij}^{(1)} n_i^{(1)} n_j^{(1)} - d_{ij}^{(2)} n_i^{(2)} n_j^{(2)} \qquad (2.1.16)$$

When surface tension effects are included, this must be modified so as to include the pressure jump due to those effects. This gives

$$p^{(1)} - p^{(2)} = d_{ij}^{(1)} n_i^{(1)} n_j^{(1)} - d_{ij}^{(2)} n_i^{(2)} n_j^{(2)} + \sigma\left(\frac{1}{R_1} + \frac{1}{R_2}\right) \qquad (2.1.17)$$

This is the boundary condition on the pressure difference across an interface between two moving fluids. When the fluids are inviscid, d_{ij} is zero and (2.1.17) reduces to the Young-Laplace formula, but without requiring the interface to be at rest.

Equation (2.1.7) is significantly simplified for radial motions of small droplets and bubbles in incompressible fluids. For such motions, the deviatoric part of the stress tensor for the fluid in the particle vanishes, as pointed out. Secondly, the velocity in the external fluid has only one component, and this is along the radial direction, or $\mathbf{u}_f = \{u_f, 0, 0\}$. Thus, every component of d_{ij} for the external fluid is zero, except d_{rr}, which is given by $2\mu_f(\partial u_f / \partial r)$. Thus, if side (2) in the previous equation refers to the external fluid side of the interface, we have, for a small bubble or droplet of radius R,

$$p_{ps} - p_{fs} = -2\mu_f \left(\frac{\partial u_f}{\partial r}\right)_{r=R} + \frac{2\sigma}{R} \tag{2.1.18}$$

where p_{ps} and p_{fs} are the particle and fluid pressures, respectively, evaluated at the corresponding sides of the interface.

Thermal. In addition to the mechanical conditions listed previously, there will some times be a need to impose boundary conditions on the temperature fields. One is that the temperatures on both sides of a boundary be equal. For later purposes, we express this condition in terms of the common value at the point of contact. Thus, if T_s represents the surface temperature, we have

$$T_f(\mathbf{x}_s, t) = T_s(\mathbf{x}_s, t) \tag{2.1.19}$$

A second condition is that the heat fluxes across the boundary be equal. In the absence of radiation heat transfer, the heat flux, $\dot{\mathbf{q}}_k$ – defined as the amount of heat flowing across a unit area in a unit time – is given by Fourier's law:

$$\dot{\mathbf{q}}_k = -k_f \nabla T_f \tag{2.1.20}$$

Thus, the condition on the heat fluxes at a surface separating two heat-conducting media is

$$k^{(1)}\mathbf{n}^{(1)} \cdot \nabla T^{(1)} = k^{(2)}\mathbf{n}^{(2)} \cdot \nabla T^{(2)} \tag{2.1.21}$$

It should be noted that this does not depend on any property of the interface (e.g., surface tension).

PROBLEMS

1. Consider a small sphere whose surface moves along the normal in an *incompressible* fluid with a speed U_s. Show that the velocity in the fluid outside the sphere is $u_s = \{U_s(R/r)^2, 0, 0\}$, where $R(t)$ is the instantaneous radius of the sphere, and r is the distance from a point in the fluid to the center of the sphere. Determine

the components of the rate-of-strain tensor and those of the deviatoric part of the stress tensor.

2. The kinetic energy of a moving fluid may be obtained from $\frac{1}{2}\int_{V_f}\rho_f u_f^2 dV$, where V_f is the volume occupied by the fluid. For the spherically symmetric, incompressible motion in problem 1, this is equal to $2\pi\rho_f\int_R^\infty u_f^2 r^2 dr$. Show that the kinetic energy of the fluid may be written as $E_{kin} = \frac{1}{2}M_0 U_s^2$, where $M_0 = 3\rho_f v_p$.

2.2 Small Particles

We now consider the particles. Generally speaking, different points within the particles will have different velocities, pressures, and temperatures. Further, these quantities vary in time owing to changes in the fluid around them. Those changes are not usually known. Thus, the problem of determining the motion of a single particle in a fluid generally requires the solution of two sets of dynamic equations: one for the fluid and another for the material in the particle. However, it is intuitively evident that, when the particles are very small, the spatial variations of pressure, temperature, and density in them are also small – in which case they may be either disregarded or replaced with average values.

Volume Averages

We first consider temperature, pressure, and density in a particle. As stated previously, these generally vary from point to point within the particle. To incorporate these variations while retaining the simplicity of having space-independent variables, we represent each of them by their average values over the volume of the particle. Thus, for example, the average of the temperature is defined by

$$\overline{T}_p(t) = \frac{1}{v_p}\int_{v_p} T_p(\mathbf{x}, t)d\tau \tag{2.2.1}$$

where the volume of the particle, v_p, is generally time dependent. The same prescription can be used to obtain average values of other state variables within the particle (e.g., pressure). It should be noted that these averages cannot satisfy conditions that depend on position.

Kinematic Considerations

We now consider the various motions that a particle can sustain. These are of course rather numerous. However, when the particle is small, the motion can be decomposed into a small number of distinct motions. To introduce these, we first recall from mechanics that an infinitesimal displacement of a rigid particle can be represented as the sum of a translation and a rotation. In the first of these, each material point in the particle is translated by the same amount. We refer to this as a uniform translation. In the second, each point is rotated about the center of mass by the same infinitesimal angle. This will be called a uniform rotation. Compressible particles, on the other hand, can be deformed in various manners, but for a sufficiently small *material element*, the most general motion can be represented as the sum of

1. a *uniform* translation,
2. a *rigid-body* rotation,
3. a *stretching* motion that can be split into
   A. a *uniform expansion* (or contraction), and
   B. a *deformation* of the element without a change of volume.

Although this split does not apply to particles of finite size, we may expect that it remains approximately valid when the particles are very small. If so, the motion of a particle can be characterized by a uniform translational motion, where each point of the particle has a velocity $\mathbf{u}_p$, a rigid-body rotation with angular velocity $\boldsymbol{\Omega}_p$, a uniform expansion with a volume-rate change dv_p/dt, and a change of area, dA_p/dt, without a change of volume. Within some limitations, this characterization enables us to consider, separately, each of these motions. For example, the drag on a small bubble that is translating while executing small volume pulsations can be calculated by considering a bubble of fixed radius.

2.3 Conservation Equations for a Small Particle

We consider here the conservation laws of mass and momentum of a particle so small that the aforementioned kinematic theorem holds. These laws are considerably simpler than those applicable to arbitrary particles because the fields within the particles do not appear in them.

Mass

By definition, the particles considered in this work have constant mass. We may take advantage of this constancy to relate the average density to the particle volume. Thus, since

$$m_p = \int_{v_p} \rho_p d\tau = \overline{\rho}_p v_p \tag{2.3.1}$$

we have $\overline{\rho}_p/\rho_{p0} = v_{p0}/v_p$, from which we get $\dot{v}_p/v_p = -\dot{\overline{\rho}}_p/\overline{\rho}_p$. It is useful to derive the last expression by considering the equation of continuity for the material inside the particle. This can be written in the same form as that for the fluid, (2.1.2), or as $D\rho_p/Dt = -\rho_p \nabla \cdot \mathbf{u}_p$. Integrating this over the volume of the particle, we obtain

$$\int_{v_p} \frac{D\rho_p}{Dt} d\tau = -\int_{v_p} \rho_p \nabla \cdot \mathbf{u}_p d\tau \tag{2.3.2}$$

Owing to the assumed smallness of the particles, $D\rho_p/Dt$ and ρ_p may be taken outside the integrals and replaced with $\dot{\overline{\rho}}_p$ and $\overline{\rho}_p$, respectively. This gives $\dot{\overline{\rho}}_p/\overline{\rho}_p = -\overline{\Delta}_p$, where $\overline{\Delta}_p$ is the spatial average of the rate of expansion, $\nabla \cdot \mathbf{u}_p$. The result may be also expressed in terms of the displacement of points on the surface of the particle. Thus, changing the volume integral into an integral over the surface of the particle, and making use of the boundary condition on the velocity, (2.1.10), we obtain

$$\frac{\dot{\overline{\rho}}_p}{\overline{\rho}_p} = -\frac{\dot{v}_p}{v_p} = -\frac{1}{v_p}\int_{A_p} \mathbf{U}_s \cdot \mathbf{n}\,dS \tag{2.3.3}$$

where A_p is the surface area of the particle.

Linear Momentum

The linear momentum of a particle is $\int_{v_p} \rho_p \mathbf{u}_p d\tau$, with both the velocity and the density generally depending on time and position within the particle. Since the particle is very small, we may take each point to translate with the same velocity, which for simplicity we denote as $\mathbf{u}_p(t)$, keeping in mind that it refers only to the translational motion. Thus, $\mathbf{u}_p(t)$ can be taken outside the integral. The remaining integral is the mass, m_p, of the particle. Hence,

$$\int_{v_p} \rho_p \mathbf{u_p} d\tau = m_p \mathbf{u_p}$$

By Newton's second law, the rate of change of the linear momentum of a particle is equal to the sum of all the forces acting on the particle. It is convenient to divide these into forces resulting from the action of the fluid around the particle and into forces that are due to external fields. The fluid forces act only on the surface of the particle and are given by the local force, $\mathbf{\Sigma}^{(f)}$, integrated over the surface of the particle. Denoting this force by $\mathbf{F}_p$, we have

$$\mathbf{F}_p = \int_{A_p} \mathbf{\Sigma}^{(f)} dS \tag{2.3.4}$$

This is the force with which the fluid acts on translating particles; it plays a central role in the dynamics of suspensions.

The external-field forces may be divided into body and surface forces. Since the particle is very small, the body force per unit mass may be considered uniform through-out the particle; furthermore, it is equal to body force per unit mass of fluid, $\mathbf{f}_{Bf}$. Since the mass of the particle is m_p, the total body force on it is $m_p \mathbf{f}_{Bf}$. In suspensions, the primary example of body force is, of course, gravity.

Now consider external forces that act on the surface of the particle. The primary example of a such a force is the electrical force that results when an electrically charged particle is placed in an electrical field. To account for such forces, we denote the surface force per unit area by $\mathbf{f}_{Sp}$. Then, Newton's second law for the particle can be written as

$$\frac{d}{dt}\int_{v_p}\rho_p\mathbf{u}_p d\tau = \mathbf{F}_p + \int_{v_p}\rho_p\mathbf{f}_B d\tau + \int_{A_p}\mathbf{f}_{Sp}dS$$

Thus,

$$m_p\frac{d\mathbf{u}_p}{dt} = \mathbf{F} \tag{2.3.5}$$

where $\mathbf{F}$ is the sum of all forces acting on the particle. This shows that the linear momentum of a small particle would be conserved if $\mathbf{F}$ vanishes, but this occurs only in strict equilibrium conditions between fluid and particle. Generally, this also requires that the external forces be absent.

Angular Momentum

We now consider angular momentum due to rigid-body rotation, that is, the angular momentum of a rigid particle rotating about its center of mass, with an instantaneous angular velocity $\mathbf{\Omega}_p$. Let $\xi = \{\xi_1, \xi_2, \xi_3\}$ be the position vector of an arbitrary point within the particle, relative to the center of mass. Then, the angular momentum of the particle is $\mathbf{L} = \int_{v_p}\xi \times \mathbf{w}_p\rho_p d\tau$, where $\mathbf{w}_p = \mathbf{\Omega}_p \times \xi$ is the velocity at a point ξ, due to the rotation alone. Since the angular velocity is constant, it can be taken outside the integral. For this purpose, we express the ith component of $\xi \times \mathbf{w}_p$ as $(\xi \times \mathbf{w}_p)_i = \varepsilon_{ijk}\varepsilon_{klm}\xi_j\Omega_{pl}\xi_m$, where ε_{ijk} is the alternating symbol tensor, which is equal to $+1$ if the indices are taken in cyclic order, -1 if they are not, and 0 if any index is repeated. Thus, if $\varepsilon_{123} = 1$, then $\varepsilon_{321} = 1$, $\varepsilon_{132} = -1$, and $\varepsilon_{122} = 0$. Using the identity $\varepsilon_{ijk}\varepsilon_{klm} = \delta_{il}\delta_{jm} - \delta_{im}\delta_{jl}$, we can write

$$L_i = I_{il}\Omega_{pl} \tag{2.3.6}$$

where $I_{il} = \int_{v_p}\rho_p(\xi_j\xi_j\delta_{il} - \xi_i\xi_l)d\tau$ is the inertia tensor of the particle and Ω_{pl} is the lth component of $\mathbf{\Omega}_p$. It is noted that all of the components of I_{ij} are constant. As with other second-order Cartesian tensors, I_{il}, may be expressed in terms of its principal axes, in which case its off-diagonal elements vanish. That is, we may write $I_{il} = I^{(i)}\delta_{il}$ (no summation over the index i is implied here), where the quantities $I^{(i)}$, $i = 1 - 3$, are called the *principal* moments of inertia. In those axes, the components of $\mathbf{L}$ are $L_1 = I^{(1)}\Omega_{p1}$, $L_2 = I^{(2)}\Omega_{p2}$, $L_3 = I^{(3)}\Omega_{p3}$. Additional simplifications are possible when the particle is symmetric with respect to one or more axes. For a sphere, all three principal moments of inertia are equal, so that we may write (2.3.6) as

$$\mathbf{L} = I\mathbf{\Omega}_p \tag{2.3.7}$$

where $I = {}^2\!/_{15} m_p R^2$ is the sum of the three principal moments of inertia for a sphere of radius R. The conservation equation for the angular momentum of a rigid particle is

$$I_{ij} \frac{d\Omega_{pj}}{dt} = N_{pi} \qquad (2.3.8)$$

where N_{pi} is the ith component of the torque applied to the particle. Since the body force is uniform over the volume occupied by the particle, the torque is given by

$$\mathbf{N}_p = \int_{A_p} \boldsymbol{\xi}_s \times \left(\boldsymbol{\Sigma}^{(f)} + \mathbf{f}_{Sp} \right) dS \qquad (2.3.9)$$

where $\boldsymbol{\xi}_s$ is the position vector of a point on the particle surface, relative to the center of mass. Thus, the angular momentum of a particle is conserved if the total applied torque vanishes.

Uniform Expansion/Compression

Although no general conservation principle exists for uniform expansions or contractions, we may take advantage of their uniformity to express the equation of mass conservation for the particle in terms of the velocity of points on its surface. If that velocity is denoted by $\mathbf{U}_s$, the rate of change of the volume of a particle is given by (2.3.3). But for a uniform expansion/contraction that velocity is uniform. Hence,

$$\dot{v}_p = \mathbf{U}_s A_p \qquad (2.3.10)$$

It must be pointed out that uniform expansions/compressions are generally possible only if the particles are spherical. Thus, if R is the instantaneous radius of the particle, we simply have $\dot{v}_p = 4\pi R^2 \dot{R}$. The corresponding average of the rate of expansion in the sphere is $\overline{\Delta}_p, = 3\dot{R}/R$.

Associated Energies

Each of the particle motions considered contributes a certain amount of energy to the total energy of the particle. These contributions are independent of one another in the sense that each motion may exist without the others. For example, a particle may rotate without translation and vice-versa. Below, we give expressions for each of the energies associated with those motions. The rigid-body motions are simpler and are considered first.

Translational Kinetic Energy

$$E_{kin} = \frac{1}{2} m_p \mathbf{u}_p^2 \qquad (2.3.11)$$

where $\mathbf{u}_p$ is the translational velocity of the particle and $\mathbf{u}_p^2 = \mathbf{u}_p \cdot \mathbf{u}_p$. Multiplying (2.3.5) by $\mathbf{u}_p$, we obtain

$$\frac{dE_{kin}}{dt} = \mathbf{F} \cdot \mathbf{u}_p \tag{2.3.12}$$

Thus, the kinetic energy is conserved if the applied force is zero, or if it acts in a direction perpendicular to the translational velocity. Neither condition is likely to occur in dynamic conditions.

Rotational Kinetic Energy. If the particle is executing a solid-body rotation about its center of mass, there will be a rotational contribution to the kinetic energy of the particle which, when principal axes are used for the inertia tensor, can be written as

$$E_{rot} = \frac{1}{2}\left[I^{(1)}\Omega_{p1}^2 + I^{(2)}\Omega_{p2}^2 + I^{(3)}\Omega_{p3}^2\right] \tag{2.3.13}$$

where $\Omega_p = \{\Omega_{p1}, \Omega_{p2}, \Omega_{p3}\}$. For a spherical particle, the principal moments of inertia are equal and

$$E_{rot} = \frac{1}{2}I\Omega_p^2 \tag{2.3.14}$$

where $\Omega_p^2 = \Omega_p \cdot \Omega_p$. Thus, the rate of change of the kinetic energy of rotation is

$$\frac{dE_{rot}}{dt} = \mathbf{N}_p \cdot \Omega_p \tag{2.3.15}$$

Pulsational Energy. Consider now particle volume changes that occur without shape deformations. As pointed out previously, such particles are generally spheres moving radially with velocity $\dot{R}$. In analogy with the translational and rotational motions, we may define a pulsational energy by means of

$$E_{pul} = \frac{1}{2}M_{pul}\dot{R}^2 \tag{2.3.16}$$

where the quantity M_{pul} has the dimensions of a mass. However, the moving surface is the interface between particle and fluid, which, by definition, has no mass. Hence, this energy is not the "kinetic energy" of the surface. A closer inspection shows that the fluid surrounding the particle moves in response to the pulsating motion and is thus associated with a kinetic energy. As shown in Problem 2.1.2 for an incompressible fluid, this kinetic energy is equal to $\sqrt[3]{2}\rho_f v_p \dot{R}^2$. Thus, for incompressible fluids, M_{pul} is given by $4\pi\rho_f R^3$ so that

$$dE_{pul}/dt = 2\pi\rho_f R^3(2\ddot{R} + 3\dot{R}^2/R)\dot{R} \tag{2.3.17}$$

Therefore, in the analogy with the translational and rotational motions, we may antic-ipate that the derivative of the pulsational energy can be written as

$$\frac{dE_{pul}}{dt} = \mathbf{F}_r \cdot \dot{\mathbf{R}} \qquad (2.3.18)$$

where $\mathbf{F}_r$ is the radial force acting on the sphere. Comparison with (2.3.17) shows that

$$F_r = 4\pi\rho_f R^3 \left(\ddot{R} + \tfrac{3}{2}\dot{R}^2/R\right) \qquad (2.3.19)$$

This result is obtained in Section 6.2 on the basis of the equations of motion.

2.4 Internal Energy

Having defined volume averages for the pressure, temperature and density of a particle, we turn to the internal energy. This has been defined at any point in terms of the temperature and the density at that point, and like those quantities, it depends on position. However, under some conditions, it is possible to represent that energy in terms of $\overline{T}_p$, $\overline{P}_p$ and $\overline{\rho}_p$. Equally important, it may be possible to express the first law of thermodynamics in terms of those quantities. Ignoring surface tension and rotation that law is

$$\frac{dE_p}{dt} = \dot{W}_p + \dot{Q}_p \qquad (2.4.1)$$

where

$$E_p(t) = \int_{v_p} \rho_p(\mathbf{x}, t)e_p(\mathbf{x}, t)d\tau \qquad (2.4.2)$$

$\dot{Q}_p$ is the heat transfer rate to the particle, and $\dot{W}_p$ is the rate at which the *external* field does work on the particle. The heat transfer rate *to* the particle may be determined from the external temperature field by means of

$$\dot{Q}_p = k_f \int_{A_p} \mathbf{n} \cdot \nabla T_f dS \qquad (2.4.3)$$

This shows that $\dot{Q}_p > 0$ if the temperature in the fluid increases as we move away from the particle. To compute the work, we need to evaluate the rate at which the external stress field does on the particle. But, in view of our assumptions, the particle can only execute rigid-particle translations and uniform expansions. It is therefore advantageous to consider the two motions separately, noting that the internal energy can be expressed as $E_p(t) = \overline{\rho_p e_p}\, v_p$ for both.

Rigid Particles. Here, the density is constant and the internal energy depends only on the temperature. Furthermore, the volume of integration is fixed because the particle is rigid. Hence, the time derivative of (2.4.2) may be expressed as

$$\frac{dE_p}{dt} = \rho_{p0} \int_{v_{p0}} c_{vp}(T_p) \frac{dT_p}{dt} d\tau \qquad (2.4.4)$$

where $c_{vp} = de_p/dT_p$ is the constant-volume specific heat for the particle material. Since the range of variation of T_p is normally small in suspensions, we take the specific heat to be constant. Further, since for rigid materials there is no difference between the constant-pressure or constant-volume specific heats, we will write the specific heat in (2.4.4) as c_{pp}. This produces no harm here and lets us write (2.4.4) in a form that is compatible with the corresponding compressible case. Thus, remembering that the volume of integration is fixed, so that the time derivative can be taken outside the integral, we obtain

$$\frac{dE_p}{dt} = \rho_{p0} v_{p0} c_{pp} \frac{d\overline{T}_p}{dt} \qquad (2.4.5)$$

Now consider the external work. This is associated with the translational motion and has two components. The first is related to the changes of kinetic energy of the particle and should therefore not appear here. The second is related to the heating produced by the relative motion, but this is negligible for the small relative velocities usually found in suspensions. Thus, for small, rigid particles, the first law of thermodynamics reduces to

$$m_p c_{pp} \frac{d\overline{T}_p}{dt} = \dot{Q}_p \qquad (2.4.6)$$

Compressible Particles. When the density of the particle changes, it is not generally possible to obtain an equally simple result as (2.4.6). Nevertheless, an expression equivalent to it can obtained when the density and temperature in the particle depart only slightly from their equilibrium values ρ_{p0} and $T_{p0} = T_0$. Then, we may expand $\rho_p e_p$, as a function of T_p and ρ_p, in Taylor series about T_0 and ρ_{p0}, retaining only leading terms. This gives

$$\rho_p e_p = \rho_{p0} e_{p0} + \rho_{p0} c_{vp} (T_p - T_0) + \rho_{p0} (\partial e_p / \partial \rho_p)_{T_0} (\rho_p - \rho_{p0})$$
$$+ e_{p0} (\rho_p - \rho_{p0}) + \cdots \qquad (2.4.7)$$

where the ellipses represent terms of higher order in $(\rho_p - \rho_{p0})$ and $(T_p - T_0)$. The thermodynamic derivative appearing here is $(\partial e_p / \partial \rho_p)_{T_0} = (p_0 - \rho_{p0} c_{Tp}^2 \beta_p T_0)/\rho_{p0}^2$. Thus, integrating (2.4.7) over v_{p0}, the mean value of the volume of the particle, and taking the time derivative of the result (2.4.4) gives

$$\frac{dE_p}{dt} = m_p c_{vp} \frac{d\overline{T}_p}{dt} + m_p \left[(p_0/\rho_{p0} - c_T^2 \beta_p T_0)/\rho_{p0} \right] \frac{d\overline{\rho}_p}{dt} + \cdots \qquad (2.4.8)$$

Consider now the external work, $\dot{W}_p$. It can include only the work of compression done by the external field. This includes the pressure as well as the normal viscous stresses in the external fluid, but the work done by the viscous stresses is of second order and can thus be neglected. There remains only the work done by the external pressure, $\int_{A_p} p_f \mathbf{U_s} \cdot \mathbf{n} \, dS$. To leading order, this is equal to $p_0 \int_{A_p} \mathbf{U_s} \cdot \mathbf{n} \, dS$, where p_0 is the ambient value of the pressure. But, on account of (2.3.10), this reduces to $-p_0(dv_p/dt)$. Thus, using $v_p = m_p/\rho_p$ and linearizing the result, we obtain $\dot{W}_p \approx -m_p(p_0/\rho_{p0}^2)(d\bar{\rho}_p/dt)$. Substituting this result and dE_p/dt from (2.4.8) in (2.4.1) yields

$$m_p c_{vp} \frac{d\overline{T}_p}{dt} - \left(m_p c_{Tp}^2 \beta_p T_0 / \rho_{p0}\right) \frac{d\bar{\rho}_p}{dt} = \dot{Q}_p \qquad (2.4.9)$$

This can be put in terms of the temperature and the pressure by expanding $p_p = p_p(T_p, \rho_p)$ and averaging the result over the volume of the particle. This gives

$$\overline{P}_p = c_{Tp}^2(\bar{\rho}_p - \rho_{p0}) + \rho_{p0} c_{Tp}^2 \beta_p (\overline{T}_p - T_0) \qquad (2.4.10)$$

Finally, using $c_{Tp}^2 \beta_p^2 T_0 = c_{pp} - c_{vp}$, (2.4.9) becomes

$$\frac{d\overline{T}_p}{dt} - \frac{\beta_p T_0}{\rho_{p0} c_{pp}} \frac{d\overline{p}_p}{dt} = \frac{\dot{Q}_p}{m_p c_{pp}} \qquad (2.4.11)$$

This applies to particles of all types. For rigid particles, $\beta_p = 0$, and we recover (2.4.6). Another simplification occurs for perfect gas bubbles since, for them, $\beta_p T_0 = 1$.

PROBLEMS

1. The Reynolds transport theorem relates the rate of change of some property in a material element to the local rate of change and the flux of that property out of the volume under consideration. Thus, if θ represents such property per unit volume, then for a particle of volume v_p enclosed by surface area A_p, we have

$$\frac{d}{dt} \int_{v_p} \theta \, d\tau = \int_{v_p} \frac{D\theta}{Dt} d\tau + \int_{A_p} \theta \mathbf{u} \cdot \mathbf{n} \, dS. \quad \text{Use this to obtain (2.3.12).}$$

2. Consider the equation of state for the material in particle [e.g., $p_p = p_p(T_p, \rho_p)$]. Show that for small departures from the ambient conditions, $\overline{p} = p_0 + \rho_0 c_{T0}^2[(\bar{\rho}_p - \rho_{p0})/\rho_{p0} + \beta_0(\overline{T}_p - T_0)] + \cdots$.

2.5 Energy Dissipation

The appearance in the equations of motion of the viscosity and thermal conductivity implies that the motions of particles in fluids are generally accompanied by energy dissipation (i.e., by the conversion of useful energy into heat). By useful energy, we mean energy available in a form that could be used to produce useful work. For example, the sound emitted by an expanding or contracting bubble could be regarded as useful work. If the motion takes place in ideal conditions, no dissipation occurs. In general, however, the irreversible actions of viscosity and heat conductivity produce some energy losses.

The rate of decrease of the energy associated with particle motions is of practical interest in a variety of situations (e.g., in determining the power needed to maintain a specified motion). Other examples may also be envisioned, but they will share some basic features. In each, a particle has acquired one or more types of motion through the action of an external agent. If that action were to stop, the motions would decay, indicating the effects of dissipation. It is evident that to sustain the motion, the external agent would have to supply energy at a rate equal to that at which energy is being dissipated. That rate can, in principle, be obtained by solving the complete set of equations, but when the motions represent small departures from equilibrium, a simpler formulation is possible. This is given next.

Let us denote the generic departure from equilibrium by $x(t)$. Then, the energy associated with each of the motions is given by an equation of the form

$$E_x = \tfrac{1}{2} M_x \dot{x}^2 \tag{2.5.1}$$

where M_x is a mass and $\dot{x}$ a velocity. Further, the conservation law for each of these energies is given by an equation of the form

$$\frac{dE_x}{dt} = F_x \dot{x} \tag{2.5.2}$$

Here, F_x is the "generalized force" corresponding to the "generalized" velocity $\dot{x}$. Now, the generalized force F_x depends on the generalized velocity $\dot{x}$ and possibly also on $\ddot{x}$ and higher order time derivatives. But when the variations with time are not very rapid, we may assume that the force depends only on the velocity [i.e., $F_x = f(\dot{x})$]. Thus, expanding this in a Taylor series about $\dot{x} = 0$, we have

$$F_x = f(0) + f'(0)\dot{x} + \tfrac{1}{2} f''(0)\dot{x}^2 + \cdots \tag{2.5.3}$$

The first term in this expansion vanishes by definition, and the third one is negligible, compared with the second when the velocities are very small. Hence,

$$\frac{dE_x}{dt} = f'(0)\dot{x}^2 \tag{2.5.4}$$

This shows the rate at which the energy changes is proportional to the square of the amplitude of the velocity. But when dissipation exists, we expect the energy to decrease in time. Since $\dot{x}^2 \geq 0$ and $dE_x/dt < 0$, we can write the right-hand side of (2.5.4) as $-\left| f'(0) \right| \dot{x}^2$. Also, in view of (2.5.1), this can be written as $-2\beta E_x(t)$, where $\beta = |f'(0)|/M_x \geq 0$. Substituting the last form into (2.5.4), and integrating the resulting differential equation, we obtain

$$E_x = E_{x0} e^{-2\beta(t-t_0)} \tag{2.5.5}$$

where E_{x0} is the value of the energy at time $t = t_0$. Thus, the energy decreases exponentially in time. Since the energy is proportional to the square of the displacement, we see that the displacement amplitude decreases as $\exp[-\beta(t - t_0)]$.

Damping Coefficient

The quantity β, which defines the rapidity of the decrease is called a *damping coefficient*. Its value is peculiar to each motion, but the following procedure, which is limited to situations where the decay rate is small, can be used to determine it for any smallamplitude motion. Formally, the damping coefficient is given by $\beta = |f'(0)|/M_x$. However, an exact determination of $f'(0)$ for a given motion usually requires the solution of the differential equation describing the motion, together with the initial conditions that exist at $t = t_0$. This amounts to solving the complete dynamical problem associated with the corresponding velocity $\dot{x}$. This is, indeed, necessary in some cases, for example when a knowledge of $\dot{x}$ is needed at every instant, or when the magnitude of displacements from equilibrium is large, in which case additional terms in (2.5.4) must be used, thus invalidating (2.5.5). But when neither of those two reasons exist, it is possible to obtain β without solving the complete problem. To show this, we first write (2.5.4) as

$$\dot{E}_x = -2\beta E_x \tag{2.5.6}$$

The left-hand side of this gives the rate at which the energy decreases in a unit time. But conservation of energy requires that this decrease be equal in magnitude, and opposite in sign, to the energy dissipation rate, $\dot{E}_x = -|(\dot{E}_x)_{lost}|$. Thus, $|(\dot{E}_x)_{lost}| = 2\beta E_x$. Let us average this over a period Δt, which does not have to be specified at this time. Then, denoting the time average by angle brackets, we have

$$\beta = \frac{\langle|(\dot{E}_x)_{lost}|\rangle}{2\langle E_x \rangle} \tag{2.5.7}$$

To obtain β from this, we need both quantities on the right-hand side. Consider $\langle E_x \rangle$ first. Using (2.5.5), we have

$$\langle E_x \rangle = E_{x0}\frac{1}{\Delta t}\int_{t_0}^{t_0+\Delta t} e^{-2\beta t}\,dt = E_{x0}\frac{1 - e^{-2\beta\Delta t}}{2\beta\Delta t} \tag{2.5.8}$$

This shows that $\langle E_x \rangle$ generally depends on β, making the procedure useless. However, if $2\beta\Delta t \ll 1$, the right-hand side of (2.5.8) reduces to E_{x0}, a known quantity that is independent of β. Thus,

$$\beta = \frac{\langle|(\dot{E}_x)_{lost}|\rangle}{2E_{x0}} \tag{2.5.9}$$

Now consider $(\dot{E}_x)_{lost}$, the rate at which energy is dissipated. If it can be *separately* calculated, then that result can be used in (2.5.9) to obtain β. This is often possible because, when dissipation is small, the energy losses can be determined on the basis of the ideal, dissipationless fields. A second method to estimate $(\dot{E}_x)_{lost}$ is as follows. The motions we are considering are due to some external device, doing work on the fluid containing the particle. That is, the device adds energy to the fluid containing the particle at a rate equal to $\dot{E}_{inp}$. Although there might be a time delay between the instant the energy is added and that when it is dissipated, it should be clear that, on the average, conservation of energy requires that the energy input rate be equal to the energy dissipation rate, so that the determination of β through (2.5.9) is again possible.

The previous procedure will be used in several contexts in some of the following chapters. The applications included there relate to both single particles and to suspensions of rigid or compressible particles.

2.6 Nondimensional Parameters

Let us now return to the main topic of the chapter, namely the basic equations of motion. In anticipation of our study of the motion of single particles in fluids, we introduce here some of the parameters that play a role in that study. To do this, we consider the fluid equations of motion and recast them in terms of nondimensional variables, using scales for distance, time, velocity, etc., that are relevant to the motions of single particles. Because these scales depend on the type of flow under study, we will concentrate on those parameters that have wide applicability. We therefore take the fluid to be incompressible, assume that there are no external forces acting on it, and that the particle is not rotating. Then, the equations of motion in a frame of reference fixed on the particle center of mass, which is translating with velocity $\mathbf{u}_p$, are

$$\nabla \cdot \mathbf{u}_f = 0 \tag{2.6.1}$$

$$\frac{\partial \mathbf{u}_f}{\partial t} + \mathbf{u}_f \cdot \nabla \mathbf{u}_f - \mathbf{u}_p \cdot \nabla \mathbf{u}_f = -\frac{1}{\rho_f}\nabla p_f + \frac{\mu_f}{\rho_f}\nabla^2 \mathbf{u}_f \tag{2.6.2}$$

$$\rho_f c_{pf}\left(\frac{\partial T_f}{\partial t} + \mathbf{u}_f \cdot \nabla T_f - \mathbf{u}_p \cdot \nabla T_f\right) = \rho_f \Phi + k_f \nabla^2 T_f \tag{2.6.3}$$

We would like to write this set of equations in such a manner that the temporal and spatial derivatives appearing in them are of comparable order of magnitude. To do this, we note that, in the vicinity of a particle, the flow changes significantly over distances which are of the order of the particle size D. Hence, we introduce the following nondimensional position vector

$$\mathbf{X} = \frac{\mathbf{x}}{D} \tag{2.6.4}$$

To scale the fluid velocity, we note that the particle and fluid velocity are equal at the surface of the particle, but that at some distance away, the fluid velocity changes. These changes are, in part, due to changes in the particle velocity. Thus, if the order of magnitude of the change in the velocity of the particle, *relative* to the fluid at some distance from it, is denoted by U_r, we can introduce a suitable nondimensional fluid velocity by means of

$$\mathbf{V}_f = \frac{\mathbf{u}_f}{U_r} \qquad (2.6.5)$$

Similarly, we nondimensionalize the particle velocity with U_r so that

$$\mathbf{V}_p = \frac{\mathbf{u}_p}{U_r} \qquad (2.6.6)$$

A time scale is also needed to obtain a nondimensional acceleration of a fluid element as it moves in the vicinity of the particle. One can be introduced in terms of the time required for the fluid to travel the distance D. But in suspensions, the particles are free to move and accelerate. As a result of this motion, the velocity of a fluid element changes in a time t by an amount which is of the order of $\dot{U}_r t$. Thus, a suitable nondimensional time is

$$t' = \frac{\dot{U}_r t}{U_r} \qquad (2.6.7)$$

There remains the pressure and the temperature in the fluid. These play a passive role in the sense that they are largely determined by the velocity field in the fluid. Therefore, the magnitude of the changes of pressure and temperature cannot be estimated before hand, and for convenience we scale them as follows:

$$\pi_f = \frac{p_f - p_{f0}}{\rho_f U_r^2}, \qquad (2.6.8a)$$

$$\theta_f = \frac{T_f - T_{f0}}{T_r} \qquad (2.6.8b)$$

where p_{f0} and T_{f0} are the (uniform) mean pressure and temperature, and T_r is the relative temperature difference. In terms of these variables, we have

$$\nabla \cdot \mathbf{V}_f = 0 \qquad (2.6.9)$$

$$A_n \frac{\partial \mathbf{V}_f}{\partial \tau} + \mathbf{V}_f \cdot \nabla \mathbf{V}_f - \mathbf{V}_p \cdot \nabla \mathbf{V}_f = \frac{1}{\mathrm{Re}} \nabla^2 \mathbf{V}_f - \nabla \pi_f \qquad (2.6.10)$$

$$A_n \frac{\partial \theta_f}{\partial t} + \mathbf{V}_f \cdot \nabla \theta_f - \mathbf{V}_p \cdot \nabla \theta_f = \frac{1}{\mathrm{Re}} \left[2\mathrm{Ec}\Phi' + \frac{1}{\mathrm{Pr}} \nabla^2 \theta_f \right] \qquad (2.6.11)$$

where Φ' is the viscous dissipation function, expressed in terms of nondimensional variables.

Thus, the scaling of the equations of motion results in the following nondimensional parameters:

Reynolds number, Re

$$Re = \frac{\rho_f U_r D}{\mu_f}$$
                                                                                        (2.6.12)

Acceleration number, An

$$A_n = \frac{D \dot{U}_r}{U_r^2}$$
                                                                                        (2.6.13)

Eckert number, Ec

$$Ec = \frac{U_r^2}{c_{pf} T_r}$$
                                                                                        (2.6.14)

Prandtl number, Pr

$$Pr = \frac{\mu_f c_{pf}}{k_f}$$
                                                                                        (2.6.15)

The physical meaning of these parameters is evident from their definition. Thus, Re is a measure of the inertia force, due to the convective acceleration, felt on a material element of fluid as it moves from one point to another, relative to the viscous forces acting on it. Similarly, the acceleration number is a measure of the effects of the local acceleration, relative to the convective acceleration. The Eckert number measures the temperature increase due to dynamic effects relative to thermal effects, and the Prandtl number is the ratio of the viscous to thermal diffusivities for the fluid.

In the case of shape deformations, other parameters appear that provide a measure of the deformation. One such parameter is the *Weber number*, We, defined as the ratio of the dynamic pressure, attempting to deform a particle, to the surface tension forces, attempting to resist it. When the dynamic pressure is estimated from the inviscid-flow equations, the Weber number becomes

$$We = \frac{\rho_f D U_r^2}{\sigma}$$
                                                                                        (2.6.16)

When viscosity is taken into account, other estimates of the dynamic pressure apply, as we will see later.

The previous list does not include motions within the particle. This is because those motions do not significantly affect the fluid. Of course, the motions within the particles can be associated with a similar set of nondimensional parameters, and some will be introduced later. More important to the response of the particles are their physical properties relative to the fluid. Thus, we may expect that some nondimensional property ratios will appear when we discuss that response in the following chapters.

2.7 Motion at Re → 0

The small size of the particles in many suspensions usually results in typical Reynolds numbers, which are very small. This smallness is mathematically advantageous because the equations of motion can then be linearized without incurring in substantial error, at least in the limit Re → 0. The equations that result when the equations of motion for an incompressible fluid are approximated for Re → 0 are known as the *Stokes equations*. To obtain them, it is useful to consider the nondimensional set derived earlier. The first is the continuity equation and is not affected by the Reynolds number. Consider now the momentum equation, written as

$$A_n \frac{\partial \mathbf{V}_f}{\partial \tau} + \nabla \pi_f = \frac{1}{\mathrm{Re}} \nabla^2 \mathbf{V}_f - \mathbf{V}_f \cdot \nabla \mathbf{V}_f + \mathbf{V}_p \cdot \nabla \mathbf{V}_f \qquad (2.7.1)$$

Because of our choice of length and velocity scales, the spatial derivatives appearing on the right-hand side are of the order of one at distances that are not far from the particle. Thus, when Re → 0, the magnitude of the first term on the right-hand side is much larger than the other terms there, so that, at such distances, we may neglect them. Nothing can be said, however, about the magnitude of the terms in the left-hand side. Thus, when Re → 0, the momentum equation becomes, approximately,

$$A_n \frac{\partial \mathbf{V}_f}{\partial \tau} + \nabla \pi_f = \frac{1}{\mathrm{Re}} \nabla^2 \mathbf{V}_f \qquad (2.7.2)$$

Consider now the nondimensional energy equation, (2.6.11). The first term on its left-hand side, $A_n(\partial \theta_f / \partial t')$, is of unknown magnitude. The second, $(\mathbf{V}_f - \mathbf{V}_p) \cdot \nabla \theta_f$ is much smaller than the sum on the right-hand side and can thus be neglected. Further, first term inside the square brackets on the right-hand side is of the order the Eckert number, which for the typical relative velocities encountered in suspensions is very small. Therefore, when the Reynolds number is very small, we obtain, approximately,

$$A_n \frac{\partial \theta_f}{\partial t'} = \frac{1}{\mathrm{Pr}\,\mathrm{Re}} \nabla^2 \theta_f \qquad (2.7.3)$$

In the original, dimensional variables, the reduced set of equations becomes

$$\nabla \cdot \mathbf{u}_f = 0 \qquad (2.7.4)$$

$$\mu_f \nabla^2 \mathbf{u}_f - \rho_f \frac{\partial \mathbf{u}_f}{\partial t} = \nabla p_f \qquad (2.7.5)$$

$$\rho_f c_{pf} \frac{\partial T_f}{\partial t} = k_f \nabla^2 T_f \qquad (2.7.6)$$

We will refer to this set of equations as the Stokes equations, although the name is more commonly used to denote only the first two. Keeping in mind that the equations apply only to incompressible fluids, we note that the pressure may be eliminated from

(2.7.5). Therefore, if the boundary conditions on the velocity are independent of the temperature, the velocity and temperature fields are decoupled. It is then possible to study them separately. This will be done in the next few chapters. While the translational velocity problem may be regarded as being the more important of the two because it determines the force on a particle, we start our discussion by considering the temperature problem because it is also important and is the simplest of the two.

3

Rigid-Particle Heat Transfer at Re ≪ 1

3.1 Introduction

Each one of the basic particle motions listed in Section 2.3 can be accompanied by temperature changes taking place in and out the particle. Particularly important are the translational and pulsational motions. In these, thermal effects can be dominant in some suspensions. The pulsational motion is considered in Chapter 6. Here, we consider rigid particles immersed in slowly moving fluids. As shown in Section 2.4, the thermal response of a very small, rigid particle can be assessed in terms of its average temperature, $\overline{T}_p$, which satisfies

$$m_p c_{pp} \frac{d\overline{T}_p}{dt} = \dot{Q}_p, \tag{3.1.1}$$

where $\dot{Q}_p$ is the heat transfer rate to the particle. This quantity is given by (2.4.3). Therefore, a knowledge of T_f is sufficient to determine the particle temperature. However, the fluid temperature is determined by (2.1.7),

$$\rho_f c_{pf} \frac{DT_f}{Dt} - \beta_f T_f \frac{Dp_f}{Dt} = \rho_f \Phi_f + k_f \nabla^2 T_f$$

that is, by an equation that includes the pressure and velocity fields. Thus, the determination of the temperature of a rigid particle in a viscous, compressible fluid requires those fields. But, for slowly moving fluids, the squares of the velocity gradients may be neglected, a fact that amounts to neglecting viscous heating. Thus, $\Phi_f = 0$. Similarly, we may also neglect the convective terms on the left-hand side of (2.1.7). The next simplification involves the pressure field in the fluid. This is important for compressible fluids; but, for slow motions, the fluid may be taken as incompressible, in which case $\beta_f T_f \to 0$. Hence, we find that the external fluid temperature is described by Fourier's heat conduction equation

$$\frac{\partial T_f}{\partial t} = \kappa_f \nabla^2 T_f \tag{3.1.2}$$

where $\kappa_f = k_f/\rho_f c_{pf}$ is the thermal diffusivity of the fluid. Thus, when the amplitudes and gradients of the fluid velocity are small, its temperature is independent of the velocity field that might exist in it. While seemingly very restrictive, these conditions apply in several situation of interest to suspensions.

In the following sections, we use (3.1.1) and (3.1.2) to study the thermal problem for rigid particles so small that their temperature is either uniform or is well represented by an average value. The nonuniform temperature distribution will be studied in Section 3.5. Because the equations that describe the temperature variations are linear, it is advantageous to first consider situations where the time dependence is harmonic. These are important in a variety of contexts (e.g., in a sound wave). Furthermore, the temperature and heat transfer results obtained for them can be regarded as the Fourier transform of such quantities when the time dependence is more general.

3.2 Harmonic Temperature Variations

We consider first a small, rigid *sphere* of radius a immersed in an incompressible fluid of infinite extent which, far from the sphere, is at a fixed temperature T_0. The particle's temperature, T_p, is either uniform or is well represented by its volume average. For simplicity, we will denote either temperature by T_p. We will also assume that it oscillates harmonically in time about T_0. That is, $T_p = T_0 + \theta_{p0} \cos \omega t$, where ω is the circular frequency, and θ_{p0} is the amplitude of the oscillations – assumed to be very small, compared with T_0. It is convenient to write the time factor as the real part of $\exp(-i\omega t)$, where $i = \sqrt{-1}$, so that $T_p = T_0 + \Re\{\theta_{p0}e^{-i\omega t}\}$, where the notation $\Re\{\tilde{f}\}$ means the real part of the complex quantity $\tilde{f}$. Alternatively, we may simply write

$$\tilde{T}_p = T_0 + \theta_{p0}e^{-i\omega t} \tag{3.2.1}$$

on the understanding that only the real part of $\tilde{T}_p$ is significant.

An advantage of writing the temperature (and other quantities) in complex form is that certain mathematical operations become very simple to carry out when they are so expressed. For example, the time derivative of a quantity that depends harmonically in time is simply equal to that quantity multiplied by $-i\omega$. However, it should be remembered that the complex representation of real quantities in mathematical operations is appropriate only when those operations are linear. Thus, for example, if $\tilde{F}$ is used to represent the real quantity F, it is obvious that $F = \Re\{\tilde{F}\}$, and that $F^2 \neq \Re\{\tilde{F}^2\}$. Instead, of course, $F^2 = \tilde{F}\tilde{F}^*$, where the asterisk is used to represent the complex conjugate of $\tilde{F}$.

Let us now return to the present problem. It is clear that owing to the heat transfer from the particle, the fluid temperature will change. However, because the sphere's temperature fluctuates harmonically in time, we expect that long after the motion has started, the fluid temperature will also fluctuate harmonically in time (at the same frequency as the sphere), but with an amplitude and phase that will generally be different from it. These differences are prescribed by (3.1.2). To solve this we write

$$\tilde{T}_f = T_0 + \tilde{\theta}_f(\mathbf{x})e^{-i\omega t} \tag{3.2.2}$$

where $\mathbf{x}$ is the position vector of a point in the fluid, relative to a set of axis fixed on the center of the sphere. The quantity $\tilde{\theta}_f(\mathbf{x})$ describes the amplitude and phase of the fluid's temperature oscillations relative to those of the sphere. Substitution into (3.1.2) gives the Helmholtz equation

$$\nabla^2\tilde{\theta}_f + K^2\tilde{\theta}_f = 0 \tag{3.2.3}$$

where

$$K^2 = i\omega/\kappa_f \tag{3.2.4}$$

Since $\sqrt{i} = \pm(1+i)/\sqrt{2}$, K has two possible values, which may be expressed as

$$K_{1,2} = \pm(1+i)/\delta_{\kappa f} \tag{3.2.5}$$

where

$$\delta_{\kappa f} = \sqrt{2\kappa_f/\omega} \tag{3.2.6}$$

This quantity has the dimensions of length and represents, as shown later, the distance over which the amplitude of the fluid temperature decreases by a significant amount from its value at the surface of the sphere. We will return to this point after the solution to (3.2.3) has been obtained. It should be noted, however, that $\delta_{\kappa f}$ depends on the frequency, becoming very large or very small as ω approaches 0 or ∞, respectively.

Now $\nabla^2\tilde{\theta}_f$ generally has three components (see Appendix C), but because the present problem has spherical symmetry around the origin, the spatial factor $\tilde{\theta}_f(\mathbf{x})$ can only depend on the radial distance, r, from the origin. Thus, (3.2.3) reduces to

$$\frac{1}{r^2}\frac{d}{dr}\left(r^2\frac{d\tilde{\theta}_f}{dr}\right) + K^2\tilde{\theta}_f = 0 \tag{3.2.7}$$

This may also be written as

$$\frac{d^2(r\tilde{\theta}_f)}{dr^2} + K^2(r\tilde{\theta}_f) = 0 \tag{3.2.8}$$

This shows that the quantity $r\tilde{\theta}_f$ satisfies the differential equation for a harmonic oscillator. Therefore, the complex temperature fluctuation in the fluid, $\tilde{T}_f' = \tilde{T}_f - T_0$ may be expressed as

$$\tilde{T}_f' = \frac{A}{r}e^{i(K_1 r - \omega t)} + \frac{B}{r}e^{i(K_2 r - \omega t)} \tag{3.2.9}$$

where A and B are constants and K_1 and K_2 are prescribed by (3.2.5). Now, because $K_2 = -(1+i)/\delta_{\kappa f}$, the second term in the previous equation becomes unbounded as $r \to \infty$. However, the boundary condition there requires that $\tilde{\theta}_f$ vanishes. Therefore, we must set B equal to zero. The value of A is obtained from the condition that the temperature of the fluid and particle be equal at $r = a$. Thus,

$$A = a\theta_{p0}e^{-iK_e a} \tag{3.2.10}$$

where we have denoted K_1 by K_e, to emphasize it refers to the external medium. Hence, the temperature field in the fluid is given by

$$\frac{\tilde{T}'_f}{\theta_{p0}} = \frac{a}{r}e^{-(r-a)/\delta_{\kappa f}+i[(r-a)/\delta_{\kappa f}-\omega t]} \tag{3.2.11}$$

The real part of this is

$$\frac{T'_f}{\theta_{p0}} = \frac{a}{r}e^{-(r-a)/\delta_{\kappa f}}\cos[\omega t - (r-a)/\delta_{\kappa f}] \tag{3.2.12}$$

This result shows several interesting features. One is that the phase of the fluctuations in the fluid, given by the argument of the cosine function, depends on both position and time, but is equal at all locations r that satisfy $(r - a)/\delta_{\kappa f} = \omega t$. Thus, the phase of the fluctuation is constant for an observer moving with a velocity, called the *phase velocity*, equal to

$$c_{\kappa f} = \sqrt{2\omega\kappa_f} \tag{3.2.13}$$

Thus, (3.2.12) represents waves emanating form the sphere and carrying thermal energy into the fluid with a phase velocity equal to $c_{\kappa f}$. We will refer to these waves as *thermal waves*. One of the actions of these waves is to eliminate temperature differences that might exist between fluid and particles. However, like all diffusive phenomena in fluids, the phase velocity of the thermal waves is small, compared with the fluid's speed of sound. For example, at 20°C and 1 atm and at a frequency equal to 10^2 Hz, the thermal wave speed is about 16 cm/sec in air and 1.3 cm/sec in water.

We also observe in (3.2.12) that the amplitude of the temperature fluctuations decreases with distance from the sphere. The decrease is due to two different effects. One is the slow decay given by the geometric $1/r$ factor resulting from the spread of the wavefront. The second, and generally more rapid decrease of the amplitude, is given by the exponential factor in (3.2.12). This shows that, at a distance $\delta_{\kappa f}$ from the sphere, the amplitude is less than 40% of its value at the surface of the sphere. Thus, the length scale $\delta_{\kappa f}$ roughly indicates how deep into the fluid do the temperature fluctuations penetrate. Because of this, $\delta_{\kappa f}$ is referred to as the *thermal penetration depth*. In the numerical example given previously, $\delta_{\kappa f}$ is about 0.025 cm in air and 0.002 cm in water. Thus, for frequencies that are considerably larger than the 10^2 Hz used in the example, the fluid around the sphere remains essentially unaffected by the temperature fluctuations of the particle. It is therefore tempting to neglect the temperature variations in the fluid completely. This is not generally possible because significant effects may take place in the thin layer of fluid around the sphere that can significantly affect the oscillations.

Finally, we note from (3.2.6) that, when the frequency is small, the penetration depth is large. In that limit, (3.2.12) shows that the temperature fluctuation in the fluid is essentially independent of the size of the particle. In that limit, the particle has become a point source of heat.

Heat Transfer Rate

We now consider the rate at which heat is transferred to and from the particle in the prior example. Because of the central symmetry, the heat transfer rate to the particle is given by

$$\dot{Q}_p = k_f \int_{A_p} (\partial T_f/\partial r) dA$$

Since the temperature gradient is uniform on the sphere's surface, it can be taken outside the integral, giving $\dot{Q}_p = 4\pi a^2 k_f (\partial T_f/\partial r)_{r=a}$. Making use of (3.2.11), this gives

$$\tilde{\dot{Q}}_p = -4\pi a k_f (1 - i K_e a)\theta_{p0} e^{-i\omega t} \tag{3.2.14}$$

Thus, the heat transfer rate fluctuates harmonically in time with the same frequency as the sphere, but with a different phase. Now, since $K_e = (1+i)/\delta_{\kappa f}$, $\dot{Q}_p$ may be expressed as

$$\tilde{\dot{Q}}_p = -4\pi a k_f (1 + z_f - i z_f)\theta_{p0} e^{-i\omega t} \tag{3.2.15}$$

where

$$z_f = \sqrt{\omega a^2/2\kappa_f} \tag{3.2.16}$$

is the ratio of particle radius to the thermal penetration depth. We may express (3.2.15) in real form using the fact that $d/dt = -i\omega$. Hence,

$$\dot{Q}_p = -4\pi a k_f (1 + z_f)(T_p - T_0) - \frac{3}{2} m_p \delta \frac{\rho_f c_{pf}}{z_f} \frac{dT_p}{dt} \tag{3.2.17}$$

where $\delta = \rho_f/\rho_p$. We thus see that $\dot{Q}_p$ depends on both the instantaneous temperature difference and the rate of change of that difference. This has important consequences when the temperature variations have an arbitrary time dependence, as it is shown in Section 3.4.

For the monochromatic case, (3.2.12) shows that different values of z_f produce different temperature fluctuations in the fluid. In particular, when $z_f \ll 1$, the size of the particle does not affect the fluid's temperature. In that limit, the heat transfer rate is given by $\dot{Q}_p = -4\pi a k_f (T_p - T_0)$. This simple formula can be interpreted as the low-frequency heat transfer rate between a particle, at instantaneous temperature T_p, and a fluid, whose temperature far from the particle is T_0. Alternatively, we could view T_0 as the temperature that the fluid would have at the location of the particle if the particle were not there. This view is superfluous in the present situation, where all temperature fluctuations are due to the particle, but will be useful in more general situations.

Low Frequencies

The simple result obtained previously by letting z_f become very small is the basis of an important result, applicable to many situations of interest. The result will be formally derived later. Thus, if *both* fluid and particle temperatures are slowly varying functions of time, not necessarily harmonic, we may anticipate that, at least in some limited range, the heat transfer rate to a sphere will be given by

$$\dot{Q}_p(t) = -4\pi a k_f (T_p - T_f) \qquad (3.2.18)$$

As shown later, this is the heat transfer rate, by conduction alone, that exists between a rigid sphere of radius a and an incompressible fluid whose temperature far from the sphere (or at the location of the sphere, in the absence of the sphere), $T_f(t)$, varies slowly in time. We shall refer to this result as the *low-frequency heat transfer*, though it is also known as the heat transfer rate due to conduction alone. It, and its counterpart for the fluid drag on a small rigid sphere, Stokes' law, are important cornerstones in the study of suspensions.

In some cases, it is necessary to include convective effects in the heat transfer rate. This is done by writing the right-hand side of (3.2.18) in terms of a heat transfer coefficient, H, as $-2\pi a H k_f (T_p - T_f)$. The coefficient is, in turn, often expressed in terms of the Nusselt number, Nu, which is equal to $2Ha/k_f$, in which case (3.2.18) becomes

$$\dot{Q}_p(t) = -2\pi a k_f \mathrm{Nu}\, (T_p - T_f)$$

In slowly moving suspensions, however, convective effects are usually small. Far more important are the effects of unsteadiness.

Let us now return to (3.2.18). Although that result is clearly a low-frequency approximation to $\dot{Q}_p$, its simplicity makes it convenient to determine the temperature of a particle in unsteady situations. This can be done by integration of (3.1.1), which we now write as

$$\frac{dT_p}{dt} = -\frac{T_p - T_f}{\tau_t} \qquad (3.2.19)$$

where

$$\tau_t = m_p c_{pp} / 4\pi k_f a \qquad (3.2.20)$$

This quantity has the dimensions of time and provides a time scale that measures the rapidity with which temperature changes occur. Since it depends on the properties of both fluids and particles, we see that the time required for those changes to take place can vary widely. This is shown below. We also note that τ_t may also be expressed as

$$\tau_t = ha^2/2\kappa_f \qquad (3.2.21)$$

where

$$h = \frac{2}{3}\frac{\rho_p c_{pp}}{\rho_f c_{pf}} \qquad (3.2.22)$$

In view of (3.1.1), we may regard $\rho_p c_{pp}$ as the thermal inertia of the particle. The quantity h is, therefore essentially equal to the ratio of thermal inertias.

Thermal Relaxation Time

Equation (3.2.19) is an example of a class of equations, called relaxation equations, that determine how rapidly time-dependent physical quantities (the particle temperature in this case) adjust to imposed changes. Among the relaxation equations, (3.2.19) is the simplest because it contains only one internal time scale, that defined by τ_t.

To understand the role of τ_t, we consider a simple example using the general solution of (3.2.19), which can be written as

$$T_p(t) = T_{p0}e^{-(t-t_0)/\tau_t} + \frac{1}{\tau_t}\int_{t_0}^{t} T_f(t')e^{(t'-t)/\tau_t}dt' \qquad (3.2.23)$$

where T_{p0} is the value of T_p at time $t = t_0$. Suppose that, for $t \leq t_0$, the particle and the fluid are at the same temperature T_0, and that at time t_0 the temperature of the fluid is raised by a constant amount, say Θ_0. Because of the particle's thermal inertia, we may anticipate that its temperature cannot adjust to the newly imposed temperature instantaneously. In fact, (3.2.23) shows that, for $t \geq t_0$, the temperature difference between fluid and particle is given by

$$\frac{T_f - T_p}{\Theta_0} = e^{-(t-t_0)/\tau_t} \qquad (3.2.24)$$

Thus, the suddenly imposed temperature difference will decrease exponentially with time. A measure of the rapidity of the decrease can be obtained by noting that in an elapsed time equal to τ_t, the temperature difference decreases from Θ_0 at $t = 0$, to a value equal to Θ_0/e. That is, temperature equalization occurs in a time that is of the same order as τ_t. The quantity τ_t is called the *thermal relaxation time* of the particle. For a given particle-fluid combination, it varies as the square of the radius and therefore increases rapidly with size, as shown in Table 3.2.1 for silica particles in both air and water.

In the example given previously, the only time scale present was τ_t. In more general situations, there may be other time scales (e.g., when a sound wave is traveling in a fluid). Here, the fluid temperature varies harmonically in time, so that the period of the waves also enters in the formulation. Whether a particle can then follow the imposed

Table 3.2.1. *Thermal relaxation time τ_t in seconds for silica particles*

Radius (μm)	In air	In water
1	2.04×10^{-5}	1.04×10^{-6}
5	5.1×10^{-4}	2.59×10^{-5}
10	2.04×10^{-3}	1.03×10^{-4}
50	5.1×10^{-2}	2.59×10^{-3}

changes depends on the ratio of its relaxation time to the time scale for the imposed temperature changes. When this ratio is very small, the particles can adjust quickly to new situations; when it is large, it cannot adjust to them at all. In the following section, we consider a simple, but important example that displays these features in some detail.

PROBLEM

Determine the temperature of a rigid sphere of radius a as a function of time when the temperature of the fluid is changed from T_0 to $T_f = T_0 + \theta_{f0} \cos(\omega t)$ at $t = 0$, if its initial temperature is also T_0.

3.3 Particle in Oscillating Temperature Field

As an application of the low-frequency heat transfer, given by (3.2.18), we consider a small sphere in a fluid whose temperature, in the absence of the sphere, oscillates harmonically in time with an amplitude θ_{f0}. As a result of the periodic heat transfer, the particle's temperature will also oscillate harmonically in time. We wish to determine the amplitude and phase of those oscillations. To do this, we may use (3.2.23), assuming that the temperature oscillations in the fluid start at some time t. This would produce in the sphere a transient temperature change that eventually dies out, leaving only a steady-state oscillation. The transient oscillation is considered in the problem above. To study the steady-state oscillation, we write the corresponding temperature fluctuations as

$$\tilde{\Theta}'_f = \theta_{f0} e^{-i\omega t} \quad \text{and} \quad \tilde{T}'_p = \tilde{\theta}_{p0} e^{-i\omega t} \tag{3.3.1}$$

where θ_{f0} is real, but where the unknown particle temperature oscillation amplitude, $\tilde{\theta}_{p0}$, is generally complex. Substituting these into (3.2.19) and solving for $\tilde{T}'_p$, we obtain the simple result

$$\mathrm{T} = (1 - i\omega\tau_t)^{-1} \tag{3.3.2}$$

where $\mathrm{T} = \tilde{T}'_p / \tilde{\Theta}'_f$ is time independent, but generally complex. The real part of this temperature ratio may be written as

$$\Re\{\mathrm{T}\} = \frac{1}{\sqrt{1 + (\omega\tau_t)^2}} \cos \eta \tag{3.3.3}$$

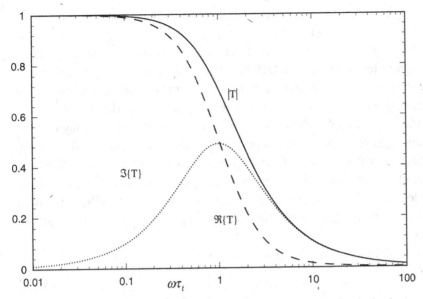

Figure 3.3.1. Temperature ratio $\tilde{T}'_p/\tilde{\Theta}'_f$ for a 100-μm diameter water droplet in air.

where $\eta = \tan^{-1}(\omega\tau_t)$. The fraction on the right-hand side expresses the amplitude of the temperature fluctuations as a function of the nondimensional product $\omega\tau_t$, as shown in Fig. 3.3.1 for a water droplet in air. This figure also shows the real and imaginary parts of the temperature ratio as a function of the frequency, or, for a fixed frequency, that ratio as a function of size. To fix ideas, we show in Fig. 3.3.2 the amplitude

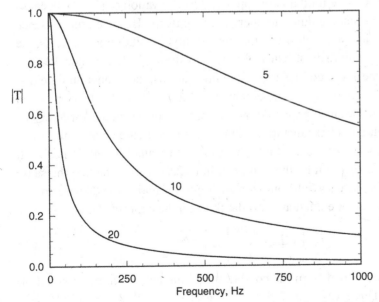

Figure 3.3.2. Amplitude of the temperature ratio for water droplets in air. Curve labels refer to the diameters of the droplets in microns.

as a function of the frequency for several particle sizes. It is seen that the size of the particle influences the amplitude in a strong manner. This is because the thermal relaxation time varies as the square of the particle radius. However, regardless of size, the particle temperature approaches that of the fluid as $\omega \to 0$. That is, at very low frequencies, where the particle time scale is much smaller than the period of the oscillation, the particle's temperature amplitude is very nearly equal to that of the fluid. Furthermore, the phase angle is also very small. We can then say that the particle and the fluid have the same temperature at all instants, so that they are in thermal equilibrium with one another. As the frequency increases, however, the ratio of the particle relaxation time and the period of the imposed oscillations increase, so that the particle is less able to follow the fluctuations in the fluid. Finally, in the limit of high frequencies, the fluctuation vanishes so that the particle's temperature remains fixed or *frozen*.

Now, from the definition of τ_t and z_f, we may write

$$\omega\tau_t = hz_f^2 \tag{3.3.4}$$

where h is given by (3.2.22). This shows that a large value of $\omega\tau_t$ generally corresponds to a large value of z_f. But such values are generally incompatible with (3.2.18), as this requires z_f to be small. The physical reason for the limitation is that (3.2.18) leaves out thermal inertia effects in the fluid. This follows from the derivation, but may also be seen by writing (3.3.4) as $z_f^2 = \omega\tau_{t,f}$, where $\tau_{t,f} = \tau_t/h$ may be regarded as the relaxation time for a fluid element having the same size as the rigid particle.

Unsteady Heat Transfer Rate

Thermal inertia effects in the fluid do not appear in the derivation of (3.3.2), because that derivation assumes that the fluid's temperature is uniform. However, the presence of a particle in the fluid usually destroys that uniformity. To study these effects, we must consider the temperature disturbance produced by the sphere's presence. We first note that because of the assumed linearity of the fluctuations, the temperature field at any point in the fluid is given by the superposition of the primary fluctuation, given by (3.2.11), and the disturbance produced by the sphere. For harmonic variations of the external temperature field, the disturbance field is similar to the temperature wave produced in a fluid by a particle whose temperature is fluctuating harmonically in time. It thus has the form given by the first term in (3.2.9), with ω having the same frequency as that in the primary field, but with an amplitude and phase that are not yet known. Thus, the temperature fluctuation in the fluid can be expressed as

$$\tilde{T}_f' = \left[1 + A\frac{e^{iK_e r}}{r}\right]\theta_{f0}e^{-i\omega t} \tag{3.3.5}$$

The constant A is determined from the condition that the temperatures be equal on the sphere's surface. This gives

$$A = (\tilde{\theta}_{p0}/\theta_{f0} - 1)\, ae^{-iK_e a} \tag{3.3.6}$$

so that

$$\tilde{T}'_f = \theta_{f0}e^{-i\omega t} + (\tilde{\theta}_{p0} - \theta_{f0})\frac{a}{r}e^{i[K_1(r-a)-\omega t]} \tag{3.3.7}$$

To determine the complex amplitude $\tilde{\theta}_{p0}$, we compute the heat transfer rate using (3.2.14) and obtain

$$\tilde{Q}_p = -4\pi a k_f(1 + z_f - iz_f)(\tilde{\theta}_{p0} - \theta_{f0})e^{-i\omega t} \tag{3.3.8}$$

But from the energy equation for the particle, (3.1.1), this rate is equal to $-i\omega m_p c_{pp}\tilde{\theta}_{p0}\exp(-i\omega t)$ so that equating the two quantities and solving for $\tilde{\theta}_{p0}$ yields, for $T = \tilde{T}'_p/\Theta'_f$,

$$T = \frac{1 + z_f - iz_f}{1 + z_f - iz_f(1 + hz_f)} \tag{3.3.9}$$

where we have used (3.3.4). Comparison with (3.3.2) shows that fluid inertia affects the temperature oscillations of the particle at all frequencies through the quantity z_f. To show this more clearly, we consider the amplitude and phase of $\tilde{\theta}_{p0} = \theta_{p0}\exp(i\eta)$. These are

$$|T| = \left[\frac{1 + 2z_f + 2z_f^2}{(1 + z_f)^2 + z_f^2(1 + hz_f)^2}\right]^{1/2} \tag{3.3.10}$$

$$\tan\eta = h(1 + z_f)z_f^2/[(1 + z_f)^2 + z_f^2(1 + hz_f)] \tag{3.3.11}$$

The temperature difference is also of interest and is

$$\frac{\tilde{T}_p - \tilde{T}_f}{\Theta'_f} = \frac{ihz_f^2}{1 + z_f - iz_f(1 + hz_f)}\Theta'_f \tag{3.3.12}$$

Figure 3.3.3 shows the amplitude and phase of T for a 100 μm diameter alumina particle in air as a function of z_f, and Fig. 3.3.4 shows $|T|$ for the same particles in water. Shown in the figures is the simple result predicted by the low-frequency heat transfer, (3.2.18), and that predicted by (3.3.9). As it is seen, both results agree well in the case of air, and this is due to the large value h has for this case. But, in the case of hydrosols (Fig. 3.3.4), significant differences occur at most frequencies.

Let us return to the heat transfer rate. This is given by (3.3.8), which we now write as

$$\tilde{Q}_p = -4\pi a k_f(1 + z_f - iz_f)(\tilde{T}_p - \tilde{T}_f) \tag{3.3.13}$$

where $\tilde{T}_p$ and $\tilde{T}_f$ are the absolute temperatures, written in complex form. To write this in real form, we first note that, for harmonic variations, $-i\omega T = dT/dt$, so that

$$\tilde{Q}_p = -4\pi a k_f(\tilde{T}_p - \tilde{T}_f) - 4\pi a k_f(1 + i)\frac{z_f}{\omega}\left(\frac{d\tilde{T}_p}{dt} - \frac{d\tilde{T}_f}{dt}\right) \tag{3.3.14}$$

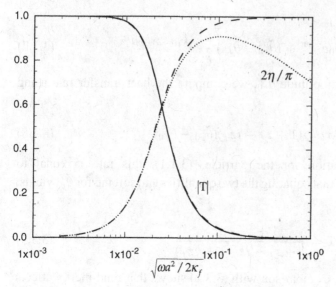

Figure 3.3.3. Amplitude and phase angle of $\tilde{T}'_p/\Theta'_f$ as a function of z_f for a 100-μm diameter alumina particle in air; dotted line (3.3.9); dashed line (3.3.2).

In real form, this can be written as

$$\dot{Q}_p = -4\pi a k_f (1 + z_f)(T_p - T_f) - 4\pi a k_f \frac{z_f}{\omega}\left(\frac{dT_p}{dt} - \frac{dT_f}{dt}\right) \quad (3.3.15)$$

This differs from low-frequency heat transfer in two ways. The first is the appearance of z_f in the factor multiplying the temperature difference; the second is the term proportional to the time derivative of that difference, which is absent in the

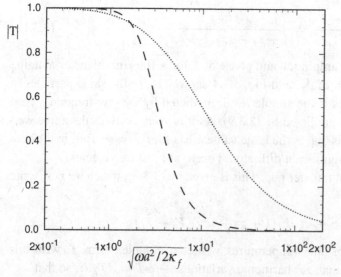

Figure 3.3.4. Amplitude of $\tilde{T}'_p/\Theta'_f$ as a function of z_f for a 100-μm diameter alumina particle in water; dotted line (3.3.9); dashed line (3.3.2).

low-frequency heat transfer. Since the time derivatives are proportional to the frequency, we see that when $z_f \to 0$, (3.2.18) is recovered. But, for finite values of z_f, the results are significantly different.

These inertia effects also change the temperature of the particle, as may be seen by writing (3.1.1) as

$$\frac{dT_p}{dt} = -\frac{T_p - T_f}{\tau_t/(1 + z_f)} - \frac{dT_p/dt - dT_f/dt}{hz_f} \tag{3.3.16}$$

The equivalent result for small frequencies is (3.2.19). Comparison between the two results shows that the simple relaxation character of the fluctuation no longer exists. One important consequence of this is that, when the temperature fluctuation in the fluid is not monochromatic, the temperature of the particle at any given instant depends on all past variations of the relative temperature rate of change. This is considered in the next section.

PROBLEM

The results given previously show that, when the fluid temperature varies harmonically in time, the particle's energy equation can be written as

$$m_p c_{pp}(1 + 1/hz_f)\frac{dT_p}{dt} = -4\pi a k_f(1 + z_f)(T_p - T_f) + (m_p c_{pp}/hz_f)\frac{dT_f}{dt}$$

indicating that the thermal inertia of the fluid increases that of the particle. Use this result to discuss the low- and high-frequency behavior of the resulting thermal inertia of the particle.

3.4 Arbitrary Time Dependence

Except for the brief discussion of transient motions in Section 3.2, all of the temperature variations studied so far have had a harmonic time dependence. We now consider rigid spheres whose temperature can change with time in a less restrictive manner. We will, however, continue to limit the discussion to particles whose temperature is either uniform or can be represented by an average value. For these, the temperature satisfies

$$m_p c_{pp}\frac{dT_p}{dt} = \dot{Q}_p \tag{3.4.1}$$

Now, as indicated in Section 3.1, the results obtained for the harmonic case, may be regarded as the Fourier transforms of the corresponding results for an arbitrary time dependence. Thus, the heat transfer rate, $\dot{Q}_p(t)$, applicable to those cases, can be obtained by inverting $\dot{Q}_p(\omega)$ by means of

$$\dot{Q}_p(t) = \int_{-\infty}^{\infty} \dot{Q}_{p\omega}e^{-i\omega t}d\omega \tag{3.4.2}$$

where we have denoted $\dot{Q}_p(\omega)$ by $\dot{Q}_{p\omega}$. As the integral implies, we need $\dot{Q}_{p\omega}$ for both positive and negative values of ω. In Section 3.3, we obtained $\dot{Q}_{p\omega}$ for the positive frequency range, and the result is given, in real form, by (3.3.12). For the present derivation, it is more convenient to use the complex form (3.3.14), which we now write as

$$
\dot{Q}_{p\omega} = -4\pi a k_f (T_{p\omega} - T_{f\omega})
$$
$$
- 4\pi a^2 \sqrt{\tfrac{1}{2}\rho_f c_{pf} k_f} \frac{1+i}{\sqrt{\omega}} \left[\left(\frac{dT_p}{dt}\right)_\omega - \left(\frac{dT_f}{dt}\right)_\omega \right] \quad \omega > 0
$$

$$(3.4.3)$$

where $T_{p\omega}$ and $T_{f\omega}$ are the Fourier transforms of T_p and T_f, respectively, and for simplicity have not used the tilde to denote complex quantities. The corresponding result for negative frequencies may be obtained following the same procedures as those used in deriving (3.3.14). That is, for negative frequencies, we would consider a time dependence equal to $\exp(i|\omega|t)$ instead of $\exp(-i\omega t)$. This would require us to select the root K_2 of K in (3.2.5), instead of K_1 as before, in order to satisfy the boundary condition at infinity. Hence, instead of the fraction $(1+i)/\sqrt{\omega}$ in the heat transfer rate, we would have $(1-i)/\sqrt{|\omega|}$. Thus, when $\omega < 0$, we would get

$$
\dot{Q}_{p\omega} = -4\pi a k_f (T_{p\omega} - T_{f\omega})
$$
$$
- 4\pi a^2 \sqrt{\tfrac{1}{2}\rho_f c_{pf} k_f} \frac{1-i}{\sqrt{|\omega|}} \left[\left(\frac{dT_p}{dt}\right)_{|\omega|} - \left(\frac{dT_f}{dt}\right)_{|\omega|} \right] \quad \omega < 0
$$

$$(3.4.4)$$

Having specified $\dot{Q}_{p\omega}$ for the entire frequency range, we now proceed to invert the result using (3.4.2). The first terms in (3.4.3) and (3.4.4) give upon inversion, $-4\pi a k_f (T_p - T_f)$. The remaining terms give

$$
-4\pi a^2 (1-i)\sqrt{\tfrac{1}{2}\rho_f c_{pf} k_f} \int_0^\infty \left[\left(\frac{dT_p}{dt}\right)_{|\omega|} - \left(\frac{dT_f}{dt}\right)_{|\omega|} \right] e^{i|\omega|t} \frac{d|\omega|}{\sqrt{|\omega|}}
$$
$$
-4\pi a^2 (1+i)\sqrt{\tfrac{1}{2}\rho_f c_{pf} k_f} \int_0^\infty \left[\left(\frac{dT_p}{dt}\right)_\omega - \left(\frac{dT_f}{dt}\right)_\omega \right] e^{-i\omega t} \frac{d\omega}{\sqrt{\omega}}
$$

These two terms are complex conjugate of one another. But the sum of two complex conjugate quantities is equal to twice the real part of either one. Hence, the sum above can be written as

$$
-4\pi a^2 \sqrt{\tfrac{1}{2}\rho_f c_{pf} k_f} \cdot 2\Re \left\{ (1+i) \int_0^\infty \left[\left(\frac{dT_p}{dt}\right)_\omega - \left(\frac{dT_f}{dt}\right)_\omega \right] e^{-i\omega t} \frac{d\omega}{\sqrt{\omega}} \right\}
$$

The integration involves two similar terms. We consider the first of them and write

$$I = 2\Re\left\{(1+i)\int_0^\infty \left(\frac{dT_p}{dt}\right)_\omega e^{-i\omega t}\frac{d\omega}{\sqrt{\omega}}\right\} \qquad (3.4.5)$$

To eliminate ω from the integrand, we express the derivative inside the integral as the Fourier transform of (dT_p/dt), that is

$$\left(\frac{dT_p}{dt}\right)_\omega = \frac{1}{2\pi}\int_{-\infty}^\infty \frac{dT_p}{dt'}e^{i\omega t'}dt' \qquad (3.4.6)$$

Hence

$$I = \frac{1}{\pi}\Re\left\{(1+i)\int_{-\infty}^\infty \int_0^\infty \frac{dT_p}{dt'}e^{i\omega(t'-t)}\frac{d\omega}{\sqrt{\omega}}dt'\right\} \qquad (3.4.7)$$

We now split the outside integral into two ranges: One from $-\infty$ to t, and the second from t to ∞. In the first, we put $\zeta = \omega(t-t')$ and in the second $\zeta = \omega(t'-t)$. These changes give

$$I = \frac{1}{\pi}\Re\left\{(1+i)\int_{-\infty}^t \frac{dT_p}{dt'}\frac{dt'}{\sqrt{t-t'}}\int_0^\infty e^{-i\zeta}\frac{d\zeta}{\sqrt{\zeta}}\right.$$
$$\left. +(1+i)\int_t^\infty \frac{dT_p}{dt'}\frac{dt'}{\sqrt{t'-t}}\int_0^\infty e^{i\zeta}\frac{d\zeta}{\sqrt{\zeta}}\right\} \qquad (3.4.8)$$

Consider now the integrals over ζ, multiplied by the $(1+i)$ factor. The first can be written as

$$(1+i)\int_0^\infty e^{-i\zeta}\frac{d\zeta}{\sqrt{\zeta}} = 2(1+i)\int_0^\infty e^{-iu^2}du$$
$$= \frac{4}{\sqrt{2}}\int_0^\infty e^{-q^2}dq = \sqrt{2\pi} \qquad (3.4.9)$$

Similarly, the second term in the sum above (3.4.7) gives $i\sqrt{2\pi}$. Therefore,

$$I = \sqrt{\frac{2}{\pi}}\Re\left\{\int_{-\infty}^t \frac{dT_p}{dt'}\frac{dt'}{\sqrt{t-t'}} + i\int_t^\infty \frac{dT_p}{dt'}\frac{dt'}{\sqrt{t'-t}}\right\} = \sqrt{\frac{2}{\pi}}\int_{-\infty}^t \frac{dT_p}{dt'}\frac{dt'}{\sqrt{t-t'}} \qquad (3.4.10)$$

A similar result is obtained with the second term inside the square bracket in (3.4.4). Collecting these separate contributions, we obtain

$$\dot{Q}_p(t) = -4\pi k_f a(T_p - T_f) - 4a^2\sqrt{\pi\rho_f c_{pf}\kappa_f}\int_{-\infty}^t \left(\frac{dT_p}{dt'} - \frac{dT_f}{dt'}\right)\frac{dt'}{\sqrt{t-t'}} \qquad (3.4.11)$$

This is the desired result. It describes the heat transfer rate, by conduction, to a spherical particle in a slowly moving, incompressible fluid whose temperature may be changing in time in an arbitrary manner.

Let us consider the right-hand side of (3.4.11). The first term is the low-frequency heat transfer and represents a heat transfer rate proportional to the instantaneous temperature difference. The second is clearly due to unsteady effects resulting from finite thermal inertia. The contribution of this term to the total heat transfer rate appears as an integral over all times prior to the current time t. That is, it depends on the complete history of the fluid and particle temperatures. The origin of this "history" term is the finite speed at which thermal waves travel. At time t, these convey to fluid elements the temperature changes that originated at the surface of the particle at earlier times.

Now, both terms on the right-hand side of (3.4.11) include the difference between the temperatures and their time derivatives. If these are specified, then the heat transfer rate can be determined by integration. But the particle's temperature is not generally known, and to obtain it we must obtain it from an integral equation,

$$m_p c_{pp} \frac{dT_p}{dt} = -4\pi k_f a (T_p - T_f) - 4a^2 \sqrt{\pi \rho_f c_{pf} \kappa_f} \int_{-\infty}^{t} \left(\frac{dT_p}{dt'} - \frac{dT_f}{dt'} \right) \frac{dt'}{\sqrt{t - t'}}$$

(3.4.12)

Because the unknown appears both inside and outside the integral, a solution to this equation cannot be generally obtained. In some instances the equation can be reduced to a differential equation of higher order, whose solution may be obtained by standard methods. Keeping in mind that the equation is of limited validity on several accounts, we shall limit the discussion here to simple cases that provide some physical insight. Further examples will be given in the next chapter, where it is shown that a nearly equal equation describes the slow, translational motion of a small, rigid sphere in a viscous fluid.

Sudden Temperature Change

As an application of the previous results, we consider the temperature of a sphere of radius a, immersed in a fluid whose temperature is suddenly changed by a small, constant amount. The example was studied in Section 3.2, where we showed that when the simple heat transfer rate is given by (3.2.18), the particle approaches the new fluid temperature exponentially – and this exponential dependence on time holds at every instant. We now consider the temperature equalization using (3.4.12). To do this, we first cast the equation in terms of temperature deviations from the initial, uniform temperature T_0, and write $T_p = T_0 + T_p'$ and $T_f = T_0 + \Theta_f'$. Second, we introduce a nondimensional time, τ, defined by

$$\tau = \frac{h}{2\pi} \frac{t}{\tau_t}$$

(3.4.13)

Thus,

$$h\frac{dT'_p}{d\tau} = -2\pi(T'_p - \Theta'_f) - \int_{-\infty}^{\tau} \frac{dT'_p/d\tau' - d\Theta'_f/d\tau'}{\sqrt{\tau - \tau'}}d\tau' \qquad (3.4.14)$$

We seek the solution of this for the following initial conditions:

$$T'_p = 0, \quad dT'_p/d\tau = 0 \quad \text{for} \quad \tau \leq \tau_0 \qquad (3.4.15a)$$

$$\Theta'_f = \Theta_{f0} \quad d\Theta'_f/d\tau = 0 \quad \text{for} \quad \tau \geq \tau_0 \qquad (3.4.15b)$$

Because the changes occur starting at a well-defined time, namely $\tau = \tau_0$, which we may take to be zero, we try a solution using Laplace transforms. Thus, the Laplace transform of a function $f(z)$ is defined by

$$\tilde{f}(s) = \int_0^{\infty} f(z)e^{-sz}dz \qquad (3.4.16)$$

We use this prescription to transform (3.4.14), term by term. Because $T'_p = 0$ at $\tau = 0$, the transform of the derivative on the left-hand side is $s\tilde{T}'_p$, where $\tilde{T}'_p$ is the transform of $T'_p(\tau)$. The transform of Θ'_f on the right-hand side is simply Θ_{f0}/s because $\Theta'_f = \Theta_{f0}$ for $\tau > 0$. For the integral, we first note that its lower limit is zero, and that it contains two terms, each of the form

$$\int_0^{\tau} f(\tau')g(\tau - \tau')d\tau' \qquad (3.4.17)$$

By the convolution theorem of Laplace transform theory, the transform of this is equal to the product of the transforms of the functions appearing in the integrand, viz. $\tilde{f}(s) \cdot \tilde{g}(s)$. Thus, the Laplace transform of the first term in the integral is

$$s\tilde{T}'_p \int_0^{\infty} \frac{1}{\sqrt{\tau - \tau'}}e^{-st}d\tau' = s\tilde{T}'_p\sqrt{\frac{\pi}{s}}$$

Similarly, we find that the transform of the second term in the integral is Θ_{f0}. This completes the term-by-term transformation of (3.4.14). Collecting the results and solving for $\tilde{T}'_p$, we obtain

$$\frac{\tilde{T}'_p}{\Theta_{f0}} = \frac{1}{s\left[1 + \sqrt{s/4\pi} + hs/2\pi\right]} \qquad (3.4.18)$$

This may be inverted exactly, with the result expressed in terms of complementary error functions of real and imaginary arguments. We will not give this complete solution, but instead use the above transform to obtain asymptotic values for the particle temperature that apply for small and large times. Such limits display the effects of the history term clearly and can be easily compared with the corresponding values predicted by the simple result (3.2.24).

To obtain these asymptotic limits, we recall that there is a one-to-one correspondence between limiting values of the nontransformed variable (the nondimensional time τ in this case) and the transformed variable s. Thus, $s \to \infty$, corresponds to

$\tau \to 0$ and vice-versa. The two limits are somewhat different in the sense that the first has a unique value, whereas the second depends on the parameter h.

1. *Short time limit.* This corresponds to $s \to \infty$. Here, (3.4.18) reduces to

$$\frac{\tilde{T}'_p}{\Theta_{f0}} = \frac{2\pi}{h} \frac{1}{s^2} + O(s^{-3/2}) \tag{3.4.19}$$

The inverse of $1/s^2$ is τ. Hence, using (3.4.13) we obtain,

$$\frac{T'_p}{\Theta_{f0}} = \frac{t}{\tau_t} + O(t/\tau_t)^{1/2} \quad t \to 0 \tag{3.4.20}$$

Thus, for very short times after the fluid temperature has been changed from 0 to Θ_{f0}, the particle temperature increases linearly with time. This limiting behavior also follows from (3.2.24). That is, history effects do not play a role in that limit, as may have been expected.

2. *Long time limit.* This follows from (3.4.18) by taking the limit $s \to 0$. The ultimate value is $\tilde{T}'_p = \Theta_{f0}/s$ and corresponds to $T'_p(t \to \infty) = \Theta_{f0}$. This applies regardless of the value of h. We would like to consider the approach to this limiting form and consider, first, the case $h = O(1)$. Here, the leading form of (3.4.18) is

$$\frac{\tilde{T}'_p}{\Theta_{f0}} \approx \frac{1}{s} - \frac{1}{\sqrt{4\pi s}} \tag{3.4.21}$$

Inverting this and expressing the result in terms of the dimensional time yields

$$\frac{T'_p}{\Theta_{f0}} = 1 - \sqrt{\frac{\tau_t/t}{2\pi h}} \quad t \to \infty \tag{3.4.22}$$

Thus, in this case, the approach to equilibrium is not exponential. Rather, the temperature of the particle approaches that of the fluid in a considerably slower manner, given by the inverse square root of time. However, when $h \gg 1$, we obtain from (3.4.18) the limiting behavior

$$\frac{T'_p}{\Theta_{f0}} = 1 - e^{-t/\tau_t} \tag{3.4.23}$$

corresponding to (3.2.24). Since $h \gg 1$ when the density ratio, $\delta = \rho_f/\rho_p$, is very small, we again see that the low-frequency heat transfer, (3.2.18), applies to solid/liquid particles in gases.

PROBLEMS

1. Show that (3.4.23) follows from (3.4.18) when $h \gg 1$.
2. The denominator in (3.4.18) may be expressed as $2\pi s(1 + \sqrt{4\pi s} + hs) = 2\pi s(\sqrt{s} + b)(\sqrt{s} + c)$, with b and c generally having both real and imaginary

parts – in which case $\overline{T}'_p(\tau)$ is determined by error functions of complex arguments. But, for certain values of h, b and c are real. Determine b and c for all
cases, but obtain $T'_p(\tau)$ only for those cases when b and c are real.

3.5 Nonuniform Particle Temperature

The results given in the preceding sections were obtained on the assumption that the
particle temperature was either uniform or was well represented by an average value.
This assumption is likely to break down when the variations in the external temperature
are very rapid. The reason for this is that temperature changes occurring at the particle's
surface are transmitted to points inside by thermal waves whose penetration depths
may be small in comparison with the radius of the particle.

To study possible temperature nonuniformities, we consider, again, a small sphere
immersed in a fluid, whose temperature, far from the sphere is oscillating harmonically in time about a mean value T_0 – this time allowing the temperature inside the
particle to depend on position. The temperature distribution in the fluid is still given
by (3.3.7), except that, now, its value at the surface of the sphere is equal to $T_p(a, t)$.
For convenience, we denote this *unknown* surface value by T_s, noting that it can also
be expressed as $T_f(a, t)$. Since T_s also varies sinusoidally, we may write it as

$$T_s = T_0 + \Re\{T'_s e^{-i\omega t}\}. \tag{3.5.1}$$

Hence, the fluid's temperature is given by

$$T'_f = \Theta'_f + \left(T'_s - \Theta'_f\right)\frac{a}{r}e^{iK_e(r-a)} \tag{3.5.2}$$

where $\Theta'_f = \theta_{f0}e^{-i\omega t}$ is the complex temperature fluctuation far from the sphere.

Consider now the particle. Its temperature is also described by Fourier's equation,
or

$$\frac{\partial T_p}{\partial t} = \kappa_p \nabla^2 T_p$$

where $\kappa_p = k_p/\rho_p c_{pp}$ is the thermal diffusivity of the particle material. For harmonic
time variations, the temperature in the particle, $T_p(r, t)$ may be expressed as

$$T_p = T_0 + \Re\{\tilde{T}'_p(r)e^{-i\omega t}\} \tag{3.5.3}$$

where the complex function $\tilde{T}'_p$ satisfies

$$\frac{1}{r^2}\frac{d}{dr}\left(r^2\frac{dT'_p}{dr}\right) + K_i^2 T'_p = 0 \tag{3.5.4}$$

and

$$K_i^2 = \frac{i\omega}{\kappa_p} \tag{3.5.5}$$

is the internal counterpart to (3.2.4). The conditions at the surface of the particle are

$$T_p' = T_s' \quad \text{and} \quad k_p \left(\frac{\partial T_p}{\partial r} \right) = k_f \left(\frac{\partial T_f}{\partial r} \right) \quad \text{at} \quad r = a \qquad (3.5.6\text{a,b})$$

The solution of (3.5.4) that remains finite at the origin is $\sin(K_i r)/r$; but to simplify the algebraic manipulations, we write it in terms of $j_0(K_i r)$, the spherical Bessel function of the first kind, of order 0 (Appendix D). Thus, the solution satisfying the first of the boundary conditions, is

$$T_p' = T_s' \frac{j_0(K_i r)}{j_0(q_i)} \qquad (3.5.7)$$

where $q_i = K_i a = (1 + i) z_p$, and

$$z_p = \sqrt{\omega a^2 / 2\kappa_p} \qquad (3.5.8)$$

is the ratio of the particle radius to the thermal penetration depth into the particle, $\delta_{\kappa p}$, defined by

$$\delta_{\kappa p} = \sqrt{2\kappa_p / \omega} \qquad (3.5.9)$$

We still need to determine the surface temperature. Evaluating the two temperature derivatives in (3.5.6b) using (3.5.2) and (3.5.7), and denoting, for simplicity, $K_e a$, by q, where $q = (1 + i) z_f$, we obtain

$$T_s' = \Theta_f' \frac{1 - iq}{1 - iq + (k_p / k_f) q_i j_0'(q_i) / j_0(q_i)} \qquad (3.5.10)$$

where the prime on j_0' represents a derivative with respect to the argument. Although algebraically involved, these results clearly show that the surface temperature generally depends on both fluid and particle temperature variations. Thus, in addition to z_f, there now appears z_p in the formulation. This is related to z_f by means of

$$z_p / z_f = \sqrt{\kappa_f / \kappa_p} = \sqrt{\tfrac{3}{2} h (k_f / k_p)} \qquad (3.5.11)$$

Since we also have $z_p / z_f = \sqrt{\delta_{\kappa f} / \delta_{\kappa p}}$, we see that the penetration depths into fluid and particle can differ significantly, owing to the generally different values of the property ratios appearing on the right-hand sides of (3.5.11).

Let us now return to (3.5.10). We wish to compare it with (3.3.9), obtained on the assumption that the particle temperature is uniform. To do this, note that when the particle is very small – compared with $\delta_{\kappa p}$ – we may expand the Bessel functions appearing in (3.5.10). Thus, for small magnitudes of their argument, we have $\zeta j_0'(\zeta)/j_0(\zeta) = -\tfrac{1}{3}\zeta^2 (1 + \zeta^2/15) + O(\zeta^6)$, so that

$$T_s' = \Theta_f' \frac{1 + z_f - iz_f}{1 + z_f - iz_f (1 + hz_f)} \qquad (3.5.12)$$

As comparison with (3.3.9) shows, this surface temperature is equal to the particle temperature obtained earlier for the uniform temperature case. However, to the same approximation, the temperature in the particle is not uniform. This may be seen by expanding the function $j_0(q_i)$ appearing in (3.5.7). For small magnitudes of the argument, we have $j_0(\zeta) = 1 - \zeta^2/6 + \zeta^4/120 + \dots$. Use of this in (3.5.7), together with the identity $q_i^2 = 2iz_p^2$, yields

$$T_p' = \tilde{T}_s'\left[1 + \frac{i}{3}\frac{a^2 - r^2}{\delta_{\kappa p}^2}\right] \qquad (3.5.13)$$

Thus, when z_p is small, the temperature distribution in the particle is parabolic, differing from T_s', by an amount which is of the order of z_p^2. Therefore, if $a \ll \delta_{\kappa p}$, we may regard the particle's temperature as uniform, as might have been expected on physical grounds. When z_p is not small, the more general result, (3.5.10), should be used. For future purposes, we write that equation as

$$T_s' = \Theta_f'\frac{k_f}{k_p}\frac{G(q_i)}{F(q, q_i)} \qquad (3.5.14)$$

where the functions $F(q, q_i)$ and $G(q_i)$ are explicitly given by

$$F(q, q_i) = \frac{1}{1 - iq} + \frac{k_f}{k_p}G(q_i) \qquad (3.5.15)$$

$$G(q_i) = j_0(q_i)/[q_i\, j_0'(q_i)]. \qquad (3.5.16)$$

Use of these two equations, we find that the particle temperature is

$$T_p' = \Theta_f'\frac{k_f}{k_p}\frac{G(q_i)}{F(q, q_i)}\frac{j_0(K_i r)}{j_0(q_i)} \qquad (3.5.17)$$

Similarly, the heat transfer rate to the particle, expressed in terms of the interior temperature field is

$$\dot{Q}_p = 4\pi a^2(\partial T_p'/\partial r)_a \qquad (3.5.18)$$

Thus, use of (3.5.7) and (3.5.16) gives

$$\dot{Q}_p = 4\pi a k_f\frac{1}{F(q, q_i)}\Theta_f'. \qquad (3.5.19)$$

This replaces the simple result given by (3.2.18), when the frequency is high. To see its effects on the particle temperature, we need to evaluate $F(q, q_i)$.

The Function $G(q_i)$

The functions $F(q, q_i)$ and $G(q_i)$ defined previously also appear in the calculation of the temperature of particles that can pulsate. Their numerical calculation is therefore

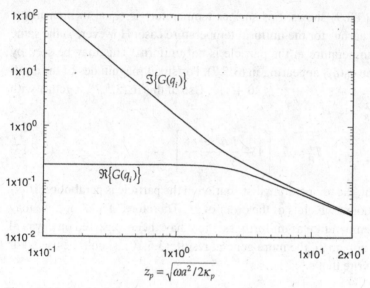

Figure 3.5.1. Real and imaginary parts of the complex function $G(q_i)$.

required several times. To help in those calculations, we give below expressions that may aid in the evaluation of $G(q_i)$. First, we use the explicit values of the Bessel functions appearing in the definition of $G(q_i)$ to write it as

$$G(q_i) = \frac{\tan(q_i)}{q_i - \tan(q_i)} = \frac{\tanh[(1-i)z_p]}{(1-i)z_p - \tanh[(1-i)z_p]} \tag{3.5.20}$$

The real and imaginary parts of $1/G(q_i)$ are

$$\Re\{1/G(q_i)\} = -1 - z_p[\sin(2z_p) + \sinh(2z_p)]/[\cos(2z_p) - \cosh(2z_p)] \tag{3.5.21}$$

$$\Im\{1/G(q_i)\} = z_p[\sinh(2z_p) - \sin(2z_p)]/[\cos(2z_p) - \cosh(2z_p)] \tag{3.5.22}$$

These can be used to compute $\Re\{G(q_i)\}$ and $\Im\{G(q_i)\}$. These two functions are shown in Fig. 3.5.1, as functions of their arguments. Their limiting forms, when $z_p \ll 1$, are $\Re\{G(q_i)\} = 1/5$ and $\Im\{G(q_i)\} = 3/2z_p^2$.

Temperature Average

Let us now return to the particle temperature field as described by the complete results, (3.5.7) and (3.5.10). Although cumbersome, these can be used at any frequency of interest to obtain the spatial variations of the temperature inside the particle. However, in most problems of interest in suspensions, these variations are of secondary importance. Nevertheless, the study of thermal effects in suspensions requires that some physically representative particle temperature be used. The question, therefore, is the identification of that quantity. One possibility is the surface temperature T_s because this is the temperature the fluid "sees." This is suitable at low frequencies because, then, the temperature inside the particle differs little from the surface value.

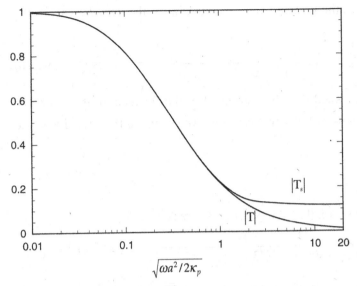

Figure 3.5.2. Magnitude of $T = \overline{T}'_p / \Theta'_f$ and $T_s = T'_s / \Theta'_f$ for a 100-μm alumina spheres in water.

But, as the frequency increases, this value becomes less representative of the particle's temperature.

A more suitable representation of the temperature field in the particle may be obtained in terms of the volume average introduced in Section 2.2. Thus, we integrate (3.5.7) using the prescription given by (2.2.1), with

$$\frac{1}{v_p} \int_{v_p} j_0(K_i r) d\tau_- = \frac{3}{q_i^3} \int_0^{q_i} u \sin(u) du = \frac{3}{q_i} j_0'(q_i) \qquad (3.5.23)$$

where we have used the identity $j_0'(\varsigma) = \cos(\varsigma)/\varsigma - \sin(\varsigma)/\varsigma^2$. This gives

$$\overline{T}'_p = \Theta'_f \frac{3}{q_i^2} \frac{k_f/k_p}{F(q, q_i)} \qquad (3.5.24)$$

where $T = \overline{T}'_p / \Theta'_f$. Using $q_i^2 = 2iz_p^2$, this may also be expressed as

$$T = \frac{i}{hz_f^2} \frac{1}{F(q, q_i)}. \qquad (3.5.25)$$

These results can be used to obtain the real and imaginary parts of the average particle temperature ratio for a wide range of frequencies. Figure 3.5.2 shows the magnitude of the complex temperature ratio as a function of z_p for a 100-μm alumina particle in water. The graph also shows the equivalent ratio for the surface temperature. It is seen that the two magnitudes are nearly equal for values of z_p as large as one. Below this value, the particle radius is smaller than the thermal penetration depth, so that temperature equalization may be expected. But beyond that value, the thermal waves are unable to reach the center of the particle within one oscillatory period, so that differences will occur. In the limit of high frequencies, the amplitude of the temperature fluctuation in the particle vanishes, but that of the surface reaches the asymptotic value

$$|T_s| \to \frac{1}{1 + (k_p/k_f)\sqrt{\kappa_p/\kappa_f}} \tag{3.5.26}$$

where $T_s = T_s'/\Theta_f'$. It is noted that the amplitude of the fluctuation of the surface temperature is larger than that of $\overline{T}_p'$. In Chapter 6, we show that the opposite is true for pulsating particles.

PROBLEMS

1. Show that when z_p is small, the surface temperature is given by
$$T_s' = \Theta_f' \frac{1 + z_f - i z_f}{1 + z_f - i z_f - (2i/3)(k_p/k_f)(a/\delta_{\kappa p})^2} + O[(a/\delta_{\kappa p})^4].$$

2. Determine the low-frequency limit of $F(q, q_i)$.
3. Derive (3.5.26) and discuss the result for various particle-fluid combinations.

3.6 Concluding Remarks

This concludes our discussion of the thermal effects that take place when a small, rigid particle is immersed in a fluid whose temperature changes in time. We have limited it to situations where the temperature differences between particles and fluid are small, and where the fluid around the particle is moving slowly but unsteadily. The corresponding problem for compressible particles is treated in Chapter 6. Convective and radiative heat transfer have not been included in this formulation. These modes of energy transfer may be important in some applications; but, in suspensions of rigid particles, the unsteady, nonuniform temperature distribution case discussed in this chapter is sufficient for most motions of interest.

4

Translational Motion at Re ≪ 1

4.1 Introduction

We now turn our attention to the first, and most basic, of the motions that very small particles can exhibit: uniform translation. When the Reynolds number is very small, this motion may be studied without taking into account other motions that the particle may also be executing (e.g., small-amplitude volume pulsations). This means that all that the particles may be regarded as rigid so far as the translational motion is concerned. Thus, the velocity of a particle undergoing translations may be obtained from Newton's second law once the forces on the particle are prescribed.

· Thus, the focus of this chapter is the particle force, that is the force with which the external fluid acts on a small, rigid particle. This force plays a central role in suspension dynamics and is considered here in some detail for the limiting case when the Reynolds number is very small. The chapter also contains some information on the forces that act on a particle at finite Re. Rotating particles are also mentioned briefly.

Basic Equations

In order to obtain the fluid force acting on a uniformly translating suspension particle, it is generally necessary to solve the equations of fluid dynamics with suitable conditions at the surface of the particle (i.e., with moving boundary conditions). Such a problem cannot yet be solved analytically for arbitrary motions or particle shapes. Fortunately, certain simplifications are possible. First, if the suspension is dilute, we may neglect the presence of other particles when computing the force on one of them. Second, small, free particles adjust themselves quickly to fluid velocity variations so that the velocity difference between particle and fluid is usually very small. This implies that the Reynolds numbers associated with particle motions are typically very small. In such cases, the nonlinear convective terms in the momentum equation can be neglected in comparison with the viscous term. In addition, if the magnitude of the fluid velocity is also small, the fluid may be taken as incompressible for most purposes. In this situation, the fluid velocity field is independent of the temperature and is determined by the Stokes equations:

$$\nabla \cdot \mathbf{u}_f = 0 \tag{4.1.1}$$

$$\rho_f \frac{\partial \mathbf{u}_f}{\partial t} - \mu_f \nabla^2 \mathbf{u}_f = -\nabla p_f \tag{4.1.2}$$

where the fluid density is now regarded as a constant. The equations are supplemented by the condition that $\mathbf{u}_f = \mathbf{U}_s$ at the surface of the particle.

A simpler set of equations applies when all motions are steady, in which case the first term in (4.1.2) is zero. That set is considered in Section 4.3. However, small particles in fluids are generally subject to constantly varying conditions, and this makes their motion unsteady. Thus, it is preferable to consider the unsteady equations from the beginning.

The previous equations can be combined to produce a single equation for the pressure or for the velocity. Thus, taking the gradient of the second equation and using the continuity equation gives,

$$\nabla^2 p_f = 0 \tag{4.1.3}$$

Similarly, taking the curl of (4.1.2) and using the fact that $\nabla \times (\nabla p_f) \equiv 0$, we obtain a single equation for the vorticity $\boldsymbol{\omega} = \nabla \times \mathbf{u}_f$:

$$\frac{\partial}{\partial t}(\nabla \times \mathbf{u}_f) = \nu_f \nabla^2 (\nabla \times \mathbf{u}_f) \tag{4.1.4}$$

Although the equation for the pressure is considerably simpler than that for the vorticity, the boundary conditions are normally expressed in terms of the velocity, thus making the last equation more useful.

But even in these simple conditions, the Stokes equations cannot be solved for arbitrary particle shapes. In fact, they can be solved only for such simple shapes that the particle boundary can be fitted by one of the several systems where Laplace's equation is separable. In what follows, we will limit ourselves to the simplest of all shapes: the sphere. Because of this selection, it will be convenient to use a polar-spherical system of coordinates, r, θ, φ, as sketched in Fig. 4.1.1. Unit vectors along those directions are $\mathbf{e}_r, \mathbf{e}_\theta, \mathbf{e}_\varphi$. The set forms a right-handed set of orthonormal vectors, that is, $\mathbf{e}_r \times \mathbf{e}_\theta = \mathbf{e}_\varphi$, $\mathbf{e}_\theta \times \mathbf{e}_\varphi = \mathbf{e}_r$ and $\mathbf{e}_\varphi \times \mathbf{e}_r = \mathbf{e}_\theta$. The elements of length along r, θ, and φ are dr, $rd\theta$, and $r \sin\theta d\varphi$, respectively, and the elements of area and volume in this system of coordinates are $dA = r^2 \sin\theta d\theta d\varphi$ and $dV = r^2 \sin\theta dr d\theta d\varphi$. Explicit expressions for the various vectorial operations required by the Stokes equations are given in Appendix C.

In the sections that follow, we will study the flow field around a spherical particle executing simple translational motions, but the main purpose of the calculations will be the determination of the fluid forces applicable in those motions. The analysis will follow the same lines as used in the last chapter to study the heat transfer problem

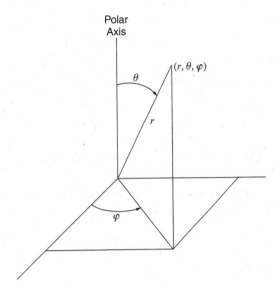

Figure 4.1.1. Spherical polar coordinates.

and will be followed by extensions of the simple results to situations where the Stokes equations no longer apply.

4.2 Translational Oscillations

We first consider translational monochromatic oscillations, without rotation, of a rigid sphere in a fluid that is otherwise at rest. The velocity of the particle and its diameter are such that the Reynolds number is very small, so that the Stokes equations apply. This particular motion is important on its own, but also its results can be used to study other motions. For example, the force that acts on more general unsteady translational motions may then be obtained from the monochromatic result by means of Fourier transform methods.

We thus take the sphere to have a translational velocity $\mathbf{u}_p(t)$ along a straight line, say one of the axes of a Cartesian system of coordinates. For purely monochromatic motions, the magnitude of $\mathbf{u}_p(t)$ is $u_p(t) = U_{p0} \cos(\omega t)$. Now consider the fluid velocity $\mathbf{u}_f$. After all transient motions have died, this is also a monochromatic function of time, having the same frequency as that of the particle. Thus, we may express both velocities as

$$u_p(t) = \Re\{U_{p0}e^{-i\omega t}\}, \qquad \mathbf{u}_f(\mathbf{x}, t) = \Re\{\mathbf{U}_f(\mathbf{x})e^{-i\omega t}\} \qquad (4.2.1a,b)$$

where the vector $\mathbf{U}_f(\mathbf{x})$ is, in general, a complex function of position that is to be determined. The vector $\mathbf{x}$ appearing here represents the position of a field point with respect to a *fixed* system of coordinates, which we place at the mean position of the sphere's center. We note in passing that, in this system, the time derivative has the usual meaning and the convective acceleration – disregarded in this analysis – is $\mathbf{u}_f \cdot \nabla \mathbf{u}_f$.

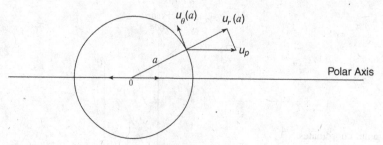

Figure 4.2.1. Geometry of flow in an axial plane.

Returning now to the differential equation for the vorticity, (4.1.4), we substitute (4.2.1b) in it and obtain

$$\nabla^2(\nabla \times \mathbf{U}_f) + K^2\nabla \times \mathbf{U}_f = 0 \qquad (4.2.2)$$

where

$$K^2 = i\omega/\nu_f \qquad (4.2.3)$$

To solve (4.2.2), we refer the motion to a polar-spherical system of coordinates with the polar axis coinciding with the direction of motion of the particle. Since the particle is spherical, the fluid motion has axial symmetry with respect to the polar axis. Therefore, the pressure and velocity fields are independent of the azimuth angle φ. Furthermore, the azimuth component of the velocity is zero, so that $\mathbf{U}_f = \mathbf{e}_r U_r + \mathbf{e}_\theta U_\theta$, where U_r and U_θ are the radial and tangential components of $\mathbf{U}_f$. Thus, on a constant φ plane, the flow would appear as sketched in Fig. 4.2.1.

The boundary condition on the fluid velocity is that $\mathbf{u}(\mathbf{x}_s, t) = \mathbf{u}_p(t)$, where $\mathbf{x}_s$ is the distance from a point on the surface of the sphere to its center and is given by $\mathbf{x}_s = \mathbf{x}_p + \mathbf{e}_r a$. Thus, the exact boundary condition requires the fluid velocity at points that are not equidistant from the fixed origin of the system of coordinates. But, if we assume that the particle displacement is very small, we may apply the boundary condition at mean position of the sphere surface (i.e., at $r = a$). Further more, since $\mathbf{u}_p$ is along the polar axis, the boundary condition can be written as

$$U_r(a) = U_{p0}\cos\theta, \qquad U_\theta(a) = -U_{p0}\sin\theta \qquad (4.2.4\text{a,b})$$

In addition, we have the physical requirement that both components should vanish as $|\mathbf{x}| \to \infty$. Now, because of the symmetry around the polar axes, the velocity components may be derived from a stream function $\psi = \psi(r, \theta)$ by means of

$$U_r = \frac{1}{r^2\sin\theta}\frac{\partial\psi}{\partial\theta}, \qquad U_\theta = -\frac{1}{r\sin\theta}\frac{\partial\psi}{\partial r} \qquad (4.2.5\text{a,b})$$

These will enable us to replace the vector equation for $\mathbf{U}_f(\mathbf{x})$ with a scalar equation for ψ. To do this, we first obtain an explicit expression for $\nabla \times \mathbf{U}_f$. Since the flow is independent of the azimuth angle, the expressions given in Appendix C show that this vector has only one nonvanishing component:

$$\nabla \times \mathbf{U}_f = \mathbf{e}_\varphi \frac{1}{r} \left[\frac{\partial(r U_\theta)}{\partial r} - \frac{\partial(U_r)}{\partial \theta} \right] \qquad (4.2.6)$$

Substituting (4.2.5) into this, we can write

$$\nabla \times \mathbf{U}_f = -\frac{\mathbf{e}_\varphi}{r \sin \theta} E^2 \psi \qquad (4.2.7)$$

where E^2 represents the following second-order differential operator

$$E^2 = \frac{\partial^2}{\partial r^2} + \frac{\sin \theta}{r^2} \frac{\partial}{\partial \theta} \left(\frac{1}{\sin \theta} \frac{\partial}{\partial \theta} \right) \qquad (4.2.8)$$

We also need $\nabla^2(\nabla \times \mathbf{U}_f)$. From the expressions given in Appendix C, we note that it, too, has only one component, and this is along the azimuthal direction. In terms of the above operator, that component is

$$[\nabla^2(\nabla \times \mathbf{U}_f)]_\varphi = \left[\nabla_\varphi^2 - \frac{1}{r^2 \sin^2 \theta} \right] \frac{E^2 \psi}{r \sin \theta} \qquad (4.2.9)$$

where ∇_φ^2 is the Laplacian operator in polar-spherical coordinates *without* azimuthal derivatives. Applying this to (4.2.7), yields

$$\nabla^2(\nabla \times \mathbf{U}_f) = -\frac{\mathbf{e}_\varphi}{r \sin \theta} E^4 \psi \qquad (4.2.10)$$

where E^4 is the operator obtained by applying E^2 to itself [i.e., $E^4 = E^2(E^2)$]. In terms of this operator, the working equation becomes

$$(E^2 + K^2)E^2 \psi = 0 \qquad (4.2.11)$$

Because of (4.2.4), ψ is subject to

$$\frac{\partial \psi}{\partial \theta} = U_{p0} a^2 \sin \theta \cos \theta, \qquad \frac{\partial \psi}{\partial r} = U_{p0} a \sin^2 \theta \qquad (4.2.12\text{a,b})$$

on $r = a$. These relationships suggest a solution of the form

$$\psi = f(r) U_{p0} \sin^2 \theta \qquad (4.2.13)$$

Using this and (4.2.8) gives

$$E^2 \psi = U_0 \sin^2 \theta \left(\frac{d^2}{dr^2} - \frac{2}{r^2} \right) f(r) \qquad (4.2.14)$$

Upon substitution of this in (4.2.11), we obtain a fourth-order, ordinary differential equation for $f(r)$

$$\left(\frac{d^2}{dr^2} + K^2 - \frac{2}{r^2} \right) \left(\frac{d^2}{dr^2} - \frac{2}{r^2} \right) f(r) = 0 \qquad (4.2.15)$$

Because the two differential operators from above are different if $K \neq 0$, the solution of this equation may be expressed as $f = f_1 + f_2$, where the functions f_1 and f_2 satisfy, respectively,

$$\left(\frac{d^2}{dr^2} - \frac{2}{r^2} \right) f_1(r) = 0, \qquad \left(\frac{d^2}{dr^2} + K^2 - \frac{2}{r^2} \right) f_2(r) = 0 \quad (4.2.16a,b)$$

When $K = 0$, that is, when the flow is steady, the two operators are equal, and the solution cannot be split in this manner. However, the differential equation (4.2.15) can then be easily solved in terms of powers or r, as shown in Section 4.3. However, for the unsteady case, (4.2.16a,b) applies. The general solution to the first of these is

$$f_1 = \frac{A}{r} + Br^2 \qquad (4.2.17)$$

but because of the boundary condition at infinity, B must be set equal to zero. The constant A can be evaluated from the boundary conditions once f_2 is determined. We note that f_1 represents a motion that is felt instantaneously everywhere in the fluid, although its magnitude decreases with distance. As we will see later, the motion represented by f_2 is not propagated instantaneously.

The solution to the second equation must also be chosen so that the boundary condition at infinity is satisfied. To do this, we first note that $K^2 = i\omega/\nu_f$ has two roots, which we write as $K_1 = (1+i)/\delta_\nu$ and $K_2 = -(1+i)/\delta_\nu$, where

$$\delta_\nu = \sqrt{2\nu_f/\omega} \qquad (4.2.18)$$

This quantity has the dimensions of length and is a measure of the distances over which viscous effects – originating at some boundary (the sphere in this case) – are significant. It is therefore called the *viscous penetration depth*. We note that this depth is large when the frequency is small, whereas at moderate and high frequencies, it is small. For example, if the fluid is air, and the frequency is 100 Hz, δ_ν is about 0.02 cm. Whether this is large or small depends on the particle size, and it is useful to measure the effects of viscosity in terms of the ratio of particle radius to viscous penetration depth. We denote the ratio of these quantities by $y = a/\delta_\nu$, or

$$y = \sqrt{\omega a^2/2\nu_f} \qquad (4.2.19)$$

Returning now to the differential equation for f_2, we note that sufficiently far from the sphere, the term $2/r^2$ becomes very small. This shows that, at such large distances, the two independent solutions of (4.2.16a,b) are proportional to $\exp(iK_1 r)$ and $\exp(iK_2 r)$, respectively. Given that K_2 is equal to $-(1+i)/\delta_\nu$, we see that the second exponential becomes unbounded as $r \to \infty$, so that the corresponding term in a solution valid at all distances has to be discarded.

Now, to obtain f_2 at all distances from the sphere, we first write (4.2.16b) in terms of the differential equation satisfied by the spherical Bessel functions of integer order.

Thus, if $g(r)$ is such a function, the equation is

$$g''(r) + \frac{2}{r}g'(r) + \left[K^2 - \frac{n(n+1)}{r^2}\right]g(r) = 0$$

whose solutions are the spherical Bessel functions of the first kind, $j_n(Kr)$, and of the second kind, $y_n(Kr)$. Sometimes, a third pair of independent solutions is used. These are the spherical Bessel functions of the third kind, $h_n^{(1)}(Kr)$ and $h_n^{(2)}(Kr)$, respectively, where

$$h_n^{(1)}(Kr) = j_n(Kr) + iy_n(Kr), \qquad h_n^{(2)}(Kr) = j_n(Kr) - iy_n(Kr)$$

To put the equation for f_2 in the above form, we put $f_2 = r^n g(r)$ in (4.2.16b) and find that it becomes identical to the standard equation if $n = 1$. Hence, the solution for $g(r)$ is given in terms of the spherical Bessel functions of integer order 1. We take the pair $h_1^{(1)}(K_1r)$, $h_1^{(2)}(K_1r)$ as our independent pair of solutions. These are explicitly given by

$$h_1^{(1)}(K_1r) = -\frac{i + K_1r}{(K_1r)^2}e^{iK_1r} \quad \text{and} \quad h_1^{(2)}(K_1r) = -\frac{K_1r - i}{(K_1r)^2}e^{-iK_1r}$$

Because $h_1^{(2)}(K_1r)$ is unbounded as $r \to \infty$, we need to retain only $h_1^{(1)}$. Thus, dropping for convenience the superscript on it, we have

$$f_2(r) = Crh_1(K_1r) \tag{4.2.20}$$

where C is a constant. If we incorporate the time factor, we see that this solution is proportional to $\exp[-r/\delta_v + i(r/\delta_v - \omega t)]$. This represents a damped wave, of viscous origins, moving away from the particle along the radial direction. As a result, different points in the fluid sense a specific change in the motion of the sphere at different times. The phase velocity of this wave is equal to

$$c_{vf} = \sqrt{2\omega v_f} \tag{4.2.21}$$

Unless the frequency is very large, this speed is low, typical of diffusion processes.

Let us now obtain the complete solution. Adding f_1 and f_2 and substituting in (4.2.13) we obtain

$$\psi = U_{p0}\left[\frac{A}{r} + Crh_1(K_1r)\right]\sin^2\theta \tag{4.2.22}$$

Using this in (4.2.5a,b) yields

$$U_r = \frac{2U_{p0}\cos\theta}{r^2}\left[\frac{A}{r} + Crh_1(K_1r)\right] \tag{4.2.23}$$

$$U_\theta = -\frac{U_{p0}\sin\theta}{r}\left\{-\frac{A}{r^2} + C\left[K_1rh_1'(K_1r) + h_1(K_1r)\right]\right\} \tag{4.2.24}$$

To determine the constants A and C, we apply the boundary conditions (4.2.4), and obtain

$$\frac{A}{a} + Cah_1(K_1a) = \frac{1}{2}a^2 \quad \text{and} \quad -\frac{A}{a^2} + C[K_1ah_1'(K_1a) + h_1(K_1a)] = a$$

$$\cdot (4.2.25a,b)$$

Solving for C, we get

$$C = \frac{3}{2}\frac{a}{K_1ah_1'(K_1a) + 2h_1(K_1a)} \tag{4.2.26}$$

Using the definition of $h_1(z)$, the denominator in this equation can be expressed as $-i\exp(iK_1a)$. Thus,

$$C = \frac{3}{2}iae^{-iK_1a} \quad \text{and} \quad A = \frac{1}{2}a^3\left[1 + 3\frac{iK_1a - 1}{(K_1a)^2}\right] \tag{4.2.27a,b}$$

Finally, using these and $K_1^2 = 2i\delta_v^{-2}$ and $K_1 = (1 + i)\delta_v^{-1}$, in the expressions for the velocity components, we obtain

$$U_r = \left(\frac{a}{r}\right)^3 U_{p0}\cos\theta - \frac{3ia\delta_v^2}{2r^3}U_{p0}\cos\theta\left[(1 - iK_1r)e^{iK_1(r-a)} - (1 - iK_1a)\right]$$

$$(4.2.28)$$

$$U_\theta = \frac{1}{2}\left(\frac{a}{r}\right)^3 U_{p0}\sin\theta + \frac{3ia\delta_v^2}{4r^3}U_{p0}\sin\theta\left[1 - iK_1a - \left(1 - iK_1r - K_1^2r^2\right)e^{iK_1(r-a)}\right]$$

$$(4.2.29)$$

These expressions describe the fluid velocity at all points in the fluid and are sufficient to obtain the pressure and the force on the sphere. However, before doing so, we consider their limiting forms, applicable to high and low frequencies. First, when $\omega \to \infty$, the second terms in both equations above vanish so that

$$U_r = U_{p0}(a/r)^3\cos\theta \quad \text{and} \quad U_\theta = U_{p0}(a/r)^3\sin\theta \tag{4.2.30a,b}$$

These results are equal to those that apply when the fluid is inviscid, as may be seen by putting $\nu_f = 0$ in (4.2.28) and (4.2.29). The reason for this is that the viscous penetration depth, δ_v, is very thin at high frequencies. Hence, the fluid can be regarded as inviscid outside this thin layer. We also see from (4.2.30) that the sphere motion is felt simultaneously at all points in the fluid, but that a few sphere radii away from the sphere, the fluid is essentially at rest while the sphere oscillates rapidly.

Let us now consider the opposite limit, namely $\omega \to 0$. Here, the sphere translates nearly steadily, with speed $u_p = U_{p0}$, but neither U_r nor U_θ have, then, limiting values that are valid everywhere. Mathematically, this occurs because the quantity $|K_1r| = \sqrt{\omega r^2/2\nu_f}$ does not have a unique limit at all values of r when $\omega \to 0$. The physical reason is that the viscous penetration depth now extends to infinity. Thus, the motion of the sphere is transmitted to the fluid at infinity by means of viscous waves having a finite speed.

We may obtain the zero-frequency limit by considering a *fixed* position r in the fluid. Since r is now fixed, it follows that $K_1 r \to 0$ as $\omega \to 0$, and

$$(1 - iK_1 r)e^{iK_1(r-a)} - (1 - iK_1 a) = \frac{1}{2}K_1^2(r^2 + a^2) + \cdots$$

Thus,

$$U_r = \frac{3}{2}\frac{a}{r}U_{p0}\left(1 - \frac{1}{3}\frac{a^2}{r^2}\right)\cos\theta, \quad \text{and} \quad U_\theta = -\frac{3}{4}\frac{a}{r}U_{p0}\left(1 + \frac{1}{3}\frac{a^2}{r^2}\right)\sin\theta$$

$$(4.2.31a,b)$$

These are the steady velocity components at some fixed point in the fluid. It is seen that the velocity field predicted by them decreases very slowly with r. It should also be emphasized that expressions (4.2.31a,b) are valid when $\omega \to 0$, but only at points that are at finite distances from the sphere. Of course, the expressions are valid when ω is exactly zero.

Fluid Pressure

We also require fluid pressure in the fluid. For a monochromatic time dependence, we can write it as

$$p(\mathbf{x}, t) = p_0 + \Re\{P_f(\mathbf{x})e^{-i\omega t}\} \tag{4.2.32}$$

where p_0 is the pressure in the fluid at rest. The complex spatial factor $P_f(\mathbf{x})$ may be computed from the velocity field by means of (4.1.2). Since the flow is incompressible, we have $\nabla^2\mathbf{u} = -\nabla \times (\nabla \times \mathbf{u})$. Hence, substitution of (4.2.32) in (4.1.2) gives

$$\nabla P_f = -\mu_f \nabla \times (\nabla \times \mathbf{U}_f) + i\omega\rho_f \mathbf{U}_f \tag{4.2.33}$$

The first term in this equation is the curl of the vorticity. As (4.2.7) shows, this has one only component, which is along the azimuth direction. Now, the curl of a vector $\mathbf{B} = \{B_r, B_\theta, B_\varphi\}$ whose only zero component, B_φ, is

$$\nabla \times \mathbf{B} = \frac{\mathbf{e}_r}{r\sin\theta}\frac{\partial(B_\varphi\sin\theta)}{\partial\theta} - \frac{\mathbf{e}_\theta}{r}\frac{\partial(rB_\varphi)}{\partial r}$$

Hence, the r and θ components of the pressure gradient are, respectively,

$$\frac{\partial P_f}{\partial r} = \frac{\mu_f}{r^2\sin\theta}\frac{\partial(E^2\psi)}{\partial\theta} + i\omega\rho_f U_r, \qquad \frac{\partial P_f}{\partial\theta} = -\frac{\mu_f}{\sin\theta}\frac{\partial(E^2\psi)}{\partial r} + i\omega\rho_f r U_\theta$$

$$(4.2.34a,b)$$

To compute $E^2\psi$, we use (4.2.14), noting that $(d^2/dr^2 - 2/r^2)f(r) = -K_1^2 f_2(r)$. Hence, $E^2\psi = -CU_{p0}\sin^2\theta K_1^2 r h_1(K_1 r)$. Substituting U_r and U_θ from (4.2.23) and

(4.2.24), and using $i\omega\rho_f = \mu_f K_1^2$, we obtain

$$\frac{\partial P_f}{\partial r} = 2\mu_f K_1^2 \frac{U_{p0}\cos\theta}{r^3} A, \qquad \frac{\partial P_f}{\partial\theta} = \mu_f K_1^2 \frac{U_{p0}\sin\theta}{r^2} A \qquad (4.2.35a,b)$$

These give, upon integration,

$$P_f = -\mu_f K_1^2 \frac{U_{p0}\cos\theta}{r^2} A \qquad (4.2.36)$$

where A is given by (4.2.27b). We thus find that the excess pressure in the fluid (i.e., the pressure due to the motion of the sphere) is proportional to the velocity of the sphere, as was to be expected given the linearity of the equations of motion. This excess pressure varies with the cosine of the angle between the position vector and the direction of motion, and decreases with distance as r^{-2}. The pressure's spatial dependence are those of a doublet, or dipole at the origin, as may be made clearer by writing (4.2.36) as

$$P_f = \frac{1}{2}\mu_f \left[(K_1 a)^2 + 3(iK_1 a - 1)\right] \mathbf{U}_{p0} \cdot \nabla\frac{a}{r} \qquad (4.2.37)$$

where $\mathbf{U}_{p0}$ is a vector along the direction of $\mathbf{u}_p$, whose magnitude is U_{p0}. To get the time-dependent pressure in real form, we reinstate the time factor and write the quantity in the square brackets in (4.2.37) as $-3(1+y) + iy(3+2y)$ so that

$$p_f - p_0 = -\frac{1}{2}\mu_f\sqrt{9(1+y)^2 + y^2(3+2y)^2}\, \mathbf{u}_p \cdot \nabla\frac{a}{r}\cos[\omega t - \eta(y)] \qquad (4.2.38)$$

where $\tan\eta = -y(3+2y)/3(1+y)$. This shows that the pressure in the fluid oscillates harmonically in time, with a phase angle, relative to that of the sphere's velocity that depends on the ratio of the sphere radius to the viscous penetration depth. However, the pressure's phase angle does not depend on position, as was the case for the velocity. In fact, we see that the fluctuations of pressure resulting from the motion of the sphere are felt simultaneously everywhere. That is, they are propagated into the fluid with an infinite speed. This unphysical result is due, of course, to our assumption that the fluid is incompressible.

Fluid Force on Sphere

We now have all the information needed to compute the fluid force on the sphere $\mathbf{F}_p$. In general, this is given by $\mathbf{F}_p = \int_{A_p} \boldsymbol{\Sigma}dA$, where $\boldsymbol{\Sigma}$ is the local traction, or surface force per unit area, expressed by (2.1.12) in terms of the components of the stress tensor, σ_{ij}. For this motion, the components $\sigma_{r\varphi}$ and $\sigma_{\theta\varphi}$ of that tensor vanish because the motion has axial symmetry. Further, the resultant force acts along the direction of motion of the particle, as sketched in Fig. 4.2.2, so that

$$F_p = \int_{A_p} [\sigma_{rr}\cos\theta - \sigma_{r\theta}\sin\theta]_{r=a}dA \qquad (4.2.39)$$

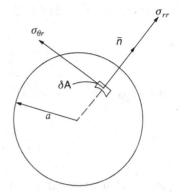

Figure 4.2.2. Fluid stresses on the particle surface.

The element of area on the surface of the sphere is $dA = a^2 \sin\theta d\theta d\varphi$. But, because of the symmetry around the polar axis, the integrand is independent of φ, so that

$$F_p = 2\pi a^2 \int_0^\pi [\sigma_{rr}\cos\theta - \sigma_{r\theta}\sin\theta]_{r=a} d\theta \qquad (4.2.40)$$

The values of σ_{rr} and $\sigma_{r\theta}$ are,

$$\sigma_{rr} = -p + 2\mu_f \frac{\partial u_r}{\partial r} \quad \text{and} \quad \sigma_{r\theta} = \mu_f r \frac{\partial}{\partial r}\left(\frac{u_\theta}{r}\right) + \frac{\mu_f}{r}\frac{\partial u_r}{\partial\theta} \qquad (4.2.41a,b)$$

It is convenient to carry out the calculation in terms of our complex pressure and velocity components. Thus, substituting (4.2.41) into (4.2.40) and noting that on $r = a$, $\partial U_r/\partial r = 0$, and $\partial U_r/\partial\theta = U_\theta$, we have

$$F_p = -2\pi a^2 \int_0^\pi \left(P_f \cos\theta + \mu_f \frac{\partial U_\theta}{\partial r}\sin\theta\right)_{r=a} d\theta \qquad (4.2.42)$$

The complex pressure is given by (4.2.36). Its value on $r = a$ is

$$P_f = \frac{3}{2}\mu_f \frac{U_{p0}\cos\theta}{a}\left[1 - iK_1 a - \frac{1}{3}(K_1 a)^2\right] \qquad (4.2.43)$$

where we have used (4.27b) for A. For the second term in (4.2.42), we have

$$\left(\frac{\partial U_\theta}{\partial r}\right)_{r=a} = -\frac{U_{p0}\sin\theta}{a}\left\{\frac{2A}{a^3} - 1 + CK_1[K_1 a h_1''(K_1 a) + 2h_1'(K_1 a)]\right\}$$

The quantity inside the square bracket in this equation is equal to $[2 - (K_1 a)^2]$ $h_1(K_1 a)/K_1 a$. Hence

$$CK_1[K_1 a h_1''(K_1 a) + 2h_1'(K_1 a)] = -\frac{3}{2}(iK_1 a - 1)\frac{2 - (K_1 a)^2}{(K_1 a)^2}$$

Also, $2A/a^3 - 1 = 3(iK_1 a - 1)/(K_1 a)^2$, so that

$$\left(\frac{\partial U_\theta}{\partial r}\right)_{r=a} = -\frac{3}{2}\frac{U_{p0}\sin\theta}{a}(iK_1 a - 1) \qquad (4.2.44)$$

Substituting this and (4.2.43) into (4.2.42), and carrying out the integration, yields

$$\frac{F_p}{2\pi a^2} = \frac{2}{3}\mu_f U_{p0} K_1^2 \frac{A}{a^2} + \frac{2\mu_f U_{p0}}{a}(iK_1 a - 1)$$

Using again the value of A, this can be written as

$$\frac{F_p}{6\pi\mu_f a U_{p0}} = -1 + iK_1 a + \frac{1}{9}(K_1 a)^2 \qquad (4.2.45)$$

It is convenient to express this in terms of the real quantity y, defined by (4.2.19). Thus,

$$F_p = -6\pi\mu_f a(1+y)U_{p0} + \frac{2}{3}\rho_f \pi a^3 \left(1 + \frac{9}{2y}\right) i\omega U_{p0} \qquad (4.2.46)$$

To obtain the time-dependent force, we multiply this by $\exp(-i\omega t)$ and then take the real part of the result. The first term will then have the time factor $U_{p0}\cos(\omega t)$, which by definition is u_p. The second term will similarly have the factor $(-i\omega U_{p0})\exp(-i\omega t)$. This is the complex acceleration of the sphere, whose real part is du_p/dt. Hence, we can express our final result in the succinct form

$$\boxed{F_p = -6\pi\mu_f a(1+y)u_p - \frac{1}{2}m_p\delta\left(1 + \frac{9}{2y}\right)\frac{du_p}{dt}} \qquad (4.2.47)$$

We have placed this result in a box because of its importance in suspension dynamics. From it, we can derive several important results applicable to all unsteady motions that have very small Reynolds numbers. We shall postpone doing that until a later section. Here, we note that the force, like the heat transfer rate to a rigid particle, has two contributions: one proportional to the velocity of the particle and another proportional to its derivative. Because of Newton's second law, we may interpret the coefficient of the second term in (4.2.47) as an addition to the mass of the sphere. This coefficient is therefore called an *added mass*. It effects will be seen later when we consider the particle's equation of motion.

Both terms in (4.2.47) involve the quantity y, which takes into account the fluid response to the sphere's unsteadiness. As comparison with (3.2.17) shows, this quantity is the viscous counterpart of z_f, the ratio of particle radius to thermal penetration depth. In fact, the analogy is close and will be exploited later on. Finally, we note that, when the frequency of oscillation approaches zero, y vanishes, and so do (du_p/dt) and $y^{-1}(du_p/dt)$. We then obtain the well-known Stokes' law:

$$\boxed{F_p = -6\pi\mu_f a u_p} \qquad (4.2.48)$$

Decay of Oscillations

Equation (4.2.53) was first obtained by Stokes in 1851. One of his reasons for study-ing this oscillatory motion was to determine the rate of decay of the oscillations of a spherical pendulum bob after the force driving the motion is removed. The results obtained herein may be used to determine that rate on the assumption that the ampli-tude of oscillation and the Reynolds number are very small. To do this, we use the energy procedure introduced in Chapter 2, Section 2.5, which is applicable when the oscillations decay slowly, as we now assume. The discussion there indicates that, after the driving force is removed, the amplitude of the oscillation decays exponentially as $\exp(-\beta t)$, where

$$\beta = \frac{\langle|\dot{E}_{lost}|\rangle}{2E_0} \tag{4.2.49}$$

Here, E_0 is the energy of the oscillation just before the driving force was withdrawn. This includes the oscillatory energy of both particle and fluid. For simplicity, we ignore the latter. Then, since the sphere is rigid, this energy is equal to kinetic energy, or $E_0 = \frac{1}{2}m_p U_{p0}^2$. Now, the energy loss rate is equal to the rate at which work is done by the force on the sphere to produce a displacement u_p in a unit time. Thus, the average work done per cycle is $\langle F_p u_p \rangle$. The force is given by (4.2.47), and the velocity is simply $u_p(t) = U_{p0}\cos(\omega t)$. Thus, the energy loss rate is given by $\langle|\dot{E}_{lost}|\rangle = 3\pi\mu_f a$ $(1 + y)U_{p0}^2$. Thus,

$$\beta = \frac{9\mu_f}{16\rho_p a^2}(1 + y) \tag{4.2.50}$$

We thus see that the decay rate decreases with the radius and density of the sphere, but increases with the frequency.

4.3 Stokes' Law

Because of its importance, it is useful to derive, again, Stokes' law, (4.2.48) – this time by considering a rigid sphere translating with a constant velocity $\mathbf{u}_p$. Rela-tive to a frame of reference fixed on the sphere, the fluid motion is unsteady, so that the momentum equation for the fluid would have a time derivative of the velocity. Since the frame of reference is moving with constant velocity, the time derivative can be expressed as $-\mathbf{u}_p \cdot \nabla\mathbf{u}_f$. Therefore, the Stokes equations are

$$\nabla \cdot \mathbf{u}_f = 0 \tag{4.3.1}$$

$$\rho_f(\mathbf{u}_f \cdot \nabla\mathbf{u}_f - \mathbf{u}_p \cdot \nabla\mathbf{u}_f) = -\nabla p_f + \mu_f \nabla^2\mathbf{u}_f \tag{4.3.2}$$

This set still contains the nonlinear convective acceleration, which contains two terms. If we neglect both on the basis that the Reynolds number is small, we obtain, upon taking the curl of the resulting equation,

$$\nabla^2(\nabla \times \mathbf{u}_f) = 0 \qquad (4.3.3)$$

This can be expressed in terms of the stream function, in terms of which (4.3.2) becomes

$$E^4\psi(r, \theta) = 0 \qquad (4.3.4)$$

Thus, if we put, as before,

$$\psi = f(r)U_{p0} \sin^2 \theta \qquad (4.3.5)$$

where $f(r)$ satisfies

$$\left(\frac{d^2}{dr^2} - \frac{2}{r^2}\right)\left(\frac{d^2}{dr^2} - \frac{2}{r^2}\right)f(r) = 0 \qquad (4.3.6)$$

To solve this equation, we seek a solution of the form $f(r) = r^q$, with q unknown. Substituting assumed solution into (4.3.6) shows that such as solution exists with q satisfying

$$[q(q - 1) - 2][(q - 2)(q - 3) - 2] = 0 \qquad (4.3.7)$$

The roots of this are $q_1 = -1$, $q_2 = 1$, $q_3 = 2$, and $q_4 = 4$. Thus, in the steady case, the function f is

$$f(r) = Ar^{-1} + Br + Cr^2 + Dr^4 \qquad (4.3.8)$$

The boundary condition at infinity requires that $C = D = 0$; those at $r = a$ yield

$$A = -\frac{1}{4}a^3 \quad \text{and} \quad B = \frac{3}{4}a \qquad (4.3.9a,b)$$

Hence, the stream function and the velocity components are

$$\psi = \frac{3}{4}U_{p0}a\left(1 - \frac{1}{3}\frac{a^2}{r^2}\right)r\sin^2\theta \qquad (4.3.10)$$

$$u_r = \frac{3}{2}U_{p0}\frac{a}{r}\left(1 - \frac{1}{3}\frac{a^2}{r^2}\right)\cos\theta, \qquad u_\theta = -\frac{3}{4}U_{p0}\frac{a}{r}\left(1 + \frac{1}{3}\frac{a^2}{r^2}\right)\sin\theta$$

$$(4.3.11a,b)$$

These velocity components were derived earlier from the corresponding results for an oscillating sphere. The excess pressure is now given by

$$p_f - p_0 = -\frac{3}{2}\mu_f\mathbf{U}_{p0}\cdot\nabla\frac{a}{r} \qquad (4.3.12)$$

At the surface of the sphere, this is equal to $3\mu_f U_{p0}\cos\theta/2a$. Thus, the excess pressure on the sphere surface is largest on the forward axial direction, where it is equal to $3\mu_f U_{p0}/2a$.

The force on the particle along the direction of motion is obtained, as before, from (4.2.39). The result is Stokes' law, which we now write as

$$F_p = -6\pi \mu_f a u_p \qquad (4.3.13)$$

We have derived this important result by considering the steady motion of a small sphere in a fluid that, in the absence of the sphere, is at rest. The same form applies when the sphere is at rest and a slow, steady flow, having a uniform velocity U_0, at infinity, flows over it. Although the streamlines are different, the force is given by (4.3.13), with u_p replaced by $-U_0$.

Let us now investigate the validity of the solution just obtained. This can be assessed by estimating the magnitude of the neglected terms in the momentum equation for the fluid, (4.3.2), using the solution obtained without them. The neglected terms were the convective acceleration terms in the equations of motion: $\mathbf{u}_f \cdot \nabla \mathbf{u}_f - \mathbf{u}_p \cdot \nabla \mathbf{u}_f$. For the solution to be applicable, both of these terms must be small in comparison with viscous term, $\nu_f \nabla^2 \mathbf{u}_r$, which was retained. In the steady case, the magnitude of this is $\mu_f |\nabla^2 \mathbf{u}_f| \approx |\nabla p_f|$. Using the solution for the pressure, we have, $\mu_f |\nabla^2 \mathbf{u}_f| \approx \mu_f a U_{p0}/r^3$. To estimate $\mathbf{u}_f \cdot \nabla \mathbf{u}_f$, we use the radial velocity component, which gives $|\mathbf{u}_f \cdot \nabla \mathbf{u}_f| \approx |u_r \partial u_r / \partial r| \approx U_{p0}^2 a^2 / r^3$. Hence,

$$\frac{\rho_f |\mathbf{u}_f \cdot \nabla \mathbf{u}_f|}{\mu_f |\nabla^2 \mathbf{u}_f|} \approx \frac{\rho_f U_{p0}^2 a^2 / r^3}{\mu_f U_{p0} a / r^3} = \mathrm{Re} \qquad (4.3.14)$$

Since, by assumption, the Reynolds number is small, we see that this neglected term is indeed small. Consider now $\mathbf{u}_p \cdot \nabla \mathbf{u}_f$. The magnitude of this is of the order of $U_{p0}^2 a / r^2$. Therefore,

$$\frac{\rho_f |\mathbf{u}_p \cdot \nabla \mathbf{u}_f|}{\mu_f |\nabla^2 \mathbf{u}_f|} \approx \frac{\rho_f U_{p0}^2 a / r^2}{\mu_f U_{p0} a / r^3} = \frac{r}{a} \mathrm{Re} \qquad (4.3.15)$$

This shows that the convective acceleration arising from the translation of the particle is small only at distances r that are not far from the surface of the sphere. At larger distances, the magnitude neglected term eventually becomes significant, so that the results are then in error. This was first pointed by Oseen, who derived a correction to Stokes' law that is applicable when the Reynolds number is small, but finite. That correction is briefly discussed in Section 4.5.

4.4 Slowly Changing Motions

Although Stokes' law applies to a small sphere moving steadily in a fluid, or to a small sphere held fixed in a slow, viscous flow, such motions are not common. The main example of the first occurs when a small sphere falls steadily at $\mathrm{Re} \ll 1$. Another steady motion occurs, of course, if a particle is completely entrained by a steadily translating fluid. However, in this case, the fluid exerts no force on the particle because there is no relative motion.

Table 4.4.1. *Dynamic relaxation time τ_d in seconds for a silica particle*

Radius (μm)	In air	In water
1	2.74×10^{-5}	4.29×10^{-7}
5	6.85×10^{-4}	1.07×10^{-5}
10	2.74×10^{-3}	4.29×10^{-5}
50	6.85×10^{-2}	1.07×10^{-3}

In fact, in most situations, the motions of free particles are unsteady and, in the absence of external forces, they are induced by changes in the velocity of the fluid around them. We will consider the effects of these changes on the particle force later. Here, we proceed as we did for the heat transfer calculation and first assume that the force acting on a sphere at rest in a steadily moving fluid would, for Re $\ll$ 1, be given by $F_p = 6\pi\mu_f a u_f$, where u_f is the fluid velocity far from the sphere. Now suppose the particle is not at rest but is moving with a steady velocity u_p, then to a first approximation we may expect that the force is given by Stokes' law, with the instantaneous relative velocity appearing in the formula. That is,

$$F_p = -6\pi\mu_f a(u_p - u_f) \qquad (4.4.1)$$

Of course, this situation is not tenable. Any relative motion between fluid and particle would produce viscous stresses on the particle surface, whose effects would be to reduce the relative velocity – that is to make the motion unsteady. As our discussion of the force on an oscillating sphere shows, unsteadiness produces viscous waves at the surface of the sphere that travel into the fluid with a finite speed. These effects are not taken into account by (4.4.1). However, we anticipate that when the relative velocity changes slowly, these effects are unimportant and take (4.4.1) to be the force on a sphere whose translational velocity relative to the fluid at infinity is $u_p(t) - u_f(t)$. As we show below, (4.4.1) produces correct results for slow motions of aerosol particles.

If (4.4.1) applies, the particle's equation of motion can be written as

$$\frac{du_p}{dt} = -\frac{u_p - u_f}{\tau_d} \qquad (4.4.2)$$

The quantity τ_d appearing in this equation plays a very important role in the dynamics of rigid particles. It is called the dynamic relaxation time of the particle, and is given by $m_p/6\pi\mu_f a$, or

$$\tau_d = \frac{2a^2}{9\nu_f\delta} \qquad (4.4.3)$$

Comparison with (3.2.19) shows that τ_d plays the same role for the particle velocity that τ_t does for the particle temperature when the heat transfer rate is given by (3.2.18).

Furthermore, since both time scales depend on size in the same manner, we see that their ratio depends only on the properties of fluid and particle – that is,

$$\frac{\tau_t}{\tau_d} = \frac{3}{2} \frac{Pr_f}{c_{pp}/c_{pf}} \tag{4.4.4}$$

Thus, for common particle-fluid material combinations, the time required by a particle to achieve thermal and dynamic equilibrium are of the same order of magnitude.

Let us now consider the meaning of the various terms appearing in (4.4.2). Thus, du_p/dt, is the acceleration of the particle as seen by a fixed observer. Sometimes, however, the particle velocity is expressed in terms of its displacement $x_p(t)$ by means of $u_p = dx_p/dt$. This also has a well-defined meaning. On the other hand, the fluid velocity has been vaguely defined as the velocity that the fluid has far from the particle. But the fluid velocity "far" from the sphere does not have a unique meaning, because it can vary with position around the sphere.

To avoid ambiguities, we take u_f to be the velocity that the fluid would have at the location of the particle if the particle were not there. In strictly one-dimensional motions, there is no distinction between the two definitions, but in less simple situations the second definition is more useful. Thus, in the one-dimensional particle equation given by (4.4.2), the fluid velocity can be expressed as $u_f = u_f[x_p(t), t]$. Thus, if u_f is known as a function of time and position in the field, then the position of the particle at each instant can be obtained by means of (4.4.2) and $u_p = \dot{x}_p$. In the next section, we consider two one-dimensional motions that are important in several contexts.

Sphere in Oscillating Fluid

Consider first a sphere immersed in a fluid that is made to execute, as a whole, small-amplitude oscillations along some straight direction at a frequency ω. The fluid's velocity is given by $u_f = U_{f0} \cos \omega t$ at every point in the fluid. Thus, we can place the origin of our system of reference anywhere in the fluid and for simplicity we place it at the center of the sphere's rest position. If we ignore the transient motion that occurs immediately after the fluid oscillations start, we expect the particle to settle into an oscillatory motion whose frequency is equal to that of the fluid. To obtain the particle velocity, we write the fluid's velocity in complex form, $u_f = U_{f0} \Re[\exp(-i\omega t)]$. This gives

$$V^{(S)} = \frac{1}{1 - i\omega\tau_d} \tag{4.4.5}$$

where $V = u_p/u_f$ and the superscript in $V^{(S)}$ is used to remind us that Stokes' law was used in deriving (4.4.5). This result shows that the sphere's amplitude of oscillation is generally smaller than that of the fluid and that the sphere lags behind the fluid. To quantify these statements, we write the sphere's velocity as $V = |V| \cos \eta$, where

$$|V^{(S)}| = \frac{1}{\sqrt{1 + \omega^2 \tau_d^2}} \quad \text{and} \quad \eta = \tan^{-1} \omega\tau_d \tag{4.4.6a,b}$$

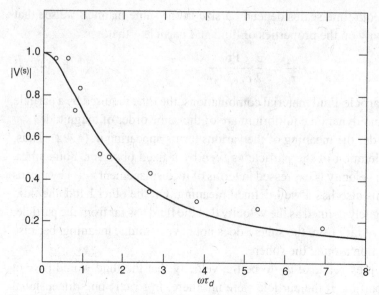

Figure 4.4.1. Magnitude of the nondimensional velocity ratio of a particle in a sound wave. Solid line, (4.4.6a); circle, experimental results of Gucker and Doyle (1956). [Reprinted with permission from Temkin (2001a) © 2001, Acoustical Society of America.]

The sphere's displacement amplitude, $|x_p| = X_{p0}$, can be determined from the velocity, but since the fluid displacement amplitude is U_{f0}/ω, it is evident that $X_{p0}/(U_{f0}/\omega)$ is also given by (4.4.6a). The ratio $|V| = X_{p0}\omega/U_{f0}$ is called the *entrainment ratio* and is shown as the solid line in Fig. 4.4.1 as a function of $\omega\tau_d$. As with $\omega\tau_t$, $\omega\tau_d$ gives the variations with frequency for a given particle size or with size for a given frequency. In either case, the figure shows that when $\omega\tau_d \ll 1$, the particle moves as though it is locked to the fluid. As the frequency increases, the particles are less able to follow the motion of the fluid, and this inability increases with particle size, as (4.4.6) shows. In the high-frequency limit, when $\omega\tau_d \gg 1$, the particles are basically at rest.

The figure also shows experimental results reported by Gucker and Doyle (1956), who measured the displacement of nonvolatile droplets entrained in plane sound waves in air. These direct measurements of the particle displacement – as well as other, less direct methods – show the validity of (4.4.6). However, despite the good agreement between experiments and theory shown in the figure, it should be reiterated that (4.4.6) is based on the heuristic arguments leading to (4.4.1). We shall return to this point in Section 4.6.

Terminal Velocity

As a second example of the use of steady forces in unsteady situations, we consider the time-dependent velocity of a sphere released from rest in a viscous fluid. Of course, the terminal velocity follows directly from a force balance once the drag law is specified.

It is nevertheless useful to see how this terminal velocity is reached after the sphere has been released from rest. Thus, if it is assumed that the viscous force is given, at all instants after release, by Stokes' law, then the equation of motion for a small sphere moving vertically in a fluid under the effects of gravity becomes

$$\frac{du_p}{dt} = (1 - \delta)g - u_p/\tau_d \qquad (4.4.7)$$

where $u_p(t)$ is positive along the direction of $\mathbf{g}$, and $\delta = \rho_f/\rho_p$. The solution to this equation, applicable when the motion starts from rest is

$$u_p(t) = (1 - \delta)g\tau_d \left(1 - e^{-t/\tau_d}\right) \qquad (4.4.8)$$

Thus, the particle velocity increases with time, eventually reaching a time-independent value – the *terminal velocity* – that is given by

$$u_{p\infty} = (1 - \delta)g\tau_d \qquad (4.4.9)$$

Thus, the particle falls or rises depending on the value of the density ratio. Particles for which $\delta = 1$ are said to be neutrally buoyant.

These results for the terminal velocity are of course applicable only for single spheres falling at Re $\ll 1$. When the Reynolds number is not very small, departures from (4.4.9) occur, and these are the basis for the measurement of the particle force. Some of these measurements will be shown in an upcoming section.

In nondilute suspensions of particles denser than the fluid, particle interaction effects result in a slower terminal velocity than applicable to single spheres. The decrease is referred to as hindered sedimentation. For additional information on this important effect, the reader is referred to the Bibliography.

4.5 Extensions to Stokes' Law

As the previous examples show, there may be situations in which small particles move in fluids at small Reynolds, but in conditions that do not exactly match those used in deriving Stokes' law. Consequently, a considerable amount of work has been done to remove some of the restrictions implied by its derivation. The following list gives some of those extensions. In most of these, the fluid force is entirely due to viscous effects and acts along the direction of the relative velocity. To avoid confusion with the force $\mathbf{F}_p$, which may include other effects, we shall use the symbol $\mathbf{F}_D$ for the viscous drag.

Very Small Particles

As the size of the sphere is decreased, the molecular nature of the fluid around it starts to play a role, particularly in gases, where the molecules are normally widely separated. A measure of the importance of molecular effects in gases is usually given in terms of a nondimensional parameter known as the *Knudsen* number. This defined as the ratio of the molecular mean free path to the size of the particle. Thus, $Kn = \ell/a$, where ℓ is the mean free path. In air at 20°C and 1 atm, this distance is of the order of 0.1 μm. Hence, sub-micron particles can be significantly affected by molecular collisions.

A correction factor for Stokes' law that takes into account this effect was proposed, independently, by Millikan and by Cunningham. Millikan's quest to determine the electrical charge of an electron showed that only the first-order term in a power series expansion in terms of Kn was sufficient to account for the molecular nature of fluids. In terms of the drag force, his correction can, for values of Kn as large as 0.4, be expressed as

$$F_D = -6\pi\mu_f a u_p/(1 + A\text{Kn}) \tag{4.5.1}$$

where $A \approx 0.87$. Millikan's experiments are detailed in a small book called *The Electron* (Millikan, 1917). In addition to the importance of its contents, the book gives a good idea of the difficulties associated with measurements involving small particles.

When the Knudsen number is large, the fluid force cannot be computed with the equations of fluid mechanics, because these apply only to continuous media. Indeed, the particles are then so small that they move in response to single molecular collisions with them. These collisions occur randomly and result in random particle displacements. This random motion is called *Brownian* motion in honor of its discoverer.

Spherical Bubbles and Droplets

Although bubbles and droplets cannot strictly be regarded as rigid, the fluid force acting on them during translational motions at small Reynolds numbers, is given by a formula that can be considered an extension of Stokes' law. The formula, obtained by Hadamard in 1911, is

$$F_D = -4\pi\mu_f a u_p \frac{1 + \frac{3}{2}(\mu_p/\mu_f)}{1 + \mu_p/\mu_f} \tag{4.5.2}$$

where μ_p is the viscosity of the fluid in the particle. For rigid particles, $\mu_p/\mu_f \to \infty$, so that this reduces to Stokes' law. Nearly the same result applies to liquid drops in gases. But for gas bubbles in liquids, where $\mu_p/\mu_f \gg 1$, (4.5.2) gives $F_D = -4\pi\mu_f a u_p$. However, experimental measurements with small bubbles show that the drag force is closer to the Stokes' value. The explanation for this is that impurities, normally present in the liquid, collect on the surface of the bubble, thus making the bubble act as a rigid sphere. Finally, the viscosity ratio for immiscible droplets in water is of order 1, so that the drag also differs form Stokes' law. However, even in these cases, Stokes' law may be assumed to hold on account of the unavoidable surface impurities.

Nonuniform Flows

Stokes' law was derived by considering the steady motion of a sphere in a fluid at rest at infinity. This situation is equivalent to one in that the sphere is held at rest in a uniform flow. In many cases, however, spheres are found in flows that cannot be considered uniform. For example, a one-dimensional fluid velocity along a tube is not uniform across the tube. From the perspective of a small particle, such nonuniformities are small, and it is possible to obtain a correction to Stokes' law that takes them into account. This was done by Faxen (1921), who obtained

$$\mathbf{F}_D = 6\pi\mu_f a \left[\mathbf{u}_p - \mathbf{u}_f(\mathbf{x}_p) - \tfrac{1}{6}\nabla^2\mathbf{u}_f(\mathbf{x}_p)\right] \tag{4.5.3}$$

Equation (4.5.3) is referred to as Faxen's first law. The second refers to the torque on a spinning particles. A derivation of the first will be given in Section 4.7.

Nonspherical Particles

Few particles are completely spherical, and much work has been carried out to obtain the equivalent formula for nonspherical particles. We will not discuss this important topic here; a very detailed account of it may be found in the book of Happel and Brenner for those shapes that can be fitted by the coordinates in a system in which Laplace's equation is separable.

More irregular particle shapes can be simulated in terms of the fundamental solutions of the Stokes equations. These are mathematical representations of the pressure and velocity produced by a spherical particle in the limit when the particle radius shrinks to a point. A glimpse to this approach is given in Section 4.7, where we introduce the first of these singular solutions, termed the *stokeslet*. It is evident, however, that irregular particles are generally acted on by torques, so that their translational motion is generally accompanied by rotations. These, also, can be simulated by singularity solutions of the Stokes' equations.

Small but Finite Reynolds Numbers

It was seen in the previous section that Stokes' solution is not valid at points in the fluid that are far from the sphere. This was first pointed out by Oseen in 1910, who obtained a first-order correction to Stokes' law. His result can be written as

$$F_D^{(O)} = -6\pi\mu_f u_p \left(1 + \tfrac{3}{16}\text{Re}\right) \tag{4.5.4}$$

Higher-order corrections have been obtained since then by many investigators, but accurate experimental data obtained by Maxworthy (1965), partially shown in Fig. 4.5.1, show that the Oseen drag is as accurate as any for Reynolds numbers below 0.8. It

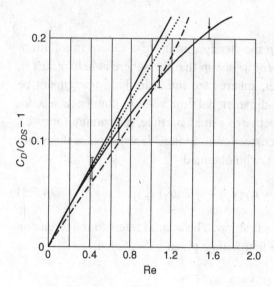

Figure 4.5.1. Nondimensional deviation of the drag coefficient from Stokes' law. Solid line Oseen; dotted line, Goldstein (1929); dash-dot line, Proudman and Pearson (1957); ⊥ Maxworthy's experimental data. [Adapted with permission from Maxworthy (1965). © 1965, Cambridge University Press.]

should be noted that, in the original article, the Reynolds number is defined in terms of the particle radius and not the diameter as used here.

Lift Force

It is evident that, for spherical particles translating without rotation in a uniform flow field, the fluid force acts along the direction of motion (i.e., there are no lift forces). However, small nonuniformities that may exist in the fluid velocity will result in lateral forces, as implied by the first of the Faxen laws. It is also evident that such nonuniformities are generally present, so that spherical particles usually rotate as they translate. But so long as Re → 0, the two motions are not coupled.

This is not the case when the Reynolds number is finite. Here, the rotational and translational motions are coupled and give rise to a lifting force even when the fluid motion is uniform. The lift force acting on a rigid particle, rotating with angular velocity $\mathbf{\Omega}_p$ – while translating at constant velocity $\mathbf{u}_p$ – has been calculated by Rubinow and Keller (1961) in conditions corresponding to Oseen flow. Their result is

$$\mathbf{F}_L = \pi a^3 \rho_f \mathbf{\Omega}_p \times \mathbf{u}_p \tag{4.5.5}$$

Rubinow and Keller also show that, to this order in the Reynolds number, the drag and the torque on the sphere are not affected by the rotation of the particles. Thus, the drag is still given by the Oseen result, and the torque by

$$\mathbf{N} = -8\pi \mu a^3 \mathbf{\Omega}_p \tag{4.5.6}$$

When the fluid velocity is not uniform, the torque is given by Faxen's second law (see Section 4.7).

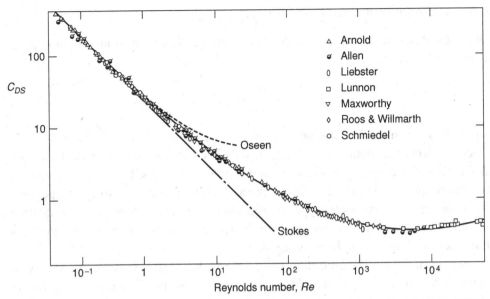

Figure 4.5.2. Steady drag coefficient for solid spheres. [Adapted with permission from Burrows (1983). © 1983, Royal Society of London.]

Empirical Forms at Finite Re

There exists a large body of experimental data that give the viscous drag in non-dimensional form, for Reynolds numbers up to a few thousand. In most experimental works, the viscous resistance is given in terms of a drag coefficient C_D, defined by

$$C_D = \frac{F_D}{\frac{1}{2}\rho_f U_r^2 \pi a^2} \tag{4.5.7}$$

Thus, for example, the Stokes and Oseen drag coefficients are given by

$$C_D^{(S)} = \frac{24}{\mathrm{Re}} \quad \text{and} \quad C_D^{(O)} = \frac{24}{\mathrm{Re}}\left(1 + \frac{3}{16}\mathrm{Re}\right) \tag{4.5.8a,b}$$

where $\mathrm{Re} = 2U_r a/\nu_f$. The experimental data falls somewhere in between these two results, as shown in Fig. 4.5.2. Most of the data shown in the figure were obtained from measurements of the terminal velocities of rigid spheres in viscous fluids under control conditions. Collectively, the data are known as the standard drag, although steady drag is a more appropriate name.

For numerical work, experimental data have been fitted by means of polynomials of high order in Re that cover the entire Re range, or by low-order polynomials, each covering a limited range of the Reynolds number. In some instances, the drag coefficient is modeled by means of simple equations that allow aspects of the motion to be studied analytically. But, in general, when the Reynolds number is finite, numerical methods are required.

4.6 Curvilinear Motion at Finite Re

When the Reynolds number is not very small, the particle's equation of motion is, of course, still given by

$$m_p \frac{d\mathbf{u}_p}{dt} = \mathbf{F} \tag{4.6.1}$$

where $\mathbf{F}$ is the sum of all the forces acting on the particle. Let us consider the fluid force, $\mathbf{F}_p$, when the motion is rectilinear and the Reynolds number is small. This force is known and it is given by (4.2.47). As that equation shows, $\mathbf{F}_p$ contains a viscous forces proportional to the velocity and the acceleration, and a viscosity-independent term, due to added mass effects, also proportional to the acceleration. The equation applies when the fluid velocity far from the sphere is at rest. A result that includes fluid motions is presented later for Re $\ll$ 1.

But when Re is finite, the fluid force is unknown, and it becomes necessary to make certain simplifying assumptions. We first note that the fluid force is not always along the direction of the relative velocity, $\mathbf{U}_r = \mathbf{u}_f - \mathbf{u}_p$. We thus decompose it into a drag force, $\mathbf{F}_D$, acting along that direction, and a lift force, $\mathbf{F}_L$, normal to it (i.e., $\mathbf{F}_p = \mathbf{F}_D + \mathbf{F}_L$). Not much can be said about $\mathbf{F}_L$ beyond what was said previously; but for spherical particles undergoing unsteady motions, it is often assumed that $\mathbf{F}_D$ can be represented by an instantaneous force in terms of the steady drag at the Reynolds number corresponding to that instant. That is, $\mathbf{F}_D(t) = \mathbf{F}_D[C_{DS}(\text{Re})]$, where C_{DS} is the steady drag and

$$\text{Re} = \frac{2a|\mathbf{u}_f - \mathbf{u}_p|}{\nu_f} \tag{4.6.2}$$

For steady motions, this representation of the drag force requires no assumption; but, as pointed out earlier, the vast majority of particle motions are unsteady.

In addition to the representation in terms of C_{DS}, there exist a number of other correlations for C_D that have been used for specific motions. For example, a drag coefficient correlation that is often used for rapidly accelerating particles is

$$C_D = 27/\text{Re}^{0.84} \tag{4.6.3}$$

(Ingebo, 1956). More inclusive lists of such correlations may be found in the review articles of Torobin and Gauvin (1957) and Rudinger (1969), and in the book by Clift, Grace, and Weber (1978).

Now, regardless of the specific correlation used, we need to express $\mathbf{F}_D$ in terms of C_D. To do this, we make use of the fact that it acts along $\mathbf{U}_r$, so that we can write $\mathbf{F}_D = F_D \mathbf{U}_r/|\mathbf{U}_r|$. Hence, using the operational definition of the drag coefficient, (4.5.7), we obtain

$$\mathbf{F}_D = \tfrac{1}{2}\rho_f \pi a^2 C_D(\text{Re})|\mathbf{u}_f - \mathbf{u}_p|(\mathbf{u}_f - \mathbf{u}_p) \tag{4.6.4}$$

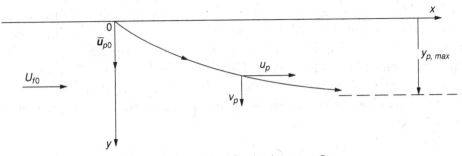

Figure 4.6.1. Schematic particle trajectory after injection in a cross-flow.

If $\mathbf{u}_f$ and $\mathbf{u}_p$ have components $\{u_f, v_f, w_f\}$ and $\{u_p, v_p, w_p\}$, respectively, along the three axes of a Cartesian system of coordinates, the magnitude of the relative velocity is

$$|\mathbf{u}_f - \mathbf{u}_p| = \sqrt{(u_f - u_p)^2 + (v_f - v_p)^2 + (w_f - w_p)^2}$$

As specified above, the drag force is proportional to the square of the relative velocity amplitude. This nonlinear dependence couples the various components of the particle velocity – a fact that does not occur when Re $\ll 1$.

While, in general, this procedure makes it necessary to resort to numerical methods, it is sometimes possible to obtain analytical solutions of the particle's equation of motion, particularly if the Reynolds number applicable to the motion does not change appreciably. The following example (Rudinger, 1974) examines one of those situations and illustrates the coupling that takes place between velocity components when the Reynolds number is finite.

Particle Injection in a Cross-Flow. Consider a one-dimensional flow of gas with uniform velocity U_{f0} along one of the directions of a rectangular coordinate system, which we take as the x-axis. At some point in the flow, rigid particles are injected in a direction perpendicular to the flow with a velocity v_{p0}, so that their initial momentum has only a vertical component. To study this motion, we assume that there is no gravity, that the particles do not rotate, and that the drag coefficient is given by

$$C_D = \frac{24}{\text{Re}} \left(1 + \tfrac{1}{6}\text{Re}^{2/3}\right) \tag{4.6.5}$$

If the initial Reynolds number is very small, the drag force is given by Stokes' law, which shows that the particle will continue to move along the vertical with the same velocity, while acquiring a velocity along the x-axis as a result of the cross-wind. What happens when the Reynolds number is not very small? To answer this, we place the origin of our system of coordinates at the point of injection, with the vertical direction coinciding with the y-axis, as sketched in Fig. 4.6.1. The velocity of the particle has, in general, components $\{u_p, v_p, 0\}$, but the fluid has only one component and this is along the x-axis. We note that the introduction of a particle into the fluid will, in the vicinity of the particle, produce a velocity perturbation that is three-dimensional. However, as pointed out previously, the fluid velocity $\mathbf{u}_f$ appearing in the drag force is the velocity that the fluid would have at the location of the particle if the particle were not there. That is, $\mathbf{u}_f = \{U_{f0}, 0, 0\}$. The motion clearly takes place on the (x, y) plane, and the magnitude of the relative velocity is $|\mathbf{u}_f - \mathbf{u}_p| = \sqrt{(U_{f0} - u_p)^2 + v_p^2}$. Therefore, the particle acceleration has only two components, and these are prescribed by (4.6.1). Thus,

$$m_p \frac{du_p}{dt} = \tfrac{1}{2}\rho_f \pi a^2 C_D |\mathbf{u}_f - \mathbf{u}_p|(U_{f0} - u_p) \tag{4.6.6}$$

$$m_p \frac{dv_p}{dt} = -\tfrac{1}{2}\rho_f \pi a^2 C_D |\mathbf{u}_p - \mathbf{u}_f| v_p \tag{4.6.7}$$

Dividing the first of these by the second and integrating the result between t_0, when $u_p = 0$, $v_p = v_{p0}$, and t, when their values are u_p and v_p, respectively, gives

$$U_{f0} - u_p = U_{f0}\frac{v_p}{v_{p0}} \tag{4.6.8}$$

This is applicable for all values of C_D and shows that, as $u_p \to U_{f0}$, the vertical component of the particle velocity approaches zero. Thus, contrary to the zero Re case, the particles can penetrate into the fluid only a finite vertical distance.

Let us determine this maximum penetration. First, we use (4.6.5) and (4.6.8) to write (4.6.7) as

$$\frac{dv_p}{dt} = -\frac{1 + \tfrac{1}{6}(2av_{p0}/v_f)^{2/3}(v_p/v_{p0})^{2/3}[1 + (U_{f0}/v_{p0})]^{1/3}}{\tau_d} v_p \tag{4.6.9}$$

If the vertical distance of the particle from the point of injection is denoted by y_p, we have $dv_p/dt = v_p(dv_p/dy_p)$, so that

$$\frac{dv_p}{dy_p} = -\frac{1 + \alpha(v_p/v_{p0})^{2/3}}{\tau_d} \tag{4.6.10}$$

where

$$\alpha = \frac{1}{6}(2av_{p0}/v_f)^{2/3}\left[1 + (U_{f0}/v_{p0})^2\right]^{1/3} \tag{4.6.11}$$

Putting $\zeta = (v_p/v_{p0})^{1/3}$ in the equation for v_p gives, after integration

$$\frac{y_p}{v_{p0}\tau_d} = \frac{3}{\alpha}\left[1 - \zeta + \frac{\tan^{-1}(\zeta\sqrt{\alpha})}{\sqrt{\alpha}} - \frac{\tan^{-1}\sqrt{\alpha}}{\sqrt{\alpha}}\right] \tag{4.6.12}$$

This gives the vertical position of the particle as a function of the of the parameter α and of the nondimensional vertical velocity ζ. This is unity at the time of injection, but decreases with time, eventually vanishing when the particle has reached the maximum penetration $y_{p,\max}$. Thus, letting $\zeta \to 0$, we obtain

$$y_{p,\max}/v_{p0}\tau_d = (3/\alpha)[1 - \tan^{-1}\sqrt{\alpha}/\sqrt{\alpha}] \tag{4.6.13}$$

The distance $v_{p0}\tau_d$ appearing here is the particle's penetration in a time equal to τ_d, when Re $\to 0$. This also follows from (4.6.13) by letting $\alpha \to 0$. Now, $[1 - \tan^{-1}\sqrt{\alpha}/\sqrt{\alpha}]$ ranges between $\alpha/3$ when $\alpha \to 0$, and unity, when $\alpha \to \infty$. Hence, we always have $y_{p,\max} \leq v_{p0}\tau_d$. Equation (4.6.13) also shows that when the Reynolds number is finite, there is coupling between the two components of the equation of motion, which allows for the transfer of momentum between the two directions.

PROBLEMS

1. Consider a rigid sphere of radius a in a suddenly imposed flow field having a constant velocity U_{f0}. The sphere is initially at rest, but accelerates rapidly in time. Assuming that the forces on it are well described by $C_D = 27/\text{Re}^{5/6}$, show that the time required by the particle to reach a fractional velocity u_p/U_{f0} is (Hoenig, 1957)

$$\Delta t = \frac{16}{27}\frac{a}{U_{f0}}\text{Re}^{5/6}\left[\frac{1}{(1 - u_p/U_{f0})^{1/6}} - 1\right]$$

2. Consider a rigid sphere released from rest in a viscous fluid, also at rest. Show that when the Oseen drag is used, its terminal velocity is given $v_{p\infty}^{(O)}/v_{p\infty}^{(S)} = [\sqrt{1+4\beta} - 1]/2\beta$, where $\beta = 3v_{p\infty}^{(S)}a/8v_f < 1$, and where $v_{p\infty}^{(S)} = (1 - \delta)g\tau_d$ is the terminal velocity of the particle under Stokes' law conditions.

3. The equation of motion for the previous problem can be expressed as $dw_p/dt' = 1 - w_p - \beta w_p^2$, where $w_p = v_p/v_{p\infty}$ and $t' = t/\tau_d$. The right-hand side of the equation can also be written as $(w_p - w_1)(w_p - w_2)$. Hence, show that the vertical velocity under Oseen' drag is given, at all times, by $v_p^{(O)}/v_{p\infty}^{(S)} = \{\sqrt{1+4\beta}\tanh[\frac{t'}{2}\sqrt{1+4\beta} + \tanh^{-1}(1 + 4\beta)^{-1/2}] - 1\}/2\beta$.

4.7 The Stokeslet

The Stokes solution for the flow past a sphere owes its existence to the linearity of the equations and to the simplicity of the geometry of the problem. But, for more irregular geometries, recourse must be made to numerical techniques. One such technique is the boundary integral method. Here, the surfaces of the particles and the walls around them are simulated in terms of distributions of the fundamental solutions of the Stokes equations. Such methods are well known in fluid mechanics, where it is shown that mass point sources, dipoles, and quadrupoles can be used to simulate a variety of flows.

In rigid-particle fluid mechanics, the mass point source has zero strength, because the particles have constant volume, so that the lowest order singularity-type solution that is relevant is the dipole. This is made of a source-sink pair and represents a force acting at a point. In the case of incompressible fluid motions at low Reynolds numbers, this singularity solution is known as the *stokeslet*.

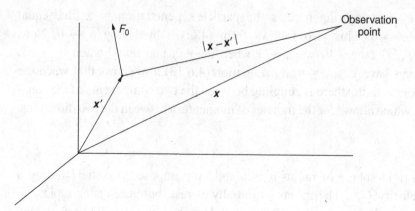

Figure 4.7.1. Stokeslet at $\mathbf{x}'$.

To introduce the stokeslet, we consider a rigid particle moving slowly in a viscous, incompressible fluid. The particle induces a fluid velocity whose magnitude decreases slowly with distance. Far from the particle, the particle appears as a point so that we may regard that velocity as being produced by a point force. But, as we get closer to the particle, its finite size becomes apparent, and it is evident that each element of area on its surface exerts a force on the fluid. Stated differently, each point on the surface can be regarded as a point force. Thus, we may simulate rigid boundaries of any shape in terms of point forces distributed over those portions of the fluid that correspond to points on the surfaces of the boundaries being simulated. Furthermore, since the Stokes equations are linear, we can obtain the flow corresponding to the actual particle by adding the separate contributions due to each point force. Of course, for the method to be successful, we need the magnitude and direction of each point force.

In order to use this approach, we first note that, in the derivation of the equation for the conservation of momentum of the fluid, it was assumed that the momentum of a material volume could change only by the application of external forces. But when forces exist within the region, the momentum equation must be modified so as to include them. Thus, if $\mathbf{f}(\mathbf{x}, t)$ represents the point force *per unit volume* acting at location $\mathbf{x}$ at time t, then, in the absence of external body forces, the linearized momentum equation would be

$$\rho_f \frac{\partial \mathbf{u}_f}{\partial t} + \nabla p_f = \mu_f \nabla^2 \mathbf{u}_f + \mathbf{f}(\mathbf{x}, t) \qquad (4.7.1)$$

Taking the gradient of this, we obtain

$$\nabla^2 p_f = \nabla \cdot \mathbf{f} \qquad (4.7.2)$$

whether $\mathbf{f}$ is time dependent or not. In what follows, we consider only the steady case, and proceed to obtain a solution to this equation, assuming first that there is only one point force in the region. Thus, we assume that at some point $\mathbf{x}'$ in the fluid, a force $\mathbf{F}_0$ acts on the fluid, as shown in Fig. 4.7.1. Then, the force per unit volume $\mathbf{f}$ may be expressed as $\mathbf{f} = \mathbf{F}_0 \delta(\mathbf{x} - \mathbf{x}')$, where $\delta(\mathbf{x} - \mathbf{x}')$ is the three-dimensional Dirac delta function, given by $\delta(x - x')\delta(y - y')\delta(z - z')$. This gives

$$\nabla^2 p_f = \mathbf{F}_0 \cdot \nabla \delta(\mathbf{x} - \mathbf{x}') \tag{4.7.3}$$

Now, $\delta(\mathbf{x} - \mathbf{x}')$ is zero everywhere, except at $\mathbf{x} = \mathbf{x}'$, where it tends to infinity in such a way that for some function $g(\mathbf{x})$

$$\int_V g(\mathbf{x})\delta(\mathbf{x} - \mathbf{x}')dV(\mathbf{x}) = g(\mathbf{x}')$$

provided the point $\mathbf{x} = \mathbf{x}'$ is included in the volume of integration. Thus, for a single point force, the Laplacian of the pressure vanishes everywhere, except at the location of the force, where it is undefined.

To obtain the solution of (4.7.3), we first show that

$$\nabla^2 \frac{1}{|\mathbf{x} - \mathbf{x}'|} = -4\pi \delta(\mathbf{x} - \mathbf{x}') \tag{4.7.4}$$

According to this, the Laplacian of $1/|\mathbf{x} - \mathbf{x}'|$ is equal to zero everywhere, except at the point $\mathbf{x} = \mathbf{x}'$, where it becomes infinity in such a manner that an integral over an infinitesimal volume surrounding that point is equal to -4π.

To prove these statements, we let $\mathbf{s} = \mathbf{x} - \mathbf{x}'$ and $s = |\mathbf{x} - \mathbf{x}'|$. Then, if $\mathbf{s} \neq 0$, $\nabla s^{-1} = -\mathbf{s}/s^3$ and $\nabla s^{-3} = -3\mathbf{s}/s^3$. These can be used to show that $\nabla^2 |\mathbf{x} - \mathbf{x}'|^{-1} = 0$ when $\mathbf{x} \neq \mathbf{x}'$. Now, consider the point $\mathbf{x} = \mathbf{x}'$. We need to show that the volume integral of $\nabla^2 |\mathbf{x} - \mathbf{x}'|^{-1}$ around that point is equal to -4π. Using the divergence theorem, we may express this volume integral as

$$\int_V \nabla^2 \frac{1}{|\mathbf{x} - \mathbf{x}'|} dV = -\int_A \frac{\mathbf{n} \cdot (\mathbf{x} - \mathbf{x}')}{|\mathbf{x} - \mathbf{x}'|^3} dA$$

where A is the area bounding volume V and $\mathbf{n}$ is the outward unit normal vector to A. We take A to be a spherical surface of small radius ε, centered at $\mathbf{x}'$. On this surface, we have $\mathbf{x} - \mathbf{x}' = \varepsilon\mathbf{n}$ so that $\mathbf{n} \cdot (\mathbf{x} - \mathbf{x}') = \varepsilon$. The element of area on this sphere is $dA = \varepsilon^2 \sin\vartheta$ so that

$$\int_V \nabla^2 \frac{1}{|\mathbf{x} - \mathbf{x}'|} dV = -\int_0^{2\pi} d\varphi \int_0^{\pi} \sin\vartheta\, d\vartheta = -4\pi$$

as was stated.

Having proved the validity of (4.7.4), we may take that equation as another operational definition of $\delta(\mathbf{x} - \mathbf{x}')$. We may therefore use it in (4.7.3) to obtain

$$\nabla^2 \left[p_f + \frac{1}{4\pi} \mathbf{F}_0 \cdot \nabla \frac{1}{|\mathbf{x} - \mathbf{x}'|} \right] = 0$$

Without loss of generality, we may put the quantity inside the square bracket equal to zero. Thus, the excess pressure induced at point $\mathbf{x}$ in the fluid is

$$p_S(\mathbf{x}) = -\mathbf{F}_0 \cdot \nabla \frac{1}{4\pi |\mathbf{x} - \mathbf{x}'|} \qquad (4.7.5)$$

where the subindex s is used as a reminder that this is the pressure produced by a point force or stokeslet. The result may also be expressed as

$$p_S(\mathbf{x}) = \frac{\mathbf{F}_0 \cdot (\mathbf{x} - \mathbf{x}')}{4\pi |\mathbf{x} - \mathbf{x}'|^3} \qquad (4.7.6)$$

Either form shows that the pressure varies as the angle between the force and the position vector of the observation point, relative to the location of the force. That dependence is typical of dipole fields.

Velocity. To describe the flow field produced by the point force, we need to calculate the velocity corresponding to the pressure specified by (4.7.5). Thus, when there is a single, steady point force at $\mathbf{x}'$, we have

$$\nabla^2 \mathbf{u}_S = \frac{1}{\mu_f} \nabla p_S - \frac{1}{\mu_f} \mathbf{F}_0 \delta(\mathbf{x} - \mathbf{x}') \qquad (4.7.7)$$

Using the result for the pressure, the right-hand side of this can be written as

$$-\frac{1}{\mu_f} \mathbf{F}_0 \delta(\mathbf{x} - \mathbf{x}') + \nabla \frac{(\mathbf{x} - \mathbf{x}') \cdot \mathbf{F}_0}{4\pi \mu_f |\mathbf{x} - \mathbf{x}'|}$$

Because of the linearity of (4.7.7) we can split the velocity $\mathbf{u}_s$ into two parts, $\mathbf{u}_s^{(1)}$ and $\mathbf{u}_s^{(2)}$, which satisfy, respectively,

$$\nabla^2 \mathbf{u}_S^{(1)} = -\frac{1}{\mu_f} \mathbf{F}_0 \delta(\mathbf{x} - \mathbf{x}') \quad \text{and} \quad \nabla^2 \mathbf{u}_S^{(2)} = \nabla \frac{(\mathbf{x} - \mathbf{x}') \cdot \mathbf{F}_0}{4\pi \mu_f |\mathbf{x} - \mathbf{x}'|} \qquad (4.7.8\text{a,b})$$

In analogy with (4.7.3), a solution of the first of these is

$$\mathbf{u}_S^{(1)} = \frac{\mathbf{F}_0}{4\pi \mu_f |\mathbf{x} - \mathbf{x}'|} \qquad (4.7.9)$$

For the second, we first note that

$$\frac{(\mathbf{x} - \mathbf{x}') \cdot \mathbf{F}_0}{|\mathbf{x} - \mathbf{x}'|} = -\frac{1}{2} \nabla^2 \frac{(\mathbf{x} - \mathbf{x}') \cdot \mathbf{F}_0}{|\mathbf{x} - \mathbf{x}'|^3}$$

so that (4.7.8b) can be written as

$$\nabla^2 \mathbf{u}_S^{(2)} = -\nabla^2 \left[\nabla \frac{(\mathbf{x} - \mathbf{x}') \cdot \mathbf{F}_0}{8\pi \mu_f |\mathbf{x} - \mathbf{x}'|} \right]$$

Therefore, a solution for $\mathbf{u}_S^{(2)}$ is

$$\mathbf{u}_S^{(2)} = -\nabla \frac{|\mathbf{x} - \mathbf{x}'| \cdot \mathbf{F}_0}{8\pi \mu_f |\mathbf{x} - \mathbf{x}'|} \qquad (4.7.10)$$

Adding this to $\mathbf{u}_S^{(1)}$ gives the induced velocity

$$\mathbf{u}_S = \frac{1}{8\pi\mu_f}\left[\frac{\mathbf{F}_0}{|\mathbf{x}-\mathbf{x}'|} + \frac{\mathbf{F}_0\cdot(\mathbf{x}-\mathbf{x}')}{|\mathbf{x}-\mathbf{x}'|^3}(\mathbf{x}-\mathbf{x}')\right] \qquad (4.7.11)$$

where we have used the fact that $(\mathbf{F}_0\cdot\nabla)\mathbf{s} = \mathbf{F}_0$.

The quantity inside the square brackets in (4.7.11) depends on the magnitude and direction of $\mathbf{F}_0$ and on the geometry of the problem, as described by the distance between the stokeslet and the observation point. It is useful to separate the two, and for this purpose, we write (4.7.11) in component form, as

$$u_{Si}(\mathbf{x}) = \frac{1}{8\pi\mu_f}O_{ij}(\mathbf{x}-\mathbf{x}')F_{0j} \qquad (4.7.12)$$

where

$$O_{ij}(\mathbf{x}-\mathbf{x}') = \frac{\delta_{ij}}{|\mathbf{x}-\mathbf{x}'|} + \frac{(x_i - x_i')(x_j - x_j')}{|\mathbf{x}-\mathbf{x}'|^3} \qquad (4.7.13)$$

is known as *Oseen's tensor*. It follows from these equations that if the force, $\mathbf{F}_0$, and its location, $\mathbf{x}'$, are known, equations (4.7.5) and (4.7.12) describe the flowfield at $\mathbf{x}$.

Distributions of Stokeslets

We may use the previous results to obtain solutions for arbitrary distributions of stokeslets that occupy some region in a fluid. The most important application of the method refers to surface distributions, and these may be considered as a special limiting form of a volume distribution, which we consider first.

Volume Distributions. Suppose that instead of a single point force at $\mathbf{x}'$, we have point forces distributed over some volume. Then, the fluid velocity induced at some point $\mathbf{x}$ is obtained by adding their separate contributions. If the distribution is continuous, we get the velocity by integrating over the volume containing the point forces. Thus, if $\mathbf{f}(\mathbf{x}')$ represents the force per unit volume, the total force in an element of volume $dV(\mathbf{x}')$ around $\mathbf{x}'$ is $\mathbf{f}(\mathbf{x}')dV(\mathbf{x}')$, giving a total force, or strength equal to $\mathbf{F}_s = \int \mathbf{f}(\mathbf{x}')dV(\mathbf{x}')$. Similarly, the total velocity induced by the distribution is

$$u_{Si} = \frac{1}{8\pi\mu_f}\int O_{ij}(\mathbf{x}-\mathbf{x}')f_j(\mathbf{x}')dV(\mathbf{x}') \qquad (4.7.14)$$

If, in addition to this induced field, the fluid has a background velocity $\mathbf{u}_f(\mathbf{x})$, then the total fluid velocity at $\mathbf{x}$ is given by the sum of the two contributions, keeping in mind that $\mathbf{u}_f(\mathbf{x})$ is to be interpreted as the velocity that the fluid would have at the location of the particle if the particle were not there.

Surface Distributions. We now limit the results obtained to situations where the point forces are distributed only over some surface A_0 (e.g., the surface of a particle).

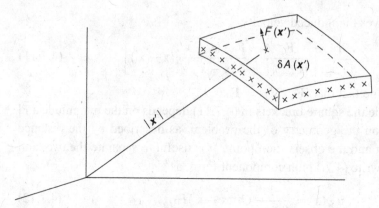

Figure 4.7.2. Surface distribution of stokeslet.

To do this, we consider a volume element enclosing part of the surface, as shown in Fig. 4.7.2, and express the volume element as $dV(\mathbf{x}') = dA(\mathbf{x}')d\ell$, where $d\ell$ is the thickness of the element. As we let this thickness go to zero, the product $\mathbf{f}(\mathbf{x}')d\ell(\mathbf{x}')$ becomes the local force per unit area acting on the fluid. By definition, this force per unit area is the local stress Σ. Hence, the fluid velocity induced at a point $\mathbf{x}$ by the surface distribution is

$$u_{Si}(\mathbf{x}) = \frac{1}{8\pi\mu_f} \int O_{ij}(\mathbf{x} - \mathbf{x}')\Sigma_j(\mathbf{x}')dA(\mathbf{x}') \qquad (4.7.15)$$

The local stress is related to the total force exerted by the distribution through

$$\mathbf{F}_s = \int \Sigma(\mathbf{x}')dA(\mathbf{x}') \qquad (4.7.16)$$

It should be noted that if the observation point $\mathbf{x}$ lies on the distribution's surface, the velocity there is influenced by other points in the distribution, including $\mathbf{x}' = \mathbf{x}$. Here, the Oseen tensor is infinite, but the induced velocity is finite, as may be verified by considering a small surface element around with center at $\mathbf{x}$.

Faxen's Laws. As an application of the representation of a particle in terms of surface distributions of stokeslets, we obtain the fluid force on a sphere of radius a, translating steadily in a fluid whose background velocity, $\mathbf{u}_f(\mathbf{x})$, is not uniform, but is a known function of position. To simulate the flow produced by the particle, we represent its surface by a uniform distribution of stokeslets, placed over the surface of a sphere of radius equal to that of the particle. These produce, at some point $\mathbf{x}$, a fluid velocity that is equal to (4.7.15). The total fluid velocity is obtained by adding this to the base flow velocity $\mathbf{u}_f$. Thus, the ith component of the total velocity is

$$U_{fi}(\mathbf{x}) = u_{fi}(\mathbf{x}) + \frac{1}{8\pi\mu_f} \int_{A_0} O_{ij}(\mathbf{x} - \mathbf{x}')\Sigma_j(\mathbf{x}')dA(\mathbf{x}') \qquad (4.7.17)$$

where A_0 is an area equal to that of the sphere.

We still need the surface stress, Σ. To obtain it, we use the fact that the total velocity $U_f(\mathbf{x})$ must satisfy on A_0 the same boundary conditions as the actual velocity. It must therefore be equal to the velocity of the particle at points on its surface. This is equal to the uniform velocity $\mathbf{u}_p$ plus a component due to the rotation of the particle, if it exists. Thus, if we denote the position vector of points on the sphere's surface by $\mathbf{x}_s$, the boundary condition can be written as $U_f(\mathbf{x}_s) = \mathbf{u}_p + \Omega_p \times (\mathbf{x}_s - \mathbf{x}_p)$, where $\mathbf{x}_p$ is the position of the center of the sphere and Ω_p is the angular velocity. Expressing the ith component of the rotational contribution to the velocity as $\varepsilon_{ijk}\Omega_{pj}(x_{sk} - x_{pk})$, where ε_{ijk} is the alternating sign tensor, we have, by (4.7.17)

$$\frac{1}{8\pi\mu_f} \int_{A_0} O_{ij}(\mathbf{x}_s - \mathbf{x}')\Sigma_j(\mathbf{x}')d A(\mathbf{x}') = u_{pi} - u_{fi}(\mathbf{x}_s) + \varepsilon_{ijk}\Omega_{pj}(x_{sk} - x_{pk})$$

(4.7.18)

This equation fixes the value of Σ. To obtain it, we multiply both sides by $d A(\mathbf{x}_s)$ and integrate over A_0. This gives, after exchanging the order of integration,

$$\int_{A_0} I_{ij}(\mathbf{x}')\Sigma_j(\mathbf{x}')d A(\mathbf{x}') = 8\pi\mu_f \int_{A_0} [u_{pi} - u_{fi}(\mathbf{x}_s) + \varepsilon_{ijk}\Omega_j(x_{sk} - x_{pk})]d A(\mathbf{x}_s)$$

(4.7.19)

where

$$I_{ij}(\mathbf{x}') = \int_{A_0} O_{ij}(\mathbf{x}_s - \mathbf{x}')d A(\mathbf{x}_s)$$
(4.7.20)

This quantity is a second-order tensor whose components depend only on the shape of the body that is being simulated. It may therefore be regarded as known. For the case of the sphere, I_{ij} can be obtained exactly. Thus, noting that $\mathbf{x}_s - \mathbf{x}_p = a\mathbf{n}$, so that $|\mathbf{x}_s - \mathbf{x}_p| = a$, we have

$$I_{ij}(\mathbf{x}') = \frac{1}{a} \int_{A_0} [\delta_{ij} + n_i n_j]d A(\mathbf{x}_s)$$
(4.7.21)

The first term contributes an amount equal to $4\pi a\delta_{ij}$, whereas the second contributes one-third of that amount. Thus, the value of I_{ij} is equal to $\frac{16}{3}\pi a\delta_{ij}$. Hence, the left-hand side of (4.7.19) vanishes, unless $i = j$. But when $i = j$, the alternating sign tensor vanishes, so that equation yields

$$\int_{A_0} \Sigma_i(\mathbf{x}')d A(\mathbf{x}') = \frac{3}{2}\mu_f \int_{A_0} [u_{pi} - u_{fi}(\mathbf{x}_s)]d A(\mathbf{x}_s)$$
(4.7.22)

Now, the left-hand side of this equation is the ith component of the force on the sphere, but, as the right-hand side shows, this force is along the direction of the relative velocity and thus represents the drag force $\mathbf{F}_D$. When the background velocity is uniform, we obtain $F_{Di} = 3(u_{pi} - u_{fi})A_0/2$, which is Stokes' law. If $\mathbf{u}_f$ is not uniform, but varies only slightly over distances that are of the order of the sphere radius, we can expand

$u_{fi}(\mathbf{x}_s)$ in Taylor series around $\mathbf{x}_p$, the center of the sphere. If the background fluid velocity there is $\mathbf{u}_f(\mathbf{x}_p)$, then the velocity at points $\mathbf{x}_s$ is, approximately,

$$u_{fi}(\mathbf{x}_s) \approx u_{fi}(\mathbf{x}_p) + (x_{sj} - x_{pj})\left(\frac{\partial u_{fi}}{\partial x_j}\right)_{\mathbf{x}_p}$$

$$+ \frac{1}{2}(x_{sj} - x_{pj})(x_{sk} - x_{pk})\left(\frac{\partial^2 u_{fi}}{\partial x_j \partial x_k}\right)_{\mathbf{x}_p}$$

We now substitute this into (4.7.22) and carry out the indicated integration term by term. The first term yields Stokes' law, as described-previously. The second vanishes because the fluid is incompressible. For the variable part in the third, we have, since $\int_{A_0} n_i n_j dA = {}^4/_3\pi a^2 \delta_{ij}$,

$$\int_{A_0} (x_{sj} - x_{pj})(x_{sk} - x_{pk})dA(\mathbf{x}_s) = \frac{4}{3}\pi a^4 \delta_{jk} \qquad (4.7.23)$$

Therefore, we obtain, using vector notation

$$\mathbf{F}_D = 6\pi\mu_f a\left[\mathbf{u}_p - \mathbf{u}_f(\mathbf{x}_p) - \frac{1}{6}\nabla^2\mathbf{u}_f(\mathbf{x}_p)\right] \qquad (4.7.24)$$

This result is known as Faxen's first law. This derivation is due to Batchelor (1972), who also obtains the torque on the particle. This is Faxen's second law and is given by

$$\mathbf{N}_p = -8\pi\mu_f a^2\left[\frac{1}{2}\nabla \times \mathbf{u}_f(\mathbf{x}_p) - \Omega_p\right] \qquad (4.7.25)$$

If the particle is torque-free, this shows that it would rotate with an angular velocity equal to one-half the vorticity of the fluid at the location of the sphere. By definition, this is the local angular velocity of the fluid. Thus, as might have been expected on physical grounds, we find that a torque-free particle rotates with the angular velocity of the fluid. The derivation also shows that, when the Reynolds number is small, the drag force and the viscous torque on a sphere are not coupled.

In addition to the stokeslet, there are other types of singularities that are used to simulate stokesian flows with particles. These techniques are described at length in the book of Kim and Karrila. The method has also been used to simulate unsteady motions (see, e.g., Pozrikidis, 1989). Earlier applications of the method to the study of macromolecules may be found in a book by Kirkwood (1967).

4.8 Unsteady Effects at Re $\ll$ 1

In Sections 4.4 and 4.5, we studied some unsteady motions of small spheres under the assumption that all changes occurred slowly. This allowed us to use Stokes' law, in which the fluid force is proportional to the instantaneous relative velocity between particle and fluid. But, in reality, that force also depends on the acceleration, as we saw for a sphere that is executing translational oscillations at a single frequency, in which case

$$F_p = -6\pi \mu_f a(1+y)u_p - \frac{1}{2}m_p\delta\left(1 + \frac{9}{2y}\right)\frac{du_p}{dt} \qquad (4.8.1)$$

In this section, we examine the effects of acceleration of both fluid and particle by considering the force on a particle that is executing translational oscillations in response to a monochromatic oscillation of the fluid, as may occur when a container filled with an incompressible fluid is made oscillate as a whole. But, before we can do that, we must evaluate the force on a small spherical particle in the oscillatory fluid. This can be obtained from the above result, provided we take into account the fluid motion. Thus, if the fluid is also executing translational oscillations having a single frequency, the force on the sphere could be obtained by solving the unsteady Stokes equations, referred to axes moving with the fluid, instead of the fixed system used to obtain (4.8.1). The analysis would be similar to that used in Section 4.2, except for two changes, which can be incorporated directly into the final result. Firstly, instead of u_p, we must use the particle velocity relative to the oscillatory fluid, $u_p - u_f$. This change gives a force

$$-6\pi \mu_f a (1+y)(u_p - u_f) - \frac{1}{2}\delta m_p\left(1 + \frac{9}{2y}\right)\frac{d(u_p - u_f)}{dt}$$

A second change is required because the frame of reference is accelerating. To obtain this, we add a fictitious force to the fluid's momentum equation. Since the fluid acceleration is du_f/dt, the fictitious force on a mass of fluid having the same volume as that of the sphere is

$$\rho_f v_p \frac{du_f}{dt} = m_p\delta\frac{du_f}{dt}$$

The need for this force may also be seen by considering a sphere moving exactly with the same velocity as the fluid. As such, it must experience the same force that a fluid particle having the same volume and shape experiences. If the fluid is at rest or is moving uniformly, this force is clearly zero; but, if the fluid is accelerating, the force on the fluid particle is given by the above amount, because of Newton's second law.

Adding the two force components, we obtain the fluid force on the particle:

$$F_p = m_p\delta\frac{du_f}{dt} - 6\pi \mu_f a (1+y)(u_p - u_f) - \frac{1}{2}m_p\delta\left(1 + \frac{9}{2y}\right)\frac{d(u_p - u_f)}{dt} \qquad (4.8.2)$$

Let us discuss the meaning of each of the terms appearing in this equation. The origin of the first term was discussed previously. The next term is the Stokes force, modified by the term $y = \sqrt{\omega a^2/2v_f}$, which as we know is the ratio of sphere radius to the viscous penetration depth, and is therefore a measure of the thickness of the viscous layer around the sphere. In a sense, this term changes the radius of the sphere from a, at zero frequencies, to $a(1 + y)$ at a finite frequency. The last term in (4.8.2) is best understood by considering first the case when the fluid's viscosity is zero. It then becomes

$$-\frac{1}{2}m_p\delta\frac{d(u_p - u_f)}{dt}$$

and represents the fluid's reaction force that acts on the particle as a result of the particle's acceleration relative to the fluid. It is called the *acceleration reaction*. Its effect, as we show herein, is to increase the effective mass of the sphere. When viscosity is included, the acceleration reaction is modified by the second term, $9/2y$, whose origin is the delayed reaction resulting from viscous effects being propagated into the fluid with a finite velocity.

Sphere in Oscillatory Fluid

We now consider the velocity of a sphere immersed in a fluid that is executing uniform translational oscillations. The situation was analyzed in Section 4.4.1, using Stokes' law. We again assume that there are no external forces, so that the sphere's equation of motion now becomes

$$m_p\left[1 + \frac{\delta}{2}\left(1 + \frac{9}{2y}\right)\right]\frac{du_p}{dt} = m_p\delta\frac{du_f}{dt} - 6\pi\mu_fa(1 + y)(u_p - u_f)$$
$$+ \frac{1}{2}m_p\delta\left(1 + \frac{9}{2y}\right)\frac{du_f}{dt} \qquad (4.8.3)$$

The coefficient of the sphere's acceleration on the left-hand side of this equation may be interpreted as the effective mass of the sphere. It consists of the actual mass, plus an amount proportional to $\frac{1}{2}\delta(1 + 9/2y)$. This quantity is therefore referred to as the coefficient of *added mass*. The first term in it is the inviscid fluid value, $\delta/2$. But, for viscous fluids, that constant value is modified by the second term, $9\delta/4y$, whose value depends on the frequency.

Now, since the sphere's motion is monochromatic, it is advantageous to write u_f as $\Re\{U_{f0}\exp(-i\omega t)\}$, with U_{f0} real. But, to save writing, we will take advantage of the linearity of the equations and simply write $u_f = U_{f0}\exp(-i\omega t)$, keeping in mind that only the real part is implied. We are interested in stationary oscillations. Here, the particle oscillates with the same frequency as the fluid. Therefore, the solution is of the form $u_p = U_{p0}\exp(-i\omega t)$, where U_{p0} is generally complex. Substitution of these complex velocities into (4.8.3) and solving for u_p/u_f yields this ratio when the fluid is viscous and incompressible. To remind us of this, we denote this ratio by $V^{(V,I)}$. Thus

$$V^{(V,I)} = 3\delta\frac{y(2y + 3) + 3i(1 + y)}{2y^2(2 + \delta) + 9y\delta + 9i\delta(1 + y)} \qquad (4.8.4)$$

where we have expressed τ_d in terms of y by means of

$$\omega\tau_d\delta = 4y^2/9 \qquad (4.8.5)$$

For future reference, we note that the relative fluid velocity is

$$u_f - u_p = \frac{4y^2(1-\delta)}{2y^2(2+\delta) + 9y\delta + 9i\delta(1+y)}u_f \qquad (4.8.6)$$

This shows, more clearly than (4.8.4), that the fluid and particle move together when the frequency is very small or when the density ratio is 1.

The amplitude of the velocity ratio is

$$\left|V^{(V,I)}\right| = 3\delta\sqrt{\frac{4y^4 + 12y^3 + 18y^2 + 18y + 9}{4(2+\delta)^2y^4 + 36\delta y^3(2+\delta) + 81\delta^2(2y^2 + 2y + 1)}}. \qquad (4.8.7)$$

Similarly, the phase angle, $\eta^{(V,I)}$, is given by

$$\tan\eta^{(V,I)} = \frac{12(1+y)y^2(1-\delta)}{4y^4(2+\delta) + 12y^3(1+2\delta) + 27\delta(2y^2 + 2y + 1)} \qquad (4.8.8)$$

As with the relative velocity, this phase angle vanishes when $\delta = 1$.

To compare these results with those obtained by means of Stokes' law, we write (4.4.6a) as

$$\left|V^{(S)}\right| = \frac{9\delta}{\sqrt{16y^4 + 81\delta^2}} \qquad (4.8.9)$$

Mathematically, this result follows from (4.8.7) by letting both δ and y approach zero in that equation, in such a manner that $y^2/\delta = O(1)$. This makes it clear that Stokes' law will yield erroneous results when the particle's density is not much larger that that of the fluid, or when the frequency is not small.

Figures (4.8.1) and (4.8.2) show the amplitude of the velocity ratio as a function of y, for two different values of the density ratio δ, as predicted by (4.8.7) and (4.8.9). The chosen values correspond to the case of silica particles in air and in water, respectively. The agreement shown in Fig. 4.8.1 between the Stokes result and (4.4.8) was to be expected because δ is equal to 5.6×10^{-4} for the case chosen for the figure. But, in Fig. 4.8.2, the value of δ is 0.45, so that inertia plays a far more important role. We also note that, in this case, the unsteady result, (4.8.7), tends to an asymptotic limit when y is large. This limiting value is

$$\left|V^{(V,I)}\right|_{y\to\infty} = \frac{3\delta}{2+\delta} \qquad (4.8.10)$$

Since this applies when $y = a/\delta_v \gg 1$, it follows that the viscous layer around the particle is then negligible. Thus, viscosity effects play no role in that limit, and if fact (4.8.10) can be obtained from (4.8.3) by neglecting viscosity altogether. Now, this limiting result predicts that, at high frequency, the velocity (and displacement) of a

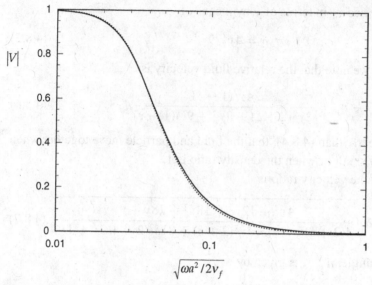

Figure 4.8.1. Particle velocity in uniformly oscillating fluid for a 100-μm diameter particle in air: solid line, (4.8.4); dotted line, (4.4.5).

particle is larger or smaller than that of the fluid depending on whether the density ratio is larger or smaller than one, respectively – that is, particles denser than the fluid move with smaller amplitudes than the fluid, whereas particles lighter than the fluid show the opposite trend. In Section 4.11, we show that this is nearly the case for a limited range of large frequencies, but that beyond these, the above limit is inapplicable because of fluid compressibility effects.

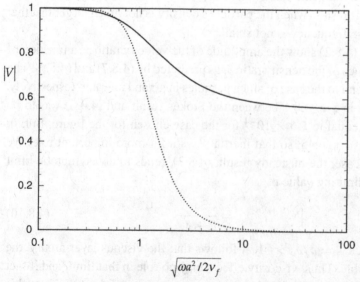

Figure 4.8.2. Particle velocity in uniformly oscillating fluid for a 100-μm diameter silica particle in water: solid line, (4.8.4); dotted line, (4.4.5).

4.9 The Basset-Boussinesq-Oseen Equation

In the pervious section, we obtained the fluid force acting on a sphere in a fluid that is executing monochromatic oscillations. That result can be used to obtain the force on the sphere for an arbitrary time dependence. This can be done by Fourier inversion, as was done with the particle heat transfer in Section 3.4. In fact, the two derivations are so similar that the equivalent force result follows from the heat transfer result by direct comparison. Thus, for example, when $\omega > 0$, the monochromatic force result, now interpreted as the Fourier transform of the fluid force, $F_p(t)$, can be written as

$$F_{p\omega} = m_p \delta \left(\frac{du_f}{dt}\right)_\omega - 6\pi \mu_f a (U_{p\omega} - U_{f\omega}) - \frac{1}{2} m_p \delta \left[\left(\frac{du_p}{dt}\right)_\omega - \left(\frac{du_f}{dt}\right)_\omega\right]$$

$$- 6\pi \rho_f a^2 (1 + i) \sqrt{\frac{\nu_f}{2\omega}} \left[\left(\frac{du_p}{dt}\right)_\omega - \left(\frac{du_f}{dt}\right)_\omega\right] \tag{4.9.1}$$

The corresponding result for the heat transfer is given by (3.4.3). Comparing the two equations, we see that the force equation contains two additional terms, namely the first and third on the right-hand side of (4.9.1), but these can be easily inverted. The remaining terms are functionally equal, and their inverse can be obtained by direct comparison. Thus, the fluid force on a sphere translating in a direction parallel to the fluid is

$$F_p(t) = m_p \delta \frac{du_f}{dt} - 6\pi \mu_f a (u_p - u_f) - \frac{1}{2} m_p \delta \left(\frac{du_p}{dt} - \frac{du_f}{dt}\right)$$

$$- 6\rho_f a^2 \sqrt{\pi \nu_f} \int_{-\infty}^t \left(\frac{du_p}{dt'} - \frac{du_f}{dt'}\right) \frac{dt'}{\sqrt{t - t'}} \tag{4.9.2}$$

a result known as the Basset-Boussinesq-Oseen equation, or the B-B-O, equation.

The meaning of the various terms on the right-hand side of the equation has been discussed previously, except the last. This term, known as the Basset, or "history" term, is due to the unsteady generation of vorticity at the surface of the sphere, that is being propagated into the fluid with a finite speed. This term produces at time t a contribution to the force that depends on the complete history of the relative acceleration.

As pointed out previously, the velocity u_f of the fluid appearing in these equations represents the velocity that the fluid would have at the location of the particle, if the particle were not there. Similarly, du_f/dt represents the acceleration that a *fluid* particle would experience at that location. In the Eulerian, or field description, this acceleration is given in terms of a local acceleration and a convective acceleration, and is usually written as Du_f/Dt. It should be emphasized, however, that in deriving (4.9.1), it was assumed that the convective acceleration in the fluid's momentum equation was negligible. Thus, self-consistency requires that, if the Eulerian description is used, the derivative should be taken to be equal to the local acceleration, $\partial u_f/\partial t$.

The B-B-O equation is limited in other ways. Thus, in addition to being limited to small Reynolds numbers, it is also limited to uniform, translational motions. The appearance of nonuniformities and/or finite Reynolds numbers are sometimes included in the B-B-O equation by the use of semiempirical coefficients. In the examples that follow, we consider uniform flows. Nonuniform motions of rigid particles at finite Reynolds numbers have been included in an extension to the equation by Maxey and Riley (1983). The equation has also been adopted to unsteady motions of droplets at small Reynolds numbers by Galindo and Gerbeth (1993) among others.

Sphere Release in Stokes Flow

To illustrate the effects of the history term in the B-B-O equation, we consider the motion of a sphere, initially held at rest in a fluid that is moving past it with a slow, unidirectional motion of magnitude U_{f0} far from the sphere. In the vicinity of the sphere, the flow is distorted by the sphere, and the distortion is initially described by Stokes flow past a sphere at rest. Intuitively, we anticipate that, if released, the sphere will accelerate, eventually acquiring a velocity equal to U_{f0}. We are interested in determining how this takes place. Now, before the sphere is released at $t_0 = 0$, the force on the sphere is $6\pi\mu_f a U_{f0}$. At larger times, it becomes

$$F_p(t) = -6\pi\mu_f a(u_p - U_{f0}) - \frac{1}{2}m_p\delta\frac{du_p}{dt} - 6\rho_f a^2\sqrt{\pi\nu_f}\int_0^t \frac{du_p}{dt'}\frac{dt'}{\sqrt{t-t'}}$$

(4.9.3)

where we have put $du_f/dt = 0$, because the fluid velocity far from the sphere is constant. In the absence of any external forces, F_p is, by Newton's second law, equal to $m_p(du_p/dt)$. Thus, introducing nondimensional time and velocity variables defined by $\tau = \nu_f t/a^2$ and $U_p = u_p/U_{f0}$, respectively, the equation of motion for the sphere becomes

$$\frac{1}{g(\delta)}\frac{dU_p}{d\tau} + U_p + \frac{1}{\sqrt{\pi}}\int_0^\tau \frac{dU_p}{dt'}\frac{d\tau'}{\sqrt{\tau-\tau'}} = 1$$

(4.9.4)

where

$$g(\delta) = \frac{9\delta/2}{1+\delta/2}$$

(4.9.5)

This quantity ranges from 0 when $\delta = 0$, to 9, when $\delta \to \infty$. Now, when the history term is dropped in (4.9.4), the solution becomes $U_p^{(S)} = 1 - \exp[-g(\delta)\tau]$, which shows that the sphere approaches the velocity of the fluid in an exponential manner, regardless of the value of δ.

Consider now the complete equation. Although the unknown appears both under integral and differential operators, its form, together with the initial condition of rest, make it suitable for treatment by means of Laplace transform methods. Thus, we transform (4.9.4), term by term, using the prescription $\tilde{f}(s) = \int_0^\infty f(t)e^{-st}dt$ to obtain the transform, $\tilde{f}(s)$, of a function of time, $f(t)$. Since the sphere velocity is initially zero, the transform of the particle acceleration is $s\tilde{U}_p(s)$. The transform of the integral in the second term can be obtained by means of the convolution theorem. This states that if

$$F(\tau) = \int_0^\tau f(t')h(\tau - t')dt'$$

then, the Laplace transform of $F(t)$ is

$$\tilde{F}(s) = \tilde{f}(s)\tilde{h}(s)$$

Hence, the Laplace transform of the nondimensional history term is $s\tilde{U}_p \cdot \sqrt{\pi/s}$, because the transform of $1/\sqrt{\tau}$ is $\sqrt{\pi/s}$. Therefore, the transformed nondimensional velocity is

$$\tilde{U}_p = \frac{g(\delta)}{s[s + g(\delta)(1 + \sqrt{s})]} \tag{4.9.6}$$

This expression may be inverted exactly, but it is useful to first obtain from it the values of U_p applicable at short and long times. These correspond, respectively, to small and large s.

1. *Small τ.* Here, we let $s \to \infty$ and obtain

$$\tilde{U}_p = \frac{g(\delta)}{s^2} - \frac{g^2(\delta)}{s^{5/2}} + \cdots$$

The inverse of the first term gives the short time velocity

$$U_p \approx g(\delta)\frac{v_f t}{a^2} \tag{4.9.7}$$

The same expression applies to $U_p^{(S)}$ when $t \to 0$, as it should since "history" effects ought to be, then, negligible.

2. *Large τ.* To obtain this limit, we must expand (4.9.6) for small value of s. However, that expression does not have a uniform limit when $s \to 0$, valid for all δ. We thus consider two separate cases:

a. $\delta \ll 1$. Here, $g(\delta) \approx 9\delta/2$, so that when $s = O(\delta) \ll 1$,

$$\tilde{U}_p = \frac{1}{s}\frac{9\delta/2}{s + 9\delta/2} + \cdots \tag{4.9.8}$$

Inversion of this gives $U_p = 1 - \exp[-g(\delta)\tau]$, which is identical to $U_p^{(S)}$. This was also to be expected because, as discussed earlier, Stokes' law applies when the density ratio is small.

b. $\delta = O(1)$ or larger. Here, $g(\delta) = O(1)$ and, for $s \to 0$, we can write (4.9.6) as

$$\tilde{U}_p \approx \frac{1}{s} - \frac{1}{\sqrt{s}} \tag{4.9.9}$$

Thus, when t is large, we obtain

$$U_p = 1 - \frac{a}{\sqrt{\pi v_f t}} \qquad (4.9.10)$$

This shows that when δ is not small, the approach to the fluid velocity is not exponential, as it is for $\delta \ll 1$. Thus, the approach to the equilibrium condition is considerably slower than predicted by Stokes' law. We may therefore expect that particles subject to a rapid sequence of fluid-velocity changes will display "memory" effects in the sense that recovery from one change has not taken place by the time the subsequent changes takes place.

Let us now consider the complete result for the transformed sphere velocity, (4.9.6). To invert it, we first write it as

$$\tilde{U}_p = \frac{g(\delta)}{\alpha - \beta} \left[\frac{1}{s(\sqrt{s} - \alpha)} - \frac{1}{s(\sqrt{s} - \beta)} \right] \qquad (4.9.11)$$

where

$$\alpha, \beta = -\frac{1}{2}g \pm \frac{1}{2}[g^2 - 4g]^{1/2} \qquad (4.9.12)$$

The roots α and β are real and different when $g > 4$, that is, when $\delta > 8/5$, and are complex conjugate otherwise. To include both cases, we put

$$\alpha = -b + ic \quad \text{and} \quad \beta = -d - ic$$

where b, c, and d are real. Thus, inverting (4.9.11), we have

$$\frac{\alpha - \beta}{g(\delta)} U_p(\tau) = \frac{1}{b - ic} \left\{ 1 - e^{(b-ic)^2 \tau} erfc[(b - ic)\tau] \right\}$$

$$- \frac{1}{d + ib} \left\{ 1 - e^{(d+ib)^2 \tau} erfc[(d + ic)\tau] \right\} \qquad (4.9.13)$$

where $erfc(z) = 1 - erf(z)$ is the complementary error function of complex argument $z = x + iy$. This equation applies for all values of δ. But when $\delta > 8/5$, a simpler form is obtained because, then, $c = 0$, so that $b = -\alpha$, $d = -\beta$. Thus,

$$U_p = 1 - \frac{b}{b - d} e^{d^2 \tau} erfc(d\sqrt{\tau}) + \frac{d}{b - d} e^{b^2 \tau} erfc(b\sqrt{\tau})$$

The error functions appearing here are real and are tabulated directly as a function of their argument. When $\delta < 8/5$, (4.9.13) has to be used. Here, the arguments are complex, and it is necessary to express the result in terms of the related function $w(z) = \exp(-z^2) erfc(-iz)$, whose values are also tabulated for various values of x and y.

Figure 4.9.1 shows the ratio of U_p to the value that is obtained on the basis of Stokes' law, $U_p^{(S)}$, for several values of the density ratio, including some below 8/5. The results are plotted as a function of the square root of the nondimensional time

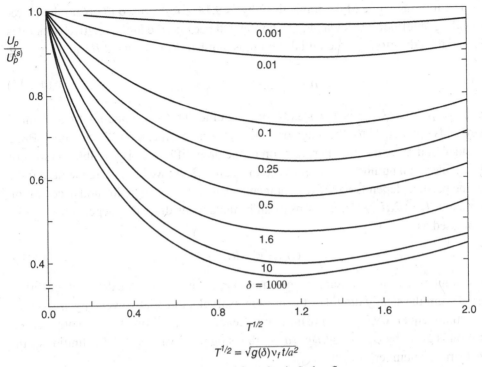

Figure 4.9.1. Sphere velocity after being released from rest in Stokes flow.

$T = g(\delta)\tau$. We see that $U_p/U_p^{(S)}$ differs from unity for all values of the density ratio, but that, as might have been anticipated, the effects of the unsteadiness on the sphere's velocity decrease with δ and nearly vanish as $\delta \to 0$.

Consider now the force on the sphere. Before the sphere is released, the force is given by the first term in (4.9.3), that is by the steady Stokes value. For $t > 0$, the other terms also contribute to the force; but, on physical grounds, we anticipate that the force is largest immediately after the sphere is released, and here the history term may be ignored. To calculate the contribution by the second term, we make use of the particle's equation of motion at $t = 0$. This gives

$$m_p(du_p/dt)_0 = 6\pi \mu_f a U_{f0}/(1 + \delta/2)$$

Hence, the second term's contribution is $-6\pi \mu_f a U_{f0}\delta/(2 + \delta)$. This is the fluid's reaction on the sphere due to the sphere's acceleration. Its magnitude is proportional to the steady Stokes drag, but its value differs from it, depending on the value of the density fraction. Thus, for very small times, the total force on the sphere, given by

$$F_p(t \to 0) \approx \frac{6\pi \mu_f a U_{f0}}{1 + \delta/2} \tag{4.9.14}$$

This is smaller than the force acting on it when it was at rest and explains the slower acceleration of the sphere, relative to the Stokes' law result.

The limiting value of the force given by (4.9.14) may be used to obtain the corresponding drag coefficient via (4.5.7), noting that because the initial relative velocity is essentially equal to U_{f0}, the initial Reynolds number, Re_0, is then $2aU_{f0}/\nu_f$, giving

$$C_D - C_{DS} = -C_{DS}\frac{\delta/2}{1 + \delta/2} \tag{4.9.15}$$

where $C_{DS} = 24/Re_0$ is the steady drag coefficient corresponding to the initial Reynolds number. Thus, the drag coefficient is seen to depend on Re_0 and on δ. But a broader, and more useful, interpretation of the drag difference is possible in terms of the acceleration number, A_n, introduced in Chapter 2. As we have seen, the fluid force on the particle depends on the instantaneous values of the Reynolds and acceleration numbers $A_n = 2a\dot{U}_r/U_r^2$. In terms of an instantaneous drag, this dependence can be expressed as

$$C_D = C_D(\text{Re}, A_n) \tag{4.9.16}$$

This implies that, for a given Reynolds number, the instantaneous drag will generally differ from the steady drag. For the present example, the relative velocity, $u_p - U_{f0}$, is initially equal to $U_{f0}[1 - g(\delta)\nu_f t/a^2]$. Hence, for small times, the relative acceleration is given by $\dot{U}_r \approx -g(\delta)\nu_f/a^2 < 0$. Using these values in the definition of the acceleration number, we get

$$A_n = -2g(\delta)/Re_0 < 0$$

Expressed in terms of A_n, the drag difference (4.9.15) is

$$C_D = C_{DS} - \tfrac{2}{3}A_n \quad (>C_{DS}) \tag{4.9.17}$$

This result applies only to a sphere released from rest in a Stokes flow and applies only when the Reynolds number is small. It nevertheless shows more clearly than (4.9.15) that unsteadiness is responsible for the departures of the drag coefficient from its steady value.

Suddenly Imposed Fluid Velocity. The motion considered here assumed that the sphere was initially held at rest. But, in most situations, the particles are free to move, and the question arises as to how they respond to changes in the fluid velocity. To get an idea about their response, we consider a sphere moving with the fluid in a steady flow. Here, the force on the particle is zero, and the streamlines describing the flow would be straight lines everywhere. Now, suppose that the fluid velocity is made to change, instantaneously, by a small amount. This could be accomplished, approximately, by the passage of a pressure pulse. Given the rapidity of propagation of the pulse and the smallness of the sphere, the sphere would find itself in a fluid that has, everywhere, a slightly different velocity. How would it respond? Of course, we anticipate that the particle would accelerate in the direction of the imposed velocity change, eventually reaching equilibrium with the fluid. This is similar to the situation in the previous

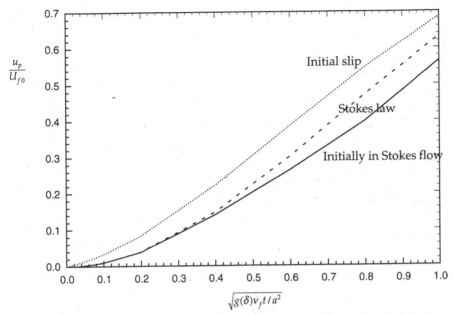

Figure 4.9.2. Initial response of a sphere after release. $\delta = 0.01$. Dashed line, Stokes' law; solid line, initially in Stokes flow; dotted line, Initially in ideal flow.

example, where the sphere was initially at rest. However, the resulting motion here must be different because the initial fluid-velocity distribution is different.

The motion is obviously equivalent to one in which both particle and fluid are initially at rest, and at some time t_0, the fluid is given a uniform, small velocity U_{f1}. Near the sphere, the streamlines must be distorted, but the pattern does not correspond to Stokes flow because Stokes flow requires a finite amount of time to be established. This follows from the fact that the thickness of the viscous layer around the sphere grow as $\sqrt{\nu_f t}$. Thus, for times that are very small, the distribution of velocity near the sphere corresponds to inviscid, incompressible flow past a sphere.

Let us look first at the particle velocity in the initial moments after the fluid's velocity was suddenly increased from zero to U_{f1}. Here, viscosity can be neglected so that all viscous forces vanish and (4.9.2) gives $F_p = \frac{3}{2}m_p\delta(du_f/dt)$. The fluid acceleration is infinite at t_0 and imparts the particle an impulse that increases its velocity by an amount equal to

$$u_p = \frac{3\delta}{2+\delta}U_{f1} \tag{4.9.18}$$

For $t > t_0$, viscosity begins to play a more significant role, and it is necessary to consider the full force equation. We shall not pose here to obtain the particle velocity for all times; but, in Fig. 4.9.2, we show the short-time variations of the nondimensional particle velocity, u_p/U_{f1}, for the case $\delta = 0.01$. Also shown in the figure are the results for the initially fixed sphere and for the Stokesian velocity, $U_p^{(S)} = 1 - \exp[-g(\delta)\tau]$. It is seen that the particle velocity increases most rapidly when the initial condition is

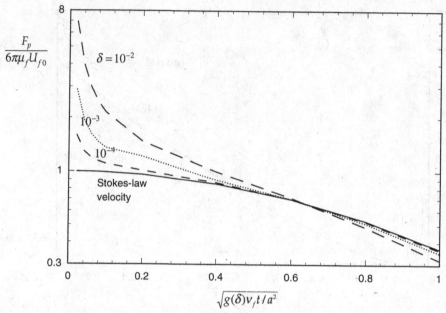

Figure 4.9.3. Particle force on a sphere initially held in an ideal flow.

one of perfect slip, on account of the much higher initial stress occurring then. Not shown in the figure is the long-time behavior. Here, all three curves approach unity, but in a different manner, with only the Stokes' law solution approaching that value exponentially.

Finally, in Fig. 4.9.3, we show the particle force divided by $6\pi\mu_f a U_{f1}$ for some small values of the density ratio. The effects of unsteadiness are quite evident, particularly at small times, when the tangential stresses are large, and for the larger values of δ, where inertia effects are more significant. Additional details about this motion may be found elsewhere (Temkin, 1972)

4.10 Unsteady Drag at Finite Re

Figure 4.9.2 also shows that the steady Stokes drag predicts a velocity ratio falling somewhere in between the two unsteady results discussed previously. This implies that when Re $\ll$ 1, the effective drag coefficient corresponding to the unsteady cases can be both larger and smaller than the steady drag and that the initial conditions play a very significant role in the motion. For finite values of Re, no exact solutions exist that describe the drag force for unsteadiness conditions. We must therefore resort to experimental investigations to guide discussion. But first we make some general observations that are based on experimental observations in both steady and unsteady conditions.

When the Reynolds number is finite, we anticipate that unsteadiness also affects the instantaneous force and, therefore, the drag coefficient. Thus, we expect that the

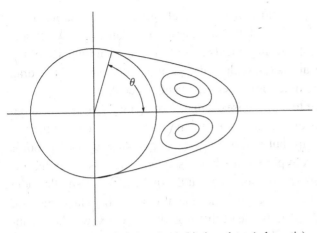

Figure 4.10.1. Recirculation region behind a sphere (schematic).

drag coefficient to depend on both the instantaneous Reynolds and acceleration numbers, or $C_D = C_D(\text{Re}, A_n)$, as already expressed by (4.9.16). Very little is known about the dependence, at finite Reynolds numbers, of C_D on A_n, but we know that if $A_n = 0$, C_D ought to be equal to the steady drag, C_{DS}, and depends on the Reynolds number. If A_n is not zero, some departures from this value are expected, so that $C_D - C_{DS}(\text{Re}) \neq 0$, as was the case in the low Reynolds number example described previously.

Let us now consider finite Reynolds numbers that can be considered steady. For them, the streamlines, as seen by an observer on the sphere, are fixed, and usually form, on the lee side of the sphere, a recirculating region that appears to remains attached for Re as large as 100. The size of that region, as measured by the angle subtended by the separation streamline is an important factor in determining the drag on the sphere (Fig. 4.10.1). A larger angle, produces a larger "form" drag, relative to the "skin-frictional" drag.

An idea of what happens when A_n is not zero may be obtained by looking at changes of the steady streamline pattern resulting from a change of the relative fluid velocity $u_f - u_p$. This can increase or decrease, depending on both the particle velocity and the fluid velocity far from the particle. As a result, A_n can be both positive and negative, but it is best to consider each case separately.

Consider first a steady flow past a fixed sphere at a Reynolds number such that a steady recirculation region exists with a subtended angle $\theta = \theta_0(\text{Re})$. Now suppose that sphere is released at some instant. In the initial stages of its motion, the absolute Reynolds number remains nearly equal to that existing before, but the relative fluid velocity experiences a deceleration. It is known from boundary layer theory that a decelerating flow is associated with an adverse pressure gradient, which induces the separation streamline to leave to sphere earlier, thus increasing the separation angle θ above the steady value θ_0. As a result, we anticipate a larger resistance force. Since the relative velocity is essentially unchanged, we also anticipate a larger drag coefficient, relative to the steady drag. We also note that the acceleration number is negative. Hence, when the Reynolds number is such that a separation region exists, we expect that $C_D > C_{DS}$ when $A_n < 0$. Similar arguments show that a sudden increase of the relative velocity will result in a positive value of A_n, and a decrease of the drag coefficient, relative to the steady drag, or $C_D < C_{DS}$ when $A_n > 0$.

Let us now consider the magnitudes of those changes, noting that A_n is proportional to $\dot{U}_r/\mathrm{Re}^2$. Suppose that we perform a sequence of experiments in which the magnitude of A_n can be changed at will, but without significantly changing the Reynolds number. This can be done, for example, by sending a sequence of pulses having different pressures. By the previous arguments, we should expect larger or smaller drag coefficients than C_{DS}, depending on the sign and magnitude of A_n. Thus, suppose $A_n < 0$. Then, as we increase the magnitude of this negative quantity, we should obtain a larger value of $C_D - C_{DS}$, but if $|A_n|$ is small, the difference can be expanded in powers series of A_n so that

$$C_D - C_{DS} = -a_1 A_n + \cdots, \qquad A_n < 0 \qquad (4.10.1)$$

where a_1 is a positive number that may depend on the fixed value of Re.

Now consider $C_D - C_{DS}$ for $A_n \geq 0$. When $A_n = 0$, that difference is, of course zero, but for $A_n > 0$ we must first have, in view of the previous argument, $C_D < C_{DS}$. Thus, for small positive values of A_n, we have

$$C_D - C_{DS} = -bA_n + \cdots, \qquad A_n > 0 \qquad (4.10.2)$$

where b_1 is positive. The decreasing trend for the drag difference indicated by this equation cannot continue indefinitely. At some point – that is for some value of A_n, $(A_n)_{\min}$ say – the drag must reach a minimum value, with the angle θ_0 having its smallest possible value (i.e., with the separation streamline moving closer and closer to the rear stagnation point). Beyond that value of A_n, $C_D - C_{DS}$ must increase relative to the minimum value, because of the larger shear in the boundary layer. Therefore, in this region, the decrease in the drag difference can be expressed as

$$C_D - C_{DS} = -c_1/A_n + \cdots \quad c_1 > 0, \qquad A_n > (A_n)_{\min} \qquad (4.10.3)$$

Experimental Results

Let us now turn to studies that have been performed recently to obtain information about the unsteady drag. Earlier experimental results were reviewed by Torobin and Gauvin (1959). A more recent review appears in the paper by Karanfilan and Kotas (1978), who also reported measurements of the drag on spheres executing translational oscillations in a fluid at rest. In these and other oscillatory motions used to study the unsteady drag, both signs of A_n occur, but published data are given in terms of the magnitude of A_n. As a result, data obtained in those studies may be regarded as giving an average value of the drag difference. For example, the results of Karanfilan and Kotas – obtained at Reynolds numbers ranging from 10^2 to 10^4, and at values of $|A_n|$ in the range $0.89 < |A_n| < 2.70$ – show that

$$C_D = C_{DS}(1 + |A_n|)^{1.2} \tag{4.10.4}$$

Below, we give the results of two unsteady drag measurements that were obtained for smaller Reynolds numbers in flows where the sign and magnitude of A_n could be controlled. In both cases, however, data were obtained with small droplets and not with rigid spheres. Nevertheless, the smallness of the droplets and the magnitude of the imposed fluid velocity were so small that the droplets remained spherical at all times. Furthermore, the effects of internal motions in the droplet on the drag could also be disregarded because of the very different viscosities of water and air. Thus, data give an idea of the effects of unsteadiness on the rigid-sphere drag.

Sphere Accelerating in Uniform Flow (Temkin and Kim, 1980). In these experiments, small water droplets were accelerated from rest by the convective flow behind a shock wave of weak strength. The imposed fluid velocity was sufficiently small that the droplets remained spherical while they accelerated. Because the relative velocity decreases in such experiments, it follows that $A_n < 0$ for them. The instantaneous drag coefficient data obtained in these experiments is shown in Fig. 4.10.2. Also shown in the figure is the steady drag data for rigid spheres in the corresponding Reynolds number range. It is seen that all data points shown in the figure have a larger drag than the steady drag. Data can be separated into groups having the same Reynolds number, but different acceleration numbers, as shown in Figure 4.10.3. Added to the data are the drag coefficients that are applicable when $A_n = 0$ at each Reynolds number. If we subtract the steady drag from each set and plot the difference $C_D - C_{DS}$, the data seem to collapse onto one line, as shown in Fig. 4.10.4. The equation of that line is given by $C_D - C_{DS} = -k_1 A_n$, where k_1 appears to be independent of the Reynolds numbers. Thus, the trend predicted by (4.10.1) seems to be followed by these data, all of which fall in the range $0 > -A_n > 0.05$.

Sphere in an Accelerating and Decelerating Flow (Temkin and Mehta, 1982). In this experiment, small spherical droplets were exposed to a pressure pulse in which the gas velocity and the excess pressure change sign, as shown in Fig. 4.10.5. As a result, the fluid velocity relative to the particle has both accelerating and decelerating components, and it is thus possible to study the effects of both positive and negative values of A_n. The results for the instantaneous drag are shown in Fig. 4.10.6. It is seen that the measured drag is larger or smaller than the steady drag, depending, respectively, on whether $A_n < 0$, or $A_n > 0$. This agrees with the heuristic arguments given previously. Data can also be separated into groups having the same Reynolds numbers and shows that the drag difference depends only on the acceleration number.

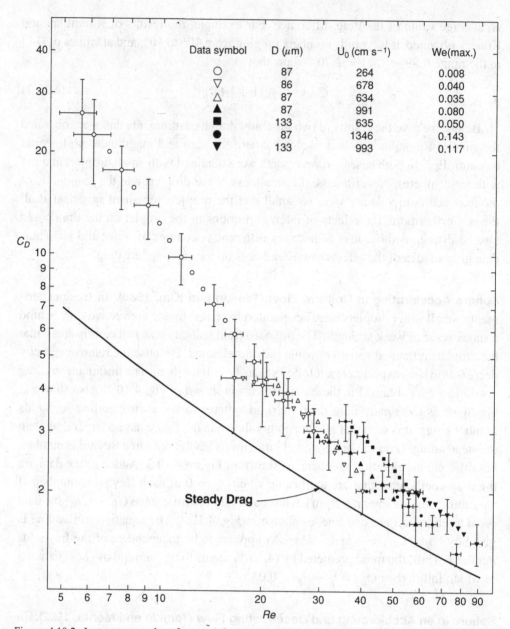

Data symbol	D (μm)	U_0 (cm s^{-1})	We(max.)
○	87	264	0.008
▽	86	678	0.040
△	87	634	0.035
▲	87	991	0.080
■	133	635	0.050
●	87	1346	0.143
▼	133	993	0.117

Figure 4.10.2. Instantaneous drag for nondeforming drops accelerating in uniform flow behind weak shock waves. $A_n < 0$. [Reprinted with permission from Temkin and Kim (1980). © 1980, Cambridge University Press.]

In Fig. 4.10.7, we show the experimental values of $C_D - C_{DS}$, as well as data fits. These are given as $C_D - C_{DS} = -a_1 A_n$ for $0 < -A_n < 0.05$ and as $C_D - C_{DS} = -a_2/A_n - a_3$ for A_n larger than about 0.005. These results appear to confirm the trends predicted by (4.10.1)–(4.10.3), except for very small positive values of A_n, where no information is available.

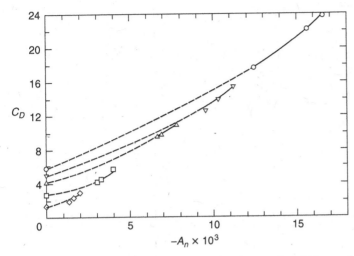

Figure 4.10.3. Instantaneous drag vs. acceleration number for several Reynolds numbers for non-deforming drops accelerating in uniform flow behind weak shock waves O, $Re = 6$; ∇, $Re = 8.1$; $\triangle$, $Re = 10.5$; $\square$, $Re = 20.5$; $\Diamond$, $Re = 51$. [Adapted with permission from Temkin and Kim (1980). © 1980, Cambridge University Press.]

4.11 Sphere in a Sound Wave

Let us now return to the translational motions at small Reynolds numbers. In Section 4.8, we studied the motion of a sphere in a fluid executing translational oscillations and obtained the particle's velocity using steady and unsteady force results, (4.4.1) and (4.8.2), respectively. These results, derived on the assumption that the fluid is incompressible, are sometimes used to study the motion of small spheres in sound waves. This might appear to be inconsistent, because sound waves owe their existence to the compressibility of the fluid. However, the wavelike motion of the fluid is not evident unless the spatial variations of the pressure and velocity are noticeable. This occurs when the length scales of interest are comparable with the acoustic wavelength λ. The length scale relevant to the force on a particle is the size of the particle. Hence, when λ is very large, compared with the size of the particles, the particle "sees" only a fluid motion that, everywhere around it, oscillates in time with the same pressure and velocity. In this situation, the forces on the sphere are well described by the incompressible results.

It follows from the above discussion that the ratio of size to wavelength gives an indication of the importance of compressibility effects. In acoustic work, it is more convenient to use the ratio of particle circumference to the wavelength, $b = 2\pi a/\lambda$. This can also be expressed in terms of the wavenumber, $k = 2\pi/\lambda$, as $b = ka$, or in terms of the frequency as $b = \omega a/c_{sf}$. The last form makes use of the ideal, equilibrium relation $k = \omega/c_{sf}$, where c_{sf} is the equilibrium sound speed for the fluid. For small particles and moderate frequencies, b is usually small, so that the incompressibility assumption is well grounded. However, since b increases with the frequency, we see

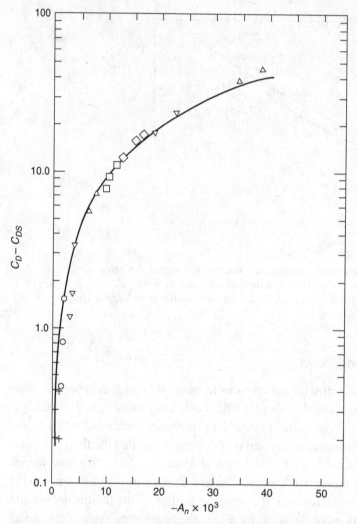

Figure 4.10.4. Difference between instantaneous and steady drag as a function of the acceleration number for nondeforming drops accelerating in uniform flow behind weak shock waves ▷, $Re = 3.2$; ◁, $Re = 4.7$; ◇, $Re = 6$; □, $Re = 8.1$; △, $Re = 10.5$; ▽, $Re = 20.5$; ○, $Re = 51$; +, $Re = 77$, [Adapted with permission from Temkin and Kim (1980). © 1980, Cambridge University Press.]

that compressibility effects will have to be included at some point. This point cannot be determined from the incompressible theory, of course, but we may anticipate that, for finite values of b, the incompressible results may display some incorrect features. One of these features, examined in the next section, is the high-frequency limit of the velocity of a particle in a sound wave. In the incompressible theory, that limit is reached for $y \gg 1$. Here, the viscous penetration depth, δ_{vf}, is much smaller than the particle radius. Hence, to a first approximation, we may study the effects of compressibility by disregarding viscous effects entirely, while retaining those due to compressibility. The results of such a study are briefly outlined in the next section; the details of the

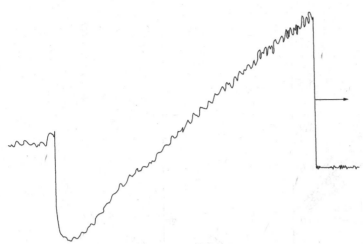

Figure 4.10.5. Pressure pulse having a gas velocity changing direction. [Reprinted with permission from Temkin and Mehta (1982). © 1982, Cambridge University Press.]

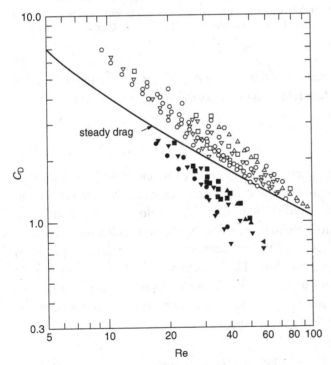

Figure 4.10.6. Instantaneous and steady drag coefficients as a function of the Reynolds number for nondeforming drops accelerating and decelerating in a pulse having both positive and negative fluid velocities. Open symbols, $dU_r/dt > 0$; filled symbols, $dU_r/dt < 0$. $\circ, \bullet, D = 115\,\mu$m; $\triangledown, \blacktriangledown, 120\,\mu$m; $\square, \blacksquare, 135\,\mu$m; $\triangle, \blacktriangle, 152\,\mu$m; $\triangleleft, \blacktriangleleft, 167\,\mu$m. [Reprinted with permission from Temkin and Mehta (1982). © 1982, Cambridge University Press.]

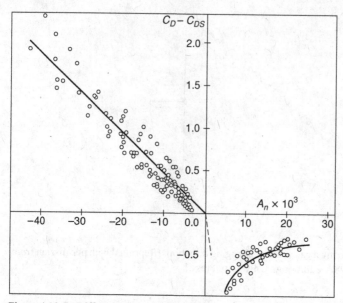

Figure 4.10.7. Difference between instantaneous and steady drag as a function of the acceleration number for nondeforming drops accelerating and decelerating in a pulse having both positive and negative fluid velocities. [Adapted with permission from Temkin and Mehta (1982). © 1982, Cambridge University Press.]

calculations may be found elsewhere (Temkin, 2001a). Later, in Section 4.11.2, we present the results of a study that retains both viscosity and compressibility.

Inviscid Fluid

Consider a freely suspended spherical particle in an inviscid fluid that sustains a sound wave. The sound wave is plane and monochromatic and will be referred to as the incident wave. Since the fluid is inviscid, the fluid velocity and pressure in it can be obtained from a velocity potential, which we write in complex form as $\phi_{in} = A \exp[i(kx - \omega t)]$. The interaction of the incident wave with the particle produces disturbances that propagate into the fluid. These disturbances are waves having the same frequency as the incident wave, but whose wave fronts are scattered into all directions around the sphere. Since the motion has axial symmetry with respect to the polar axis, the scattered wave field can be expressed as

$$\phi_{sc} = \sum_{n=0}^{\infty} i^n (2n + 1) C_n h_n^{(1)}(kr) P_n(\cos\theta) e^{-i\omega t} \qquad (4.11.1)$$

where the C_n are constants determined by the boundary conditions, $h_n^{(1)}(kr)$ is the spherical Bessel function of the third kind, and $P_n(\cos\theta)$ is the Legendre polynomial of the first kind. Each of the terms in (4.11.1) represents an axially symmetric solution of the wave equation in spherical polar coordinates.

The complete acoustic field in the fluid is given by the sum, $\phi = \phi_{in} + \phi_{sc}$, of the incident and scattered wave potentials, subject to the boundary conditions. Since the fluid is inviscid, the only boundary condition that can be imposed at the surface of the particle is that the normal component of the velocities be equal, or

$$\mathbf{u}_f \cdot \mathbf{n} = \mathbf{U}_p \cdot \mathbf{n} \quad r = a \tag{4.11.2}$$

To apply this boundary condition, we express the incident wave potential as

$$\phi_{in} = \sum_{n=0}^{\infty} i^n (2n + 1) A j_n(kr) P_n(\cos\theta) e^{-i\omega t} \tag{4.11.3}$$

Evaluating the fluid velocity from $\mathbf{u}_f = \nabla\phi$, and applying the boundary condition, we obtain, when $n \neq 1$,

$$C_n = -A \frac{j_n'(b)}{h_n'(b)}, \quad n \neq 1 \tag{4.11.4}$$

where $b = ka$. The value of C_0 is used later and can be written as

$$C_0 = iA \frac{\sin b - b \cos b}{1 - ib} e^{-ib} \tag{4.11.5}$$

For $n = 1$, the boundary condition gives

$$A j_1'(b) + C_1 h_1'(b) = U_{p0}/3ik \tag{4.11.6}$$

where U_{p0} is the amplitude, generally complex, of the sphere's velocity. Since this velocity is not known, we require an additional equation to determine C_1. This is provided by the equation of motion of the particle, which in the absence of external forces is $m_p(d\mathbf{U}_p/dt) = \mathbf{F}_p$. For an inviscid fluid, the particle force is equal to $\mathbf{F}_p = -\int_{A_p} p_f' \mathbf{n} dA$, where $p_f' = i\rho_{f0}\omega\phi$. Carrying out these calculations, we obtain a second equation for U_{p0}, namely

$$U_{p0}/3ik = \delta[A j_1(b) + C_1 h_1(b)]/b \tag{4.11.7}$$

Equating this to (4.11.6), we obtain

$$C_1 = -A \frac{b j_1'(b) - \delta j_1(b)}{b h_1'(b) - \delta h_1(b)} \tag{4.11.8}$$

Using the explicit values of $j_1(x)$ and $h_1(x)$ given in the appendices, this result can be expressed in terms of elementary functions as

$$C_1 = -iA \frac{[1 - b^2/(2 + \delta)] \sin b - b \cos b}{1 - b^2/(2 + \delta) - ib} e^{-ib} \tag{4.11.9}$$

This completes the solution. It may be used to obtain the particle velocity, as well as the amplitude of the scattered waves.

Velocity. The particle velocity can be obtained using (4.11.7) and the above value of C_1. We wish to express that velocity in terms of the fluid velocity far from the sphere. First, we note that the amplitude of the fluid in the incident wave, U_{f0}, is obtained from A by means of $A = U_{f0}/ik$. Thus, substitution of (4.11.9) into (4.11.7) yields

$$V^{(C)} = \frac{3\delta}{2+\delta} \frac{e^{-ib}}{1 - b^2/(2+\delta) - ib} \tag{4.11.10}$$

where the superscript C on $V = u_p/u_f$ is used to remind us that the result applies to compressible, but inviscid, fluids. Several features can be noted from this result. First, when $b = 0$, it reduces to $|V^{(C)}| = 3\delta/(2+\delta)$. This is equal to the viscous result obtained from (4.8.4) in the limit $y \to \infty$ [see equation (4.8.10)]. Second, we see that $|V^{(C)}| \to 0$ as $b \gg 1$. Thus, contrary to the viscous case, we see that the sphere remains at rest at very high frequencies. A corollary of this is that the fluid force on the particle vanishes in that limit. Finally, at intermediate values of b, the particle executes translational oscillations, the amplitude of which depends on the density ratio.

In figures, 4.11.1 and 4.11.2 we show the variations of V with the frequency for a 100-μm diameter silica sphere in both air and water, using the various results that have been obtained. To show the effects of compressibility more clearly, we show in the first the real part of V, and, in the second, the imaginary part. As the frequency parameter we choose the variable y.

The figures also show the viscous, incompressible theory as given by (4.8.4) and clearly show the influence of compressibility. It is also seen that both sets of results agree in a limited range of high frequencies where the viscous layer is thin, while at the same time not so high that compressibility effects are important. In addition, the figures show the velocity ratio as predicted by a theory, discussed later, that includes both compressibility and viscosity.

Scattering. The induced particle velocity is one of the effects of the particle-wave interaction; another is the scattering of acoustic energy, as represented by the infinite series, (4.11.1), of acoustic waves that the sphere produces. That energy represents a loss to the incoming energy – a loss that can be important at high frequencies. The loss is assessed in terms of a scattering cross-section, σ_s, defined as that fraction of the incident wave that transmits a power equal to the scattered power. That cross section is given by

$$\sigma_s = 4\pi a^2 \sum_{n=0}^{\infty} \frac{(2n+1)}{b^2} \frac{|C_n|^2}{|A|^2} \tag{4.11.11}$$

When b is small, only the first two terms contribute significantly to the sum and yield,

$$\frac{\sigma_s}{\pi a^2} = \frac{4}{9}\left[1 + 3\left(\frac{1-\delta}{2+\delta}\right)\right]b^4 \tag{4.11.12}$$

The second term inside the square bracket originates in the $n = 1$ term in (4.11.11). That term is proportional to the velocity of the particle. Because of this, the associated cross-section is referred to as *self-scattering*. We note that this contribution to the total cross-section vanishes when $\delta = 1$. This is as it should be, because when $\delta = 1$, there is no relative motion between fluid and particle.

Equation (4.11.12) also shows that, when $b \ll 1$, the energy scattered by the sphere is a very small fraction of the incident energy. As we show later, other kinds of losses overwhelm the scattering loss. This means that unless the frequency is very large, the scattering loss can be neglected.

Viscous, Compressible Fluid

We give below, without derivation, the velocity of a sphere in a sound wave when both the compressibility and the viscosity of the fluid are included. The derivation follows similar lines as the inviscid theory above. The main difference is that, in addition to the scalar potential $\phi = \phi_{in} + \phi_{sc}$, there is a need to consider a vector potential, $\mathbf{B}$, so that the fluid velocity is now given by $\mathbf{u}_f = \nabla\phi + \nabla \times \mathbf{B}$. Expansions of these potentials into spherical harmonics leads to the particle velocity, after the viscous boundary conditions are satisfied. In complex form, the velocity ratio can then be expressed as (Temkin and Leung, 1976; Temkin, 2001a)

$$V = -3\delta \frac{F + iG}{H + iI} e^{-ib} \tag{4.11.13}$$

where

$$F = 2y^2 + 3y + (b/y)^2(1 + y) \tag{4.11.14}$$

$$G = 3(1 + y) - 2b^2 - b^2/y \tag{4.11.15}$$

$$\begin{aligned}H = {} & 2y^2(b^2 - 2 - \delta) + y[b^2(1 + 2\delta) - 9\delta(b + 1)] \\ & + 9\delta b(2b^2/9 - 1) + 3\delta(b^2/y)(b - 1) - 3\delta(b/y)^2\end{aligned} \tag{4.11.16}$$

$$\begin{aligned}I = {} & 2y^2 b(2 + \delta) + y[9\delta(b - 1) + b^2(1 + 2\delta)] + b^2(1 + 4\delta) \\ & + 3\delta(b^2/y)(1 + b) + 3\delta b(b/y)^2 - 9\delta\end{aligned} \tag{4.11.17}$$

These may be used to obtain the real and imaginary parts of the velocity ratio, as well as its amplitude and phase when the fluid external to the sphere is both viscous and compressible.

Figures 4.11.1 and 4.11.2 display the dependence of $|V|$ for a silica particle in air and in water, respectively, as predicted by three theories: the viscous, incompressible theory, (4.8.4); the inviscid, compressible theory (4.11.10); and the viscous, compressible theory, (4.11.13). It is seen that the last set overlaps with the viscous, incompressible

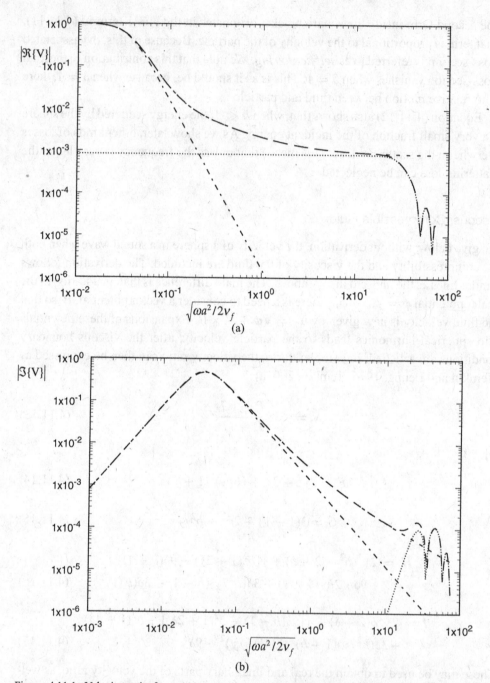

Figure 4.11.1. Velocity ratio for a 100-μm silica particle in a sound wave in air: dotted line, $V^{(C)}$, (4.11.10); dash-dot line, $V^{(S)}$, (4.4.5); short dashed line, $V^{(V,I)}$, (4.8.4); long dashed line, V, (4.11.13). (a) Real part; (b) imaginary part.

theory at low and intermediate frequencies, and with the inviscid, compressible theory at high frequencies. Thus, (4.11.13) incorporates the limited results and may be used to describe the behavior of a small, rigid sphere in a sound wave at frequencies that extend from zero to very high values.

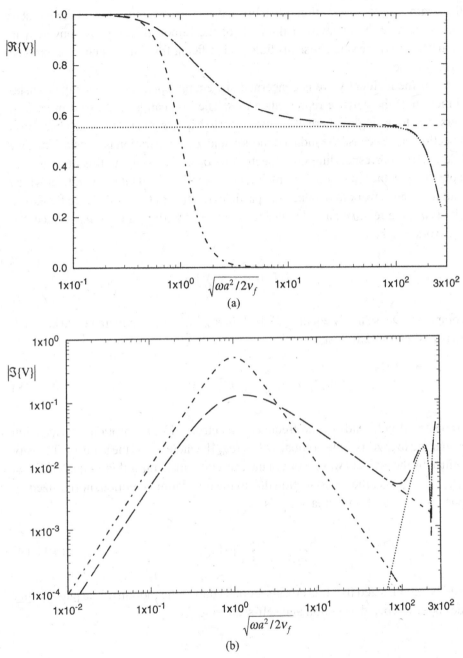

Figure 4.11.2. Velocity ratio for a 100-μm silica particle in a sound wave in water: dotted line, $V^{(C)}$, (4.11.10); dash-dot line, $V^{(S)}$, (4.4.5); short dashed line, $V^{(V,I)}$, (4.8.4); long dashed line, V, (4.11.13). (a) Real part; (b) imaginary part.

Absorption Cross-Section

The previous paragraphs describe the translational motion of a particle induced by a sound wave. Whether or not the fluid is viscous, the induced motion requires the expenditure of some energy. This energy is provided by the incident waves. Thus, once that the steady, oscillatory motion is established, energy is removed from the incident wave continuously. In an inviscid fluid, all of the removed energy is converted in scattered waves, whereas in a viscous fluid, a significant fraction is spent to overcome viscous resistance.

So far as the incident wave is concerned, the energy spent to produce the translational motion of the particle represents a loss. The loss can be expressed in various manners; but, for a single sphere, it is customary to express it in terms of an absorption cross-section, σ_a. Because the induced motion under consideration is translational, we shall denote the corresponding cross-section by $\sigma_{a,tr}$. To compute this quantity, we first note that the energy required to maintain the motion is equal to the rate at which the fluid force does work to displace the particle, relative to the fluid. Since the fluid's velocity is u_f, the relative particle displacement per unit time is $u_p - u_f$. Hence, the average work rate is

$$\frac{1}{2}\Re\{F_p(u_p - u_f)^*\}$$

The complex force is simply equal to $F_p = -i\omega m_p u_p$. Thus, the average rate at which the energy is removed, or lost, is

$$|\langle \dot{e}_{lost}\rangle| = \frac{1}{2}\omega m_p U_{f0}^2 |\Im\{V\}| \tag{4.11.18}$$

This vanishes at both ends of the frequency spectrum. The absorption cross-section corresponding to this loss rate is equal to $I_{inc}|\langle \dot{e}_{lost}\rangle|$, where I_{inc} is the acoustic intensity transmitted by the incident wave. For a plane wave in an ideal fluid, this is equal to $I_{inc} = \frac{1}{2}\rho_{f0}c_{sf}U_{f0}^2$. Hence, the cross-section due to the translational motion, normalized by the sphere's cross-sectional area, πa^2, is

$$\frac{\sigma_{a,tr}}{\pi a^2} = \frac{4}{3}\frac{b}{\delta}|\Im\{V\}| \tag{4.11.19}$$

It should be emphasized that this cross-section is due to the motion of the particle and includes both viscous dissipation and self-scattering.

PROBLEM

If in (4.11.19), we use $V^{(C)}$ from (4.11.10), we obtain the nondimensional power spent by a sound wave to move a sphere in an inviscid fluid. That power is required because the scattered waves remove energy from the incident wave. Determine that power and reduce your results for the case $b \ll 1$. Compare your results to those given by the small b expansion given by (4.11.12).

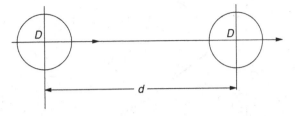

Figure 4.12.1. Two equal spheres in axial motion.

4.12 Force Reduction Due to Particle Interactions

All of the results that have been obtained so far relate to a single particle in a fluid. They are important because, as stated before, they are applicable to particles in sufficiently dilute suspensions. While this is true, at least in an approximate manner, there is a need to quantify the statement. That is, we would like to know the volume concentration range in which we may disregard particle interaction effects.

Strictly speaking, particle interactions can never be disregarded, because the magnitude of the disturbances produced by one particle decreases very slowly with distance from it. Thus, these disturbances affect the motion of neighboring particles to a certain extent, regardless of how far are they located. This means that, in a suspension in motion, each particle will interact with many others. But it is evident that the interaction with its closest neighbor is the most important. Thus, in order to estimate the importance of particle-interaction effects, we need to consider only one particle pair. Now, it is clear that the interactions can approximately be disregarded for particle-pair separations exceeding some value. A knowledge of this value can, with the aid of (1.4.1), be used to estimate the maximum value of the volume concentration. That equation shows that the volume concentration in a suspension is, approximately,

$$\phi_v \approx (D/\bar{\lambda})^{-1/3} \tag{4.12.1}$$

where $\bar{\lambda}$ is a distance between two neighboring particles and D is some mean diameter.

There remains the need to define the direction and value of the distance that enables us to use single particle results in suspensions. The most direct way of doing this is in terms of the fluid force acting on each of the particles in a given pair. Although this force is not known for all cases, some information about it exists that is sufficient for our purposes. The information exists for both small and finite Reynolds numbers, and is discussed next.

Very Small Reynolds Numbers

There are several studies available of the fluid force acting on a rigid sphere that is located near another. Among these, the work of Stimson and Jeffrey (1926) is useful because it relates to two equal spheres, moving together at a constant velocity in a straight line at Re ≪ 1. The geometry of the problem is sketched in Fig. 4.12.1. For such a geometry, Stokes' equations are can be solved using separation of variables. It

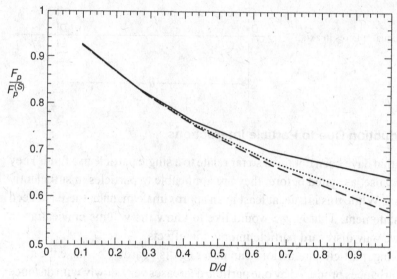

Figure 4.12.2. Sphere drag decrease as a function of particle separation. Solid line, Stimson & Jeffrey; dashed line, (4.12.2); dotted line, (4.12.3).

is thus possible to obtain an exact solution, which is what Stimson and Jeffreys did. We shall not pose here to present any of the details of their calculations and limit the presentation to showing their final result in graphical form. Thus, Fig. 4.12.2 shows the ratio of the force on either particles to the force that would apply if the particle where isolated. The variations are shown as a function of D/d, in terms of which the dilute limit is obtained for $D/d \to 0$. Also shown in the graph is a result obtained by Happel and Brenner (1965), using an approximate method. That result, also obtained by Tam (1969), can be expressed succinctly as

$$\frac{F_p}{F_p^{(S)}} = [1 + K(D/d)]^{-1} \tag{4.12.2}$$

where $K = 3/4$. A slightly different formulation was also obtained by Tam [equation (A8) of his article]. That expression is

$$\frac{F_p}{F_p^{(S)}} = [1 + K[D/d - (D/d)^3/3]^{-1} \tag{4.12.3}$$

This equation is also displayed graphically in Fig. 4.12.2. As we see, the three results are quite similar, except at close separations where some differences occur.

The figure also shows that axial distances of the order of 10 particle diameters will result in an error in the particle force of about 10%, relative to Stokes' law. In suspensions, free particles do not arrange themselves in this neat pattern, which means that the average distance between particles that produces a 10% error is somewhat larger than 10 diameters. Taking a distance of 15–20 diameters, for example, leads to an estimated maximum volume concentration to fall somewhere in the range of 10^{-3}

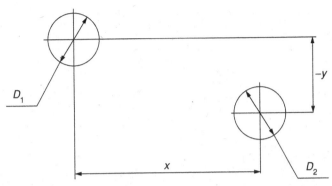

Figure 4.12.3. Particle pair on a plane passing through their centers.

to 10^{-2}. This is, of course, only an estimate value for the maximum acceptable value of ϕ_v. It is based on a particular steady flow and on Stokes' law, and assumes that the stated drag differences are acceptable.

This estimate is based on existing rigid-particle results. Liquid droplets can also be affected by interaction effects. These have been studied by Haber, Hetsroni, and Solan (1973) for a wide variety of separations and relative viscosities. Their results show that equal-sized droplets also experience a reduction in the drag and that the reduction is comparable with that experienced by rigid spheres. This is not surprising because, as discussed earlier, small, nondeforming droplets behave as rigid particles.

Finite Reynolds Numbers

Here, the information regarding particle interactions stems from experimental work or from numerical simulations. However, the available simulations are not relevant to small, freely suspended particles that are responding to externally imposed flow fields. We thus limit the discussion to experimental work; in particular, to one study of interactions between droplet pairs responding to the uniform convective flow behind a weak shock wave, while falling under the effects of gravity (Temkin and Ecker, 1989). Other experiments available include some where the interactions between droplets in vertically falling streams were considered. Because such streams are composed of a large number of droplet pairs, the data obtained with them yield relatively little information about interactions between any one pair.

To fix ideas, we show in Fig. 4.12.3 a schematic of a droplet pair in the experiments. The imposed flow is from left to right, and the droplet on the right, called the second droplet, is at some instant at position (x, y) relative to the first. However, the centers of both droplets fall on the same plane. This alignment, difficult to achieve in practice, allows the interactions to be observed in a simple, yet fairly general manner.

Now, the imposed flow is produced by the passage of a plane shock wave of weak strength and rectangular profile (i.e., the fluid velocity in it may be considered constant while the droplets accelerate). The maximum relative Reynolds number, based on

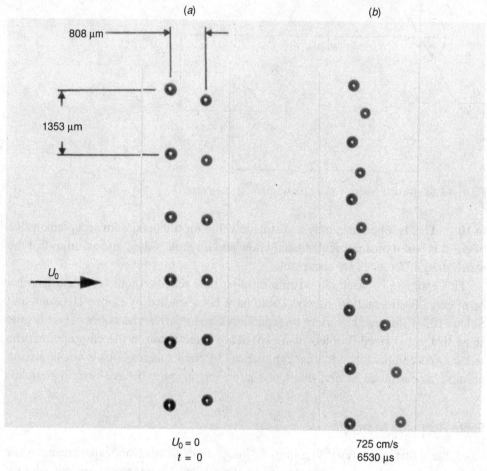

Figure 4.12.4. Nondeforming droplet streams before and during an experiment. (a) Streams before passage of shock wave. (b) Streams 6.5 msec after passage of shock wave. Re = 135; We = 0.14. [Reprinted with permission from Temkin and Ecker (1989). © 1989, Cambridge University Press]

the first droplet alone, was about 135, and the maximum velocity was such that the maximum Weber number, We, was 0.15. Previous experiments had shown that, for these Weber numbers, the droplets remain spherical (Temkin and Kim, 1980). Thus, the obtained data refer to spheres. Furthermore, given that the viscosity ratio, μ_p/μ_f is very large, these droplets may be regarded as being nearly rigid.

In the experiments, two droplet streams having controlled size, separations, and alignments were injected across a shock tube, as indicated in Fig. 4.12.4a. This shows the two streams falling vertically under the effects of gravity *before* their being exposed to the horizontal shock wave flow. Several important observations can be made regarding the droplets in the two streams. First, the diameters of the droplets in the first stream are slightly larger than those in the second. This arrangement was chosen to avoid collisions produced by purely size effects, such as might occur if a small droplet were placed in the first stream. Second, the vertical distance between adjacent

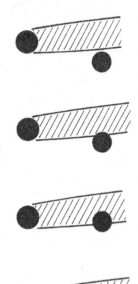

Figure 4.12.5. Schematic wakes behind droplets in first stream. [Reprinted with permission from Temkin and Ecker (1989). © 1989, Cambridge University Press]

droplets in each stream remains constant as the droplets fall. This means that their velocity is constant and that whatever interactions might exist between adjacent vertical droplets, their motion is not affected. Third, the horizontal distance between the two streams is constant even though the vertical velocities differ. This again indicates that no interactions occur at this time between the droplets in the two streams even though the distance is only a few diameters.

Now consider Fig. 4.12.4b. This shows the two streams 6.5 msec after the horizontal flow has been imposed. During the intervening time, both streams have moved slightly to the right, in response to the imposed flow. But while the droplets in the first stream remain aligned along the vertical, indicating that they have all moved horizontally by the same amount, some of those in the second stream display different horizontal displacements. Such nonuniformities in the second stream are due to dynamic interactions with droplets in the first. The reason is that soon after the horizontal flow is imposed, wakes are formed behind each droplet. Those behind the droplets in the first stream are shown, schematically, in Fig. 4.12.5. The deviations from uniform displacement in the second stream can then be visualized in terms of their position, relative to the wakes. Thus, those droplets that have not entered a wake region are displaced to the right by larger amounts. On the other hand, droplets in those wakes show smaller horizontal displacements, and therefore are now closer to the droplets in the first stream.

These differences are now simpler to explain. Droplets moving in the wake of the upstream droplet experience a reduced drag and, therefore, their horizontal displacements are decreased by amounts that depend on their positions in the wake. On the other hand, measurements of the drag on the droplets in the first stream shows that

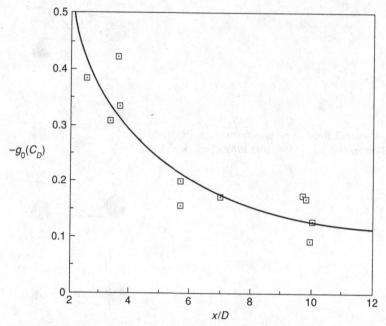

Figure 4.12.6. Drag decrease on second droplet when on wake axis. [Reprinted with permission from Temkin and Ecker (1989). © 1989, Cambridge University Press]

they not significantly affected by their own wake. Furthermore, computations of their drag coefficient shows that they are not affected by the presence of other droplets and move as though they were isolated.

Now, the reduction of drag for the droplet in the second stream is maximum, as might have been anticipated, for those droplets that are located along the axes of the wake, with larger reductions occurring for smaller separations. This is shown in Fig. 4.12.6, where the drag reduction on the second droplet, normalized by its isolated value, C_{D_∞}, is shown as a function of the nondimensional axial distance from the center of the first droplet. As the graph shows, the reduction at the shorter distances are large and are comparable with those experienced by similarly placed sphere in small Reynolds number flows (see Fig. 4.12.2). As the axial distance increases, the reduction decreases, becoming equal to about 0.1 at $D/x = 0.1$. Coincidentally, this is close to the reduction experienced at the same distance when Re $\ll$ 1.

The solid line in Fig. 4.12.6 is a best fit of the data and is given by

$$g_0(C_D) = \frac{C_{D_\infty} - C_{D_0}}{C_{D_\infty}} = -Ae^{-B(x/D)} \tag{4.12.4}$$

where C_{D_0} is the actual drag on the second droplet when the droplet is located along the axial line and C_{D_∞} is its isolated value, $A = 0.595$ and $B = 0.145$. The standard

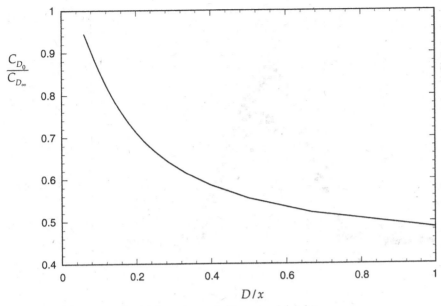

Figure 4.12.7. Fractional axial drag change for the second droplet.

deviations on A and B are, respectively, 0.047 and 0.022. Using this equation, we may write

$$\frac{C_{D_o}}{C_{D_\infty}} = 1 - Ae^{-B(x/D)} \qquad (4.12.5)$$

Figure 4.12.7 shows this drag ratio as a function of D/x so that we may compare the reductions with those shown in Fig. 4.12.2 for $Re \ll 1$. Although, as we can see in the figure, the reductions are comparable at small and large distances, significant differences occur in between, with the finite Reynolds number results showing larger decreases.

As the second droplet moves away from the axial line, the drag decreases are significantly reduced. The lateral drag reduction data collapse into a single trend when scaled with $g_0(C_D)$, as shown in Fig. 4.12.8. A best fit of the data points shown in the figure is given by

$$g(C_D) = g_0(C_D)e^{-C(y/D)^2} \qquad (4.12.6)$$

where y is the lateral distance and $C = 2.25$. Because of the manner the experiments were conducted, this lateral distance may be taken to be the radial distance from the axis to the center of a droplet. Seen in this manner, the wake region, defined as the region where the second droplet experiences a reduced drag, is a cigar-shaped, three-dimensional region, as sketched in Fig. 4.12.9.

The region of drag reduction is demarked by a paraboloid of revolution described by

$$\frac{x}{D} = \frac{C}{B}\left[1 - \left(\frac{r}{D}\right)^2\right] \qquad (4.12.7)$$

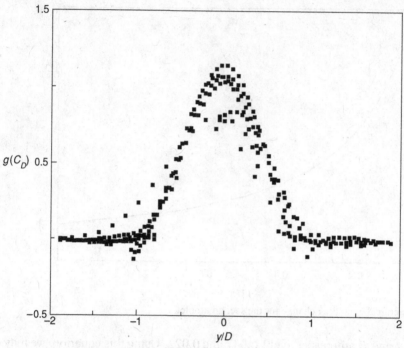

Figure 4.12.8. Drag decrease along the lateral direction. [Reprinted with permission from Temkin and Ecker (1989). © 1989, Cambridge University Press]

where r is the distance from the surface of the paraboloid to the axes and $C \approx 15$. As this equation shows, the region extends to about 15 diameters along the horizontal direction. But, at such distances, the reduction is very small. Also, droplets located outside this paraboloid experience no drag reduction. This is equivalent to stating that, at these Reynolds numbers, such particles are not affected by particle interactions.

Returning to the question posed at the beginning of this section regarding an estimate of the maximum volume concentration for which the dilute results are approximately applicable, we conclude that an axial distance of about 10 diameters might be used as the closest distance for which the single-particle results may be used without incurring into a large error. Again, in suspensions of free particles, the actual distances may be larger and, as in the previous case, we take a distance of the order of 15–20 diameters to be maximum distance that is compatible with the dilute force. Although the physical picture at small and large Reynolds numbers is different, this estimate is equal to that found for small Reynolds numbers, and also corresponds to

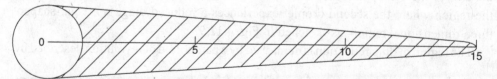

Figure 4.12.9. Region of reduced drag behind first droplet.

ϕ_v in the range of 10^{-3} to 10^{-2}. Of course, these are estimates obtained from two specialized conditions. Nevertheless, they provide an idea about the values of the volume concentrations that may be regarded as dilute.

4.13 Concluding Remarks

In this chapter, we have examined the force on a small, rigid sphere at small Reynolds numbers and have studied its response to a variety of fluid motions. The examination has allowed us to look at various effects, including fluid inertia, compressibility, and viscosity. We have also given some extensions of the basic results that allow calculations in less restrictive cases and have considered in some detail the motion of a spherical particle in a sound wave.

5

Shape Deformations

5.1 Introduction

We now turn our attention to particles that are not rigid. By definition, the relative positions of material points within these particles are not fixed, so that the application of external forces usually deforms them. It will be recalled that the deformations of small particles were divided into deformations without change of volume and uniform expansions. The first of these is briefly considered in this chapter. Volume changes are considered in Chapter 6.

Shape deformations can, in principle, be induced in all types of particles, provided the forces applied to them are sufficiently large. But, because in suspensions those forces are small, solid particles normally retain their shape. We therefore limit the present discussion to small bubbles and droplets, keeping in mind that interfacial effects now come into play.

Bubbles and droplets that are not exposed to any external force must be spherical. This follows from the Young-Laplace equation, (2.1.11), which shows that the pressure difference across a curved interface, at any point of the interface, is proportional to the curvature, $R_1^{-1} + R_2^{-1}$, of the interface at that point. If no forces are applied, the pressure difference must be the same for all points on the interface and that means that the curvature is constant over the whole surface.

But even when external fields are present, small droplets and bubbles may, for many purposes be considered spherical. An idea of the size below which particles may be regarded as spherical is obtained by considering gravitational effects. Let $2a$ represent the lateral dimension of the particle along $\mathbf{g}$. Then, the change in the hydrostatic pressure in the fluid outside the particle in that distance is of the order of $\Delta p_h \approx 2a\rho_f g$. On the other hand, the excess pressure in the particle produced by surface tension forces is of the order of $\Delta p_\sigma \approx 2\sigma/a$. Hence, the ratio of hydrostatic to surface tension pressures is of the order of

$$\frac{\Delta p_h}{\Delta p_\sigma} \approx \left(\frac{a}{\Delta_c}\right)^2 \tag{5.1.1}$$

where $\Delta_c = \sqrt{\sigma/\rho_f g}$. This quantity has the dimensions of length and is called the *capillary length*. Thus, (5.1.1) shows that hydrostatic effects may be ignored, compared with those produced by surface tension when the lateral dimension of the particle is much smaller than the capillary length. In the case of a water-air interface, Δ_c is about 0.27 cm. A typical cloud droplet has a diameter of less than 100 μm, so that the deformations produced by the hydrostatic pressure can be ignored.

As the size of a droplet (or bubble) increases, its shape becomes increasingly deformed. Thus, a typical raindrop, whose sizes, at ground level, are in the millimeter range, are not spherical. Furthermore, the surface of such drops oscillates in time as the droplets fall. For larger sizes than these the oscillations are unstable, and can shatter a drop into many drops of small, but more stable size.

A nondimensional parameter that can be used to characterize shape deformations in dynamic conditions is the Weber number, We. This parameter is a measure of the fluid-dynamic forces trying to deform the particle, relative to the surface tension forces, attempting to keep it spherical. Thus, if Δp_d represents the dynamic pressure difference acting across a single drop or bubble, the Weber number can be defined by We $= \Delta p_d/\Delta p_\sigma$. To estimate Δp_d, we consider first flows for which Re $\ll 1$. Here, we may use the pressure distribution around a sphere as predicted by the Stokes solution. This shows that $\Delta p_d \approx 3\mu_f U_r/D$. Hence,

$$\text{We} \approx \mu_f U_r/\sigma$$

A numerical example is helpful. Taking the magnitude of the relative velocity, U_r, to be equal to $\nu_f/\rho_f D$, so that Re $= 1$, we find that for a $D = 100$-μm droplet in air, the Weber number is of the order of 10^{-4}. This is sufficiently small that we may regard single droplets in air as spherical when Re $= 1$ or less. Consider now higher Reynolds numbers. Here, $\Delta p_d \approx \rho_f U_r^2$ so that the Weber number is given by (2.6.16). If we now take Re $= 100$ and $D = 100$ μm, we find that We $\approx O(1)$.

Droplet motions having We $\ll 1$ are characterized by spherical shapes because the surface tension forces overwhelm external forces attempting to deform the particle. But, when We $\approx O(1)$, the fluid dynamic forces and the surface tension forces have comparable magnitudes. One might therefore anticipate that the droplets will execute some kind of oscillation. For Weber numbers larger than these, the oscillations become unstable and may result in droplet breakup.

The processes underlying these motions are very interesting and the deformation and breakup of small droplets and bubbles are fascinating. We will, however, limit the analytical discussion to the small oscillations that a spherical droplet or bubble can sustain if its surface is perturbed lightly. The chapter also includes some information on the breakup of liquid surfaces, including droplets as well as some comments on certain other processes that are influenced by surface tension effects.

5.2 Energy Considerations

Suppose a spherical droplet or a bubble is deformed by the action of an external force, and that, at some time, the force is removed. We may, on physical grounds, expect the particle to start executing some kind of oscillatory motion. However, we do not know ahead of time whether the oscillations are stable or unstable, except for small deformations, where we anticipate that eventually the oscillations will die out and the particle's shape will again become spherical. When the deformations are finite, we can obtain some information about the equilibrium shape and stability of a given surface. To do this, we use the minimum energy principle from thermodynamics. This states that the energy of a system is a minimum when the system is in equilibrium. Stated differently, if some of the independent variables of a system are given a *virtual* displacement from their equilibrium values, those variables will change until the system reaches equilibrium again, although the new equilibrium condition may differ from the original one. In fact, the system will, from among all possible equilibrium states available to it, chose the one that has the smallest energy for the given constraints. The following example, due to Penner and Li (1967), may clarify this.

Equilibrium Pressure in a Droplet or Bubble

Consider an isothermal system consisting of a spherical droplet (or a bubble) immersed in a fluid, both enclosed by a rigid container. The pressure in the exterior fluid is p_f, and we wish to determine the pressure, p_p, in the droplet. To do so, we consider a virtual change in the system, defined as a change of the independent variables, consistent with the constraints imposed on the system. In this example, the constraints are that the temperature and total volume remain constant. Therefore, the separate volumes of the particle and fluid, v_p and V_f, respectively, can change so long as the temperature and the total volume remains constant. Thus, a change in which the volume of the particle is increased by an amount δv_p, requires a change of fluid volume equal to $\delta V_f = -\delta v_p$. Accompanying these volume changes, there is a change in the total energy of the system. This is divided as follows. First, the change of volumes requires energy expenditures equal to $-p_p \delta v_p$ and $-p_f \delta V_f$. Second, there is the work associated with stretching the interface separating the particle from the fluid, and this is $\sigma \delta A_p$ if the area of the particle increases by an amount equal to δA_p. Thus, the change of energy of the system changes by an amount equal to

$$\delta E = (p_f - p_p)\delta v_p + \sigma \delta A_p \tag{5.2.1}$$

The minimum energy principle requires this to be zero, giving $p_p = p_f + \sigma(\delta A_p/\delta v_p)$. Now, if R is the instantaneous radius of the droplet, $\delta v_p = 4\pi R^2 dR$ and $\delta A_p = 8\pi R dR$. Hence,

$$p_p = p_f + 2\sigma/R \tag{5.2.2}$$

which is the Young-Laplace equation for a spherical surface.

Now suppose that we slightly deform a spherical droplet so that it appears as an ellipsoid of revolution and then release it. The surface area of this ellipsoid is slightly larger than that of the nondeformed particle. Hence, the system would return to its initial, spherical shape because it has a smaller surface area, but the return will be via damped oscillations about the equilibrium shape. These small-amplitude oscillations involve certain shapes – or modes – that are important in several contexts. These are considered next.

5.3 Surface Vibrations of a Droplet

When the external force fields are weak, shape deformations usually have small amplitude and are controlled by surface tension effects that enable the surface of the droplet to oscillate about its equilibrium spherical shape. Because the oscillations have small amplitudes, they can be expressed in terms of the characteristic modes of oscillation. By a mode of oscillation, we mean the shapes associated with one of the normal, or characteristic, frequencies of vibrations.

Below, we obtain these frequencies and their corresponding modes of oscillation assuming that that there are no dissipative effects. This means that viscosity is to be disregarded both inside and outside the droplet, and that, in addition, the external medium should be regarded as incompressible so that energy cannot be radiated by the droplet in the form of acoustic waves. In those conditions, the vibrations will be permanent – that is, once the oscillations are started, they never decay in time.

The oscillatory motion can be studied by means of the equations of motion for ideal and incompressible fluids, applied to the fluid in the particle, and to the fluid outside it, i.e.,

$$\nabla \cdot \mathbf{u}_p = 0 \qquad \nabla \cdot \mathbf{u}_f = 0 \tag{5.3.1a,b}$$

$$\rho_p \frac{\partial \mathbf{u}_p}{\partial t} + \nabla p_p = 0 \qquad \rho_f \frac{\partial \mathbf{u}_f}{\partial t} + \nabla p_f = 0 \tag{5.3.2a,b}$$

It should be noted that the fluid outside can be a gas or another liquid. In the latter case, it will be assumed that the two liquids are immiscible.

Each of these systems of equations are to be solved, subject to boundary conditions on the surface of the particle. These state that the normal components of the velocity in and out of the particle be equal at the interface, that is that,

$$\mathbf{u}_p \cdot \mathbf{n} = \mathbf{u}_f \cdot \mathbf{n} \tag{5.3.3}$$

Similarly, the pressures on either side of the surface satisfy the jump condition given by (2.1.17). Since the fluids are taken as inviscid, this reduces

$$p_p - p_f = \sigma \left(\frac{1}{R_1} + \frac{1}{R_2} \right) \tag{5.3.4}$$

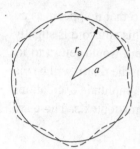

Figure 5.3.1. Surface disturbance on a plane φ = constant.

These conditions apply at every instant everywhere on the surface of the particle, which, in general is not spherical, as sketched in Fig. 5.3.1. The sketch shows an idealized cross-section of the particle at some instant. If we select a different azimuth angle, the pattern will change revealing that the disturbances generally depend on that angle. But, for simplicity, we will consider only modes of oscillation that are symmetrical about the polar axis, but that may depend on the polar angle θ.

The surface of the droplet is described by an equation relating the coordinates of points on it. Thus, if r_s is the radial distance from the undisturbed center of the particle to a point on its surface, that equation is of the form $f(r_s, \theta, \varphi, t) = 0$. Since φ is constant, this equation describes a plane, closed curve. It is then convenient to write the equation of this curve in terms of the radial distance, $\zeta = \zeta(\theta, t)$, between points on the surface of the disturbed particle from the corresponding point on the undisturbed surface:

$$f(r_s, \theta, \varphi_0, t) = r_s - a - \zeta(\theta, t) = 0 \tag{5.3.5}$$

Thus, the radial distance from the center of the particle to points on the surface are given by

$$r_s = a + \zeta(\theta, t) \tag{5.3.6}$$

Now, the boundary conditions apply at points on the particle's surface. However, for small-amplitude vibrations, $|\zeta| \ll a$, we may apply them at $r_s = a$. It is, of course, necessary to retain, to leading order, the effects of the deformation described by $\zeta(\theta, t)$. For this purpose, we need to obtain the local value of the unit normal vector $\mathbf{n}$. This can be obtained from the equation of the surface by means of $\mathbf{n} = \nabla f / |\nabla f|$. Thus, if $\mathbf{e}_r$ and $\mathbf{e}_\theta$ are unit vectors along the r and θ directions, respectively, we have

$$\mathbf{n} = \frac{\mathbf{e}_r + \mathbf{e}_\theta(\partial \zeta / \partial \theta)/r}{\sqrt{1 + (\partial \zeta / \partial \theta)^2 / r^2}} \tag{5.3.7}$$

For small disturbances, we may approximate this by

$$\mathbf{n} \approx \mathbf{e}_r + \mathbf{e}_\theta \frac{1}{r} \frac{\partial \zeta}{\partial \theta} \tag{5.3.8}$$

which shows that, in the linear approximation, the local departures from the spherical shape are proportional to the slope of the deformation.

Since the motions are irrotational, we may obtain the solutions of the two sets of equations in terms of velocity potentials. However, since the equations are of the same nature, we need solve only one set, and later adopt the solution to the other. We thus put

$$\mathbf{u} = \nabla\phi \tag{5.3.9}$$

and

$$p = p_0 - \rho_0\frac{\partial\phi}{\partial t} \tag{5.3.10}$$

where the static pressure p_0 is equal to p_{p0} inside and p_{f0} outside the drop. The velocity potential satisfies Laplace's equation $\nabla^2\phi = 0$. Keeping in mind that ϕ is independent of the azimuth angle, that equation reduces to

$$\frac{1}{r^2}\frac{\partial}{\partial r}\left(r^2\frac{\partial\phi}{\partial r}\right) + \frac{1}{r^2\sin\theta}\left(\sin^2\theta\frac{\partial\phi}{\partial\theta}\right) = 0 \tag{5.3.11}$$

This equation admits a solution by separation of variables. For the time factor, we use $\exp(-i\omega t)$ because we are searching for harmonic oscillations. It is to be noted, however, that stationary oscillations require ω to be real, because a complex value would indicate either oscillations with either a diminishing or a growing amplitude. Thus, putting

$$\phi = \Theta(\theta)R(r)e^{-i\omega t} \tag{5.3.12}$$

we obtain, upon substitution in (5.3.11),

$$\frac{1}{\sin\theta}\frac{d}{d\theta}\left(\sin\theta\frac{d\Theta}{d\theta}\right) + C\Theta = 0 \tag{5.3.13}$$

$$\frac{d}{dr}\left(r^2\frac{dR}{dr}\right) - CR = 0 \tag{5.3.14}$$

where C is the separation constant.

The equation for Θ can be changed into a standard form by putting $x = \cos\vartheta$. This gives

$$(1 - x^2)\Theta''(x) - 2x\Theta'(x) + C\Theta(x) = 0 \tag{5.3.15}$$

Solutions to this may be obtained in terms of infinite series of powers of x (i.e., x^n), where n is an integer. The series terminate at some point if C is taken to be given by

$$C = n(n+1) \quad n = 0, \pm 1, \pm 2, \ldots \tag{5.3.16}$$

This yields

$$(1 - x^2)\Theta''(x) - 2x\Theta'(x) + n(n+1)\Theta(x) = 0 \tag{5.3.17}$$

whose solutions are the Legendre's polynomials $P_n(x)$ and $Q_n(x)$. Explicit values for some of these are given in Appendix E. These show that $Q_n(x)$ is unbounded at the points $x = \pm 1$, corresponding to 0, π, respectively. Thus, we discard from the general solution of (5.3.17) the term corresponding to $Q_n(x)$, so that $\Theta = P_n(\cos\theta)$. With $C = n(n+1)$, the equation for R becomes

$$R''(r) + \frac{2}{r}R'(r) - \frac{n(n+1)}{r^2}R(r) = 0 \tag{5.3.18}$$

Assuming a solution of the form $R = r^l$, we obtain $l(l+1) = n(n+1)$, with solutions $l = n$ and $l = -(n+1)$. For positive n, the first diverges at infinity but is finite at the origin, whereas the second displays opposite trends. Thus, r^n may be used to represent the internal radial field, whereas $r^{-(n+1)}$ gives the external field's radial dependence. Hence,

$$\phi_p = e^{-i\omega t}\sum_{n=0}^{\infty} A_n r^n P_n(\cos\theta) \quad \text{and} \quad \phi_f = e^{-i\omega t}\sum_{n=0}^{\infty} B_n r^{-n-1} P_n(\cos\theta)$$
$$\tag{5.3.19a,b}$$

An expression relating A_n to B_n is provided by the boundary condition on the normal component of the velocity which applied at $r = a$ is $(\partial\phi_p/\partial r)_a = (\partial\phi_f/\partial r)_a$. This gives

$$B_n = -\frac{n}{n+1}A_n a^{2n+1}, \quad n = 0, 1, 2, \ldots \tag{5.3.20}$$

Thus, in the external region we have

$$\phi_f = e^{-i\omega t}\sum_{n=0}^{\infty}\frac{na^{2n+1}}{n+1}A_n r^{-n-1}P_n(\cos\theta) \tag{5.3.21}$$

To determine the coefficients A_n, we express the velocity of the interface in terms of the surface displacements. The distance, along the radial direction, from a point on the surface to the center of the sphere, was given by (5.3.6). The velocity of the surface along r is thus $\partial\zeta/\partial t$. On the other hand, the radial component of the velocities is given by $\partial\phi/\partial r$ for both the internal and external fluids. The two are equal to $\partial\zeta/\partial t$ at points on the surface, but because of the smallness of the displacements, we apply the conditions at $r = a$. Hence, we require that

$$\left(\frac{\partial\phi}{\partial r}\right)_{r=a} = \frac{\partial\zeta}{\partial t} \tag{5.3.22}$$

for both fluids. Now, so long as $|\zeta|$ is small, ζ can depend on θ in an arbitrary manner, and the result for the velocity potential makes it clear that a convenient expression for

ζ is as an infinite series of the P_n. Thus, we express formally ζ as

$$\zeta = e^{-i\omega t} \sum_{n=0}^{\infty} \varepsilon_n P_n(\cos\theta) \qquad (5.3.23)$$

where ε_n is the amplitude corresponding to each value of n. Although the series starts at $n = 0$, physical arguments show that the first two terms in it should be absent. Thus, the term with $n = 0$ corresponds to a uniform radial displacement of the surface, and this represents a change of volume. Similarly, the $n = 1$ term represents a translational oscillation of the particle as a whole, and this uniform translation involves no deformation. We thus omit the first two terms in (5.3.23).

Taking the time derivative of (5.3.23) and equating it to $(\partial\phi_p/\partial r)_{r=a}$, we obtain

$$A_n = -\frac{i\omega\varepsilon_n}{na^{n-1}} \qquad n = 2, 3, \ldots \qquad (5.3.24)$$

Hence,

$$\phi_p = -i\omega e^{-i\omega t} \sum_{n=2}^{\infty} \frac{\varepsilon_n}{n}\left(\frac{r}{a}\right)^n P_n(\cos\theta) \quad \text{and}$$

$$\phi_f = i\omega e^{-i\omega t} \sum_{n=2}^{\infty} \frac{\varepsilon_n}{n+1}\left(\frac{a}{r}\right)^{n+1} P_n(\cos\theta) . \qquad (5.3.25a,b)$$

The pressures inside and outside the drop can be computed from these by means of (5.3.10). We need only the pressure jump across the interface, and is given by the real part of

$$(p_p - p_f)_{r=a} = p_{p0} - p_{f0} + \omega^2 ae^{-i\omega t} \sum_{n=2}^{\infty} \left(\frac{\rho_p}{n} + \frac{\rho_f}{n+1}\right) P_n(\cos\theta)$$

$$(5.3.26)$$

On the other hand, this pressure jump can also be computed from Laplace's formula, (5.3.4), which can be expressed as

$$p_{p0} - p_{f0} = \sigma\nabla\cdot\mathbf{n} \qquad (5.3.27)$$

Here $\mathbf{n}$ is the unit outward normal to the surface, given by 5.3.7. The divergence of the normal vector $\mathbf{n}$ may be obtained from (5.3.8) and (5.3.23). Thus,

$$\nabla\cdot\mathbf{n} = \frac{2}{r} - \frac{1}{r^2}\sum_{n=0}^{\infty} \varepsilon_n n(n+1)P_n(\cos\theta)e^{-i\omega t} \qquad (5.3.28)$$

Now, to obtain the pressure jump at the surface using (5.3.27), the previous expression for $\nabla\cdot\mathbf{n}$ should be evaluated at the interface, where $r = a + \zeta$. Since ζ is small everywhere, we have $1/r \approx (1/a)(1 - \zeta/a)$. Therefore, the sum of the principal curvatures

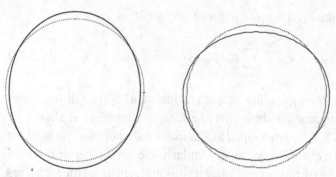

Figure 5.3.2. Sketch of the $n = 2$ shape mode. The dotted line represents the undisturbed surface.

is, to order ε_n,

$$\nabla \cdot \mathbf{n} = \frac{2}{a} + \frac{1}{a^2} \sum_{n=0}^{\infty} \varepsilon_n [n(n+1) - 2] P_n(\cos\theta) e^{-i\omega t} \qquad (5.3.29)$$

The first term on the right-hand side is the total curvature of the nondeformed drop, which, when multiplied by σ, is equal to the static pressure jump $p_{p0} - p_{f0}$. The remaining terms give

$$\sum_{n=0}^{\infty} \varepsilon_n a \{\omega^2 [\rho_p/n + \rho_f/(n+1)] - \sigma(n-1)(n+2)/a^2\} P_n(\cos\theta) = 0$$

$$(5.3.30)$$

This must hold for all values of n. Therefore, the quantity inside the braces must be zero for all values of n. This specifies the frequency corresponding to each n, viz.

$$\omega_n^2 = \sigma \frac{n(n-1)(n+1)(n+2)}{[\rho_p(n+1) + \rho_f n]a^3} \qquad n = 2, 3, \ldots \qquad (5.3.31)$$

It is noted that the frequencies defined by this are real for all values of n. Each of these values gives the frequency associated with a natural mode of oscillation, which in turn are each prescribed by the real part of (5.3.23), or

$$\zeta_n = \varepsilon_n \cos(\omega_n t) P_n(\cos\theta), \qquad n = 2, 3, \ldots \qquad (5.3.32)$$

The shapes and frequencies associated with each value of n represent different deformation modes. The lowest nonzero mode corresponds to the $n = 2$ term. In it, the displacements of points on the surface of the droplet vary as $\cos^2\theta$. Thus, the droplet surface oscillates between oblate and prolate spheroidal shapes about a mean spherical surface, as sketched in Fig. 5.3.2. Instantaneous profiles of the next two modes are shown in Fig. 5.3.3. Now, the frequencies of oscillation, given by (5.3.31), are seen to depend on various quantities, including the densities of the interior and exterior fluids. On the basis of the density differences, we can distinguish three different cases: drops in gases, bubbles in liquids, and droplets in liquids. In the first case, the density

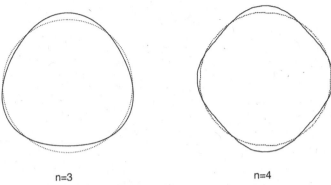

n=3   n=4

Figure 5.3.3. Sketches of the third and fourth shape modes.

of the liquid in the drop is much larger than that of the gas outside, or $\delta \ll 1$, so that (5.3.31) gives, for the lowest mode,

$$\omega_2 = \sqrt{8\sigma/\rho_p a^3} \qquad \delta \ll 1 \qquad\qquad (5.3.33)$$

When $\delta \gg 1$, corresponding to incompressible bubbles in water, the formula gives

$$\omega_2 = \sqrt{12\sigma/\rho_f a^3} \qquad \delta \gg 1 \qquad\qquad (5.3.34)$$

Finally, when $\rho_f/\rho_p \approx 1$, corresponding to immiscible droplets in liquids, the lowest frequency of oscillation is

$$\omega_2 \approx \sqrt{24\sigma/5\rho_p a^3} \qquad \delta = O(1) \qquad\qquad (5.3.35)$$

Some numerical examples might be useful. Thus, the value of ω_2 for a 1-mm diameter air bubble in water is about 2.6×10^3 sec^{-1}, whereas the corresponding value for a 1-mm water droplet in air will be about 6.1×10^4 sec^{-1}, and that for a 1-mm benzene droplet in water will be 1.5×10^3 sec^{-1}. Since $\omega_n \propto a^{-3/2}$ for all modes, we see that the corresponding frequency varies rapidly as the radius is changed.

Two more comments about these results. First, the modes described by (5.3.32) were obtained on the assumption that the deformations were symmetrical with respect to the polar axis. Nonsymmetric modes can also be sustained by the surface, but their frequencies are still given by (5.3.31). Second, the discussion above shows that the deformation is given by a series of noninteracting modes. This result owes its existence to the linear approximation used in the derivation and implies that that no energy is transferred between modes. This ceases to be valid when nonlinear effects are considered.

It should also be added that the forces on droplets that are oscillating while they translate at finite Re can be significantly different from those experienced by nondeforming drops. The drag force applicable to these is often given in terms of empirical correlations. The reader is referred to the bibliography, where a number of works dealing with this important topic are included.

Figure 5.4.1. Breakup of a liquid jet.

5.4 Breakup of Liquid Surfaces

In every one of the normal vibrations studied previously, a given point of the particle's surface oscillates about its equilibrium position, at the corresponding characteristic frequency. As noted earlier, these frequencies are all real, indicating that the oscillations are permanent. This is due to two different factors. The first is our neglect of all dissipative mechanisms, such as viscosity and acoustic radiation. The inclusion of any of these in the calculations would result, as shown in Section 2.5, in an exponential decay of the amplitudes of oscillation. The second factor resulting in permanent oscillations is the stability of the oscillations. By this we mean that if the surface of a fluid particle is deformed slightly and then released, it will execute oscillations having a decaying amplitude, eventually returning to the initial shape. The argument does not necessarily hold in for all surfaces. There exist situations where deformed shapes have smaller surface area than the nondeformed shape, in which case the deformed surface will not return to its original form after release. A well-known example of a liquid mass breaking up as a result of unstable growth of surface deformations is provided by a liquid jet. Figure 5.4.1 shows a photograph of a thin jet in the initial stages of the instability. The jet surface shows a disturbance whose amplitude increases with distance from the exit orifice at the top of the jet, not shown in the figure. The jet is unstable to such disturbances because its surface area decreases with increasing amplitude of the deformation. Thus, even if the disturbance could be removed, the jet would not return to its initial, cylindrical shape. Instead, we see that the disturbance results in thick bulges and thin troughs that later form filaments that eventually detach from the bulges. This action forms a stream of droplets having different sizes, which, owing to their different velocities, can later collide to form a stream of larger, but nearly uniform size droplets. The mechanism has been used for some time to produce droplets having controlled sizes and separation, such as those shown in Section 4.12. We shall not pose here to describe this very remarkable instability in any detail, nor the techniques that are now used to produce droplets taking advantage of it. The interested reader is referred to the very extensive literature on the subject, some of which is cited in the bibliography.

Very small droplets and bubbles in motion are able to remain spherical, or nearly so if they vibrate, because of surface tension. Unstable droplet deformations can, of course, be induced when the particles are not small, or when the applied forces are very large. Such conditions may sometimes be present in suspension motions, even when the Reynolds number is not large. For example, droplets moving in fluids in the presence of a high shear will experience significant deformation and even breakup. Closer to the main topic of this book are the deformation and breakup that can occur in intense acoustic fields. This is put to use in acoustic atomizers, which use high-frequency sound waves to rupture thin surfaces to produce mists of very small droplets.

Several other examples of droplet breakup deserve notice. One occurs during the acoustic levitation of droplets. Here, droplets of a desired size are positioned in a vertical acoustical field of high intensity. As a result of acoustic forces, the droplets can, if the field is sufficiently strong, be kept suspended against gravity. For small droplets or for moderate fields, the droplets retain a near spherical shape. Otherwise, they exhibit considerable flattening. Further increase of the acoustic intensity results in breakup.

Droplet breakup is of some importance in cloud physics, where it occurs when a falling cloud droplet has, as a result of collisions with other droplets, grown so large that it experiences strong deformation and breakup. This limits the sizes that raindrops can acquire. Hailstones, which also grow by accretion with droplets and other hailstones, do not break so easily and can become much larger than raindrops. Another situation where droplet breakup occurs is when thunder propagates in a cloud. These are sharp-front pressure waves of moderate to large amplitude, (i.e., shock waves). The encounter of such a wave with a droplet can produce a wide variety of effects; but, for a range of amplitudes and droplet sizes, the result is deformation and breakup.

In the figures given later, we show a sequence of photographs obtained in a shock tube (Kim, 1977) with 270-μm diameter water droplets exposed to shock waves. For these, the initial Reynolds number was 670, corresponding to a maximum Weber number equal to 6.3. In the experiments, single streams of droplets are vertically injected across the tube, where they are exposed to a shock wave, as shown by the photograph

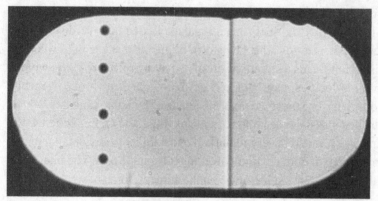

Figure 5.4.2. Droplet stream falling across shock tube 24.9 μsec after passage of shock front, appearing to the right. (Fig. 2a of Temkin and Kim, 1980) [Reprinted with permission from Temkin and Kim (1980). © 1980, Cambridge University Press.]

in Fig. 5.4.2, taken 25 μsec after that impact. In this large-field photograph, the flow is from left to right. The dark circles are the shadow cast by the droplet on the focal plane and the bright-dark line to the right of the droplets is the shock front, traveling to the right. At this time the droplet is very nearly spherical as shown in Fig. 5.4.3a.

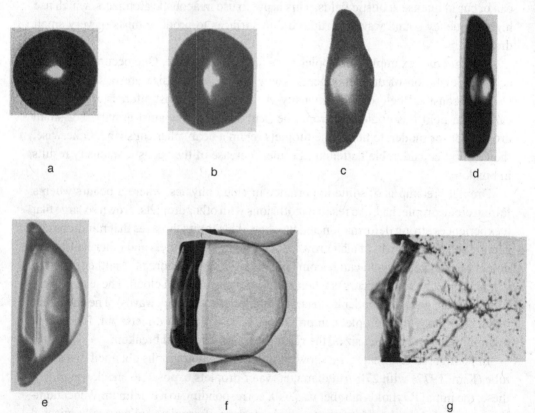

Figure 5.4.3. Lateral views of a water droplet breaking up in the bag-mode (Kim, 1977). $D = 270\,\mu$m, $U_0 = 37$ m/sec. $(We)_{max} = 6.3$, $(Re)_{max} = 670$. Elapsed times after impact: (a) 25 μsec; (b) 100 μsec; (c) 350 μsec; (d) 450 μsec; (e) 850 μsec; (f) 1,060 μsec; and (g) 1,270 μsec.

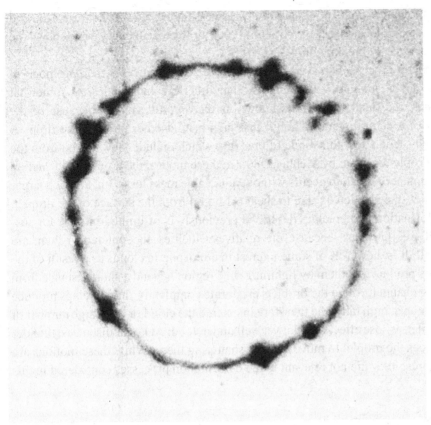

Figure 5.4.4. Liquid ring containing about 90% of the initial droplet mass in the process of breakup (Gast, 1991).

The earliest visible shape change, as seen at right angles to the shock motion, occurs later. Figure 5.4.4b shows the droplet at 100 μsec. Here, we see a flattening developing at the back side of the droplet that is probably due to the separated flow region developing there. At later times, the droplet flattening increases, as shown in the next two photographs in the sequence (c) and (d). Photographs taken from the rear of the droplet at these times (Mehta, 1980; Gast 1991) show that the central region in the front of the droplet is curved inward. This is due to the incoming air flow, which has a maximum pressure there and which has begun to push the liquid inward. That forcing motion continues and results in the formation of an incipient bag at the rear of the droplet, (e). This bag continues to grow with the lateral dimension of the droplet, so that it nearly joins with the adjacent droplets in the initial stream, while thinning out in the process (f). Eventually, the bag shatters, forming a rather large number of much smaller droplets (g).

The bulk of the initial mass of the droplet, however, remains in a liquid ring whose diameter is about three initial droplet diameters. At a slightly later time that depends on the Weber number, the ring becomes unstable, forming the equivalent of a circular jet displaying bulges and troughs. The troughs thin out and break. The ring thus results in a small number of drops of essentially two sizes. Figure 5.4.4, taken from the end of the shock tube, located 1.5 m, or about 5,000 droplet diameters away, shows the ring in the process of breaking up (Gast, 1991).

The breakup process displayed in the above figures, known as the *bag* mode of breakup, occurs at smaller Weber numbers than other breakup modes. At somewhat higher values, the breakup mode is known as the *umbrella* mode, because of the appearance of a central stem of liquid that lags behind other parts of the drop. A photographic sequence of this mode of breakup, which include views taken from the rear of the droplet, is given by Mehta (1980). In these modes, the droplet deformation up to the moment of breakup occurs without noticeable mass loss. But, at much larger flow velocities, the gas flow is able to shear off liquid from the surface of the droplet.

The deformation and breakup described previously is of limited interest for suspensions of free particles, because the relative velocities encountered in them are usually small. It is, however, of some interest in atmospheric clouds as a result of the thunderclaps produced in them by lightning. In a region beyond a short distance from the lightning channel, where the droplets evaporate completely, thunderclaps produce droplet breakup at high rates and therefore inject into the clouds a very large number of smaller droplets whose effects are not yet well understood. At larger distances, thunder simply induces the droplet to move without shattering them. While these motions are of some interest, they are not relevant to the deformation processes considered in this chapter.

5.5 Concluding Remarks

In this chapter, we have considered some of the effects of surface tension. The incursion into this field has been limited in many ways, but gives a glimpse into some fascinating examples of those effects.

6

Volume Pulsations

6.1 Introduction

We now come to the last of the basic motions of a small particle considered in this book: uniform expansions and compressions. Unlike the shape deformations treated in the previous chapter, this motion involves volume changes. In principle, these occur in all types of particles because all materials are compressible to some extent. But the best example is provided by small gas bubbles in liquids.

Now, the volume changes considered here – uniform expansions or contractions – cannot occur unless the particle is spherical. This means that they are limited to changes in the radius of the particle. Of course, if no forces are applied, a small compressible particle will have a fixed radius. This radius, called the *equilibrium radius*, is determined by surface tension effects. Thus, if the interior and exterior static pressures are p_{p0} and p_{f0}, respectively, we have, using the Laplace-Young equation

$$a = \frac{2\sigma}{p_{p0} - p_{f0}} \tag{6.1.1}$$

A change of any of the quantities on the right-hand side of this equation will result in a new equilibrium radius, but it is the external pressure that plays the dominant role. This does not mean that the internal pressure is unimportant. Quite the contrary. The internal pressure also affects the radius of the particle, but that pressure plays essentially a passive role in the sense that its value is fixed if the external pressure is constant.

However, the smallest variations in the external fluid produces volume pulsations in the particle. By this we mean those motions where every point on the surface of the particle is displaced along the radial direction by an identical amount. We are interested in several details of the pulsations, such as the flow field they produce, their natural frequencies of oscillation, and their amplitude under imposed forces. These quantities are important in a variety of contexts, and their study offers a glimpse into the fascinating world of bubble dynamics.

As with previous chapters, the discussion here is largely limited to small-amplitude motions. But even then, the pulsational problem is far from simple because of the many effects that take place when a particle pulsates. Thus, some assumptions will be required. These will be introduced as needed. One of them, to be verified later, is that the particles' internal properties are either uniform or can be represented by spatial average values. As in Chapter 2, we define the volume average of some physical quantity $f(\mathbf{x}, t)$ by means of

$$\overline{f}(t) = \frac{1}{v_{\mathrm{p}}} \int_{v_{\mathrm{p}}} f(\mathbf{x}, t) d \, V(\mathbf{x}) \qquad (6.1.2)$$

where $v_{\mathrm{p}}(t)$ is the instantaneous volume of the particle. Thus, the average density in a particle is $\overline{\rho}_p(t) = v_{\mathrm{p}}^{-1} \int_{v_{\mathrm{p}}} \rho_p(\mathbf{x}, t) dV$, so that the mass of the particle can be expressed as $m_p = \overline{\rho}_p v_{\mathrm{p}}$. Since m_p is constant, we also have

$$\dot{\overline{\rho}}_p/\overline{\rho}_p = -\dot{v}_{\mathrm{p}}/v_{\mathrm{p}} \qquad (6.1.3)$$

where the dots over $\overline{\rho}_p$ and v_p represent time derivatives. Furthermore, if the particle's instantaneous radius is $R(t)$, we also have $\dot{\overline{\rho}}_p/\overline{\rho}_p = -3\dot{R}/R$, where $\dot{R}(t)$ is the velocity of points on the particle surface. It should be stated that, while (6.1.3) makes no assumption regarding the uniformity of the density, we will either assume that to be the case or use the average value to represent the density. An exception to this occurs in Section 6.4, where the internal fields are examined.

But even if the properties inside the sphere are well represented by spatial averages, the study of particle pulsations in fluids is not straightforward because of thermal effects. These also affect the displacement of the particle's surface, owing to thermal expansion. We postpone the discussion of such effects until Section 6.6 and begin our study of particle pulsations by assuming that they take place isothermally. The restriction will enable us to study the motion in its simplest form.

6.2 Motion Produced by Pulsating Sphere

We begin by considering a sphere of equilibrium radius a, executing harmonic, radial pulsations of uniform amplitude, ε, assumed small, in a body of fluid of large extent. The motion produces in the surrounding fluid a flow field whose velocity is along the radial direction of a spherical system of coordinates fixed at the sphere's center. That is, $\mathbf{u}_f = \{u_f(r, t), 0, 0\}$. We wish to obtain this velocity as well as the accompanying pressure. As defined, the problem is linear, but it is convenient to start by writing the complete set of governing equations in their general form, taking advantage of the fact that the velocity field is irrotational, or $\nabla \times \mathbf{u}_f = 0$. Hence, $\nabla^2 \mathbf{u}_f = \nabla(\nabla \cdot \mathbf{u}_f)$, so that $\mathbf{u}_f \cdot \nabla \mathbf{u}_f = \frac{1}{2}\nabla u_f^2$, where $u_f^2 = \mathbf{u}_f \cdot \mathbf{u}_f$. Using these two relations in the momentum equation, (2.1.3), without the body force term, we can write that equation as

$$\rho_f \left(\partial \mathbf{u}_f/\partial t + \tfrac{1}{2}\nabla u_f^2\right) = -\nabla p_f + \tfrac{4}{3}\mu_f \nabla(\nabla \cdot \mathbf{u}_f) \qquad (6.2.1)$$

The remaining governing equations are, as before

$$\frac{\partial \rho_f}{\partial t} + \nabla \cdot (\rho_f \mathbf{u}_f) = 0 \tag{6.2.2}$$

$$\rho_f c_{pf} \frac{DT_f}{Dt} - \rho_f \beta_f \frac{Dp_f}{Dt} = \rho_f \Phi_f + k_f \nabla^2 T_f \tag{6.2.3}$$

$$p_f = p_f(\rho_f, T_f) \tag{6.2.4}$$

The set retains the effects of compressibility, viscosity, heat conductivity, and nonlinearity. Our main concern is the small-amplitude, linear motion. But, before reducing our system to study it in detail, it is useful to consider the external motion in the absence of compressibility and heat conductivity. As shown below, the reduced system can be solved without linearizing it, thus providing some information about nonlinear effects.

Incompressible Fluid

Here, $\rho_f = \rho_{f0}$, a constant, so that the continuity equation becomes.

$$\nabla \cdot \mathbf{u}_f = 0 \tag{6.2.5}$$

Using this in the momentum equation, we obtain

$$\rho_{f0} \left(\partial \mathbf{u}_f / \partial t + \tfrac{1}{2} \nabla u_f^2 \right) = -\nabla p_f \tag{6.2.6}$$

These equations are to be solved subject to conditions at the sphere's surface and at infinity. Thus, if $U_s(t)$ is the surface velocity, we require that $u_f = U_s$ on the surface of the sphere and that far from it, u_f vanishes. Similarly, the fluid pressure at infinity equals the pressure there. We denote this pressure by p_∞, noting that it may be a function of time. Since the only component of u_f is along r and does not depend on any other coordinate, the continuity equation reduces to

$$\partial(r^2 u_f)/\partial r = 0 \tag{6.2.7}$$

Integrating this and applying the boundary condition at the instantaneous position, $R(t)$, of the surface, we obtain

$$u_f = U_s \left(R/r \right)^2 \tag{6.2.8}$$

where $U_s = \dot{R}$ is the radial velocity of the surface. Thus, the magnitude of the fluid velocity is proportional to the velocity of the sphere, and is everywhere in phase with it; it also decays rapidly with distance. The pressure in the fluid corresponding to this velocity follows from (6.2.6). Thus, substitution of (6.2.8) gives

$$\frac{\partial p_f}{\partial r} = -\rho_{f0} \left[\frac{2R\dot{R}^2 + R^2 \ddot{R}}{r^2} + \frac{1}{2} \frac{\partial}{\partial r} \left(\frac{U_s R^2}{r^2} \right)^2 \right]$$

Integrating between ∞, where $p_f = p_{f\infty}$, and r, gives

$$p_f = p_{f\infty} + \rho_{f0} \left(\frac{2R\dot{R}^2 + R^2\ddot{R}}{r} - \frac{\dot{R}^2 R^2}{2r^4} \right) \tag{6.2.9}$$

Unlike the velocity field, which only depends on the position and velocity of the surface, we see that the pressure also depends on its acceleration. It is also seen that neither the pressure nor the velocity in the fluid depend on the viscosity of the fluid. For future reference, we note that these results have been obtained without linearizing the equations or the boundary conditions and thus apply to arbitrary surface motions.

Rayleigh's Equation

Equation (6.2.9) was obtained on the assumption that the radial displacement of the particle was known, and that what we required was the pressure produced by the motion. In reality, the situation is the opposite. The surface of small bubbles tends to pulsate in response to changes in the external pressure, but the amplitude of the pulsation is not predetermined. We wish to obtain that amplitude for arbitrary values of the external pressure. If the fluid is regarded as incompressible, as done previously we may obtain an equation for the position of the surface from (6.2.9) as follows. First, we evaluate that equation at $r = R$ to obtain the external pressure on the particle surface, p_{fs}. Thus,

$$\rho_{f0} \left(R\ddot{R} + \tfrac{3}{2}\dot{R}^2 \right) = -(p_{f\infty} - p_{fs}) \tag{6.2.10}$$

We note in passing that, if surface tension effects are ignored, p_{fs} is equal to the internal pressure. Multiplying (6.2.10) by the surface area of the particle, we obtain the incompressible fluid force on the pulsating particle. Thus,

$$F_r = 4\pi R^3 \rho_{f0} \left(\ddot{R} + \tfrac{3}{2}\dot{R}^2/R \right),$$

a result that was obtained earlier [see Equation (2.3.19)].

Bubble Collapse. Equation (6.2.10) was first obtained by Rayleigh in his study of the collapse of a spherical cavity, what we now call a cavitation bubble. These can be formed in a variety of ways. The example treated by Rayleigh refers to gas bubbles created when small pockets of gas trapped in crevices on the surface of a solid surface overcome the surface tension force keeping them there. This may occur because of a reduction of the liquid pressure outside the crevice. The released gas forms spherical gas bubbles whose radius can grow significantly, sometimes overshooting its equilibrium value. This overshoot cannot continue indefinitely. At some point, an inward motion is initiated that may damage solid surfaces that are close to the bubbles. The effect, called cavitation, was of interest to Rayleigh and continues to be of interest today.

Below is Rayleigh's analysis of the collapse of a bubble in its simplest form. Here, a bubble has attained its maximum radius, R_m, in a fluid that is at rest at infinity. First, we note that if the bubble is large the effect of surface tension may be neglected so that we may put $p_{fs} = p_p$, giving

$$R\ddot{R} + \tfrac{3}{2}\dot{R}^2 = \frac{p_p - p_{f\infty}}{\rho_{f0}}$$

To obtain the instantaneous position of the bubble's surface, that is to obtain $R(t)$, we need to integrate this equation for a prescribed value of the liquid pressure at infinity. This requires that p_p be known as a function of the radius. If the gas in the bubble can be taken as perfect, we could – since the motion is, by assumption, isothermal – use the perfect gas equation of state to write $p_p = CR^{-3}$, where C is a constant that is proportional to the mean density of the gas in the bubble. But, if the bubble is large, the mass of the gas in it can be approximately neglected, so that $p_p \approx 0$. Thus, eliminating $\ddot{R}$ from this equation by means of $\ddot{R} = [d(R^3\dot{R}^2)/dt - 3R^2\dot{R}^3]/2\dot{R}R^3$, we obtain

$$\frac{1}{2}\frac{d}{dt}\left(R^3\dot{R}^2\right) = \frac{p_{f\infty}R^2\dot{R}}{\rho_{f0}}, \tag{6.2.11}$$

This can be integrated between $t = 0$, when $R = R_m$ and $\dot{R} = 0$, and some other time t, giving

$$(R^{3/2}\dot{R})^2 = \frac{2}{3}\frac{p_{f\infty}}{\rho_{f0}}\left(R_m^3 - R^3\right)$$

Although an explicit solution of this cannot be obtained analytically, we can obtain from it the time that a bubble requires to attain a given value, R, if its value at $t = 0$ is R_m. This time is given by

$$t = \left(\frac{3\rho_{f0}}{2p_{f\infty}}\right)^{1/2}\int_{R/R_m}^{1}\frac{r^{3/2}dr}{(1 - r^3)^{1/2}}$$

where $r = R/R_m$. This result was used by Rayleigh to estimate the time required for total collapse, t_0, say, by letting $R \to 0$. The resulting integral may be expressed in terms of gamma functions which, when evaluated, give $t_0 = 0.915R_m\sqrt{\rho_{f0}/p_{f\infty}}$. According to this, a bubble having a maximum radius equal to 1 mm and immersed in water at 1 atm would collapse in about 10^{-4} sec. However, the radial velocity at the time of collapse would be infinite.

Of course the assumptions made to obtain this result are not valid as the bubble becomes very small. Thus, the mass of the gas inside the bubble cannot be neglected then. This means that the pressure in the bubble must attain rather large values, which, owing to the rapidity of the collapse, involve rather high temperatures. Hence, the assumption that the temperature is constant is also invalidated by real bubble.

In fact, during its initial stages, the inward motion appears to follow the Rayleigh collapse described briefly previously. But, in the last moments, the variations are nearly adiabatic, with a consequent large increase in the temperature, possibly reaching values in excess of 10, 000 K. It is in these final stages of the collapse that the bubble emits short-duration light pulses. This remarkable phenomenon is known as *sonoluminescence*. The emission appears to be the result of the very high temperatures that the gas in the bubble reaches in the last moments of the collapse. At such temperatures, the gas in a bubble become ionized, creating a plasma (a gas made of free electrons and other ions). As the electrons and ions recombine, light is emitted. In the process, molecular dissociation of the gas or gases in the bubble occurs. This phenomenon is referred to *sonochemistry*. Sonoluminescence and sonochemistry are fascinating topics, but are beyond the scope of this book. For additional information, the reader is referred to the references cited in the bibliography.

The Rayleigh-Plesset Equation

Let us now return to (6.2.10) to obtain p_p, this time including some of the effects left out in the Rayleigh equation, while still assuming that the pulsations are isothermal and take place in an incompressible liquid. We note that $p_{f\infty}$ is a known function of time, but that the surface pressure on the side of the fluid, p_{fs}, is unknown. To obtain it, we use the boundary condition on the normal components of the stress tensors for the interior and exterior fluids, (2.1.17). Since only the interior fluid is compressible, that condition becomes,

$$p_{fs} - 2\mu_f \left(\frac{\partial u_f}{\partial r}\right)_{r=R} = p_{ps} - \frac{2\sigma}{R} - 2\mu_p \left(\frac{\partial u_p}{\partial r} - \frac{1}{3}\nabla \cdot \mathbf{u}_p\right)_{r=R}$$

The first term on the right-hand side of this denotes the value of the particle pressure, evaluated at the particle side of the surface. By assumption, the pressure within the particle is either uniform, or well represented by its average value. Hence, $p_{ps} = p_p$. The last term on the right-hand side of this equation represents the deviatoric component of the stress tensor in the interior fluid (i.e., in the particle). Since the particle is small, and since the pulsational motion is one of uniform expansions and contractions, this term vanishes identically, as (2.1.18) indicates. On the other hand, the corresponding quantity for the exterior fluid, $2\mu_f(\partial u_f/\partial r)$, does not vanish. From (6.2.8), we obtain $(\partial u_f/\partial r)_{r=R} = -2\dot{R}/R$. Substitution into (6.2.10) and using $p_{ps} = p_p$, we obtain

$$\rho_{f0}\left(R\ddot{R} + 4\nu_f\frac{\dot{R}}{R} + \frac{3}{2}\dot{R}^2\right) = -\left(p_{f\infty} + \frac{2\sigma}{R} - p_p\right) \tag{6.2.12}$$

Equation (6.2.12) is known as the Rayleigh-Plesset equation. It describes the variations of the particle radius with time, resulting from the pressure difference between particle and fluid. But since the particle pressure may be regarded as a function of the particle radius, R, the equation prescribes the variations of that radius in response to the external pressure far from the sphere.

As derived, (6.2.12) is limited to motions in incompressible fluids. Several extensions exist that include compressibility effects, while retaining the nonlinear character of the equation. This aspect of the motion is important on its own, but is outside the scope of this work. The interested reader is referred to more detailed works on the subject, some of which are listed in the bibliography.

Compressible Fluid

We now consider the external fields produced by small-amplitude, radial motions of a pulsating sphere in a compressible fluid, but continue to regard the motions as isothermal. As in the previous case, this approach makes the energy equation unnecessary. We shall, however, retain the effects of viscosity. First, we linearize the equations, assuming that that the density and pressure deviate only slightly from their ambient values (e.g., $p_f = p_0 + p'_f$ and $\rho_f = \rho_{f0} + \rho'_f$, with $p'_f \ll p_0$ and $\rho'_f \ll \rho_{f0}$). Using these in the above system and dropping terms of second order, we obtain

$$\frac{\partial \rho'_f}{\partial t} + \rho_{f0}\nabla \cdot \mathbf{u}_f = 0 \tag{6.2.13}$$

$$\rho_{f0}(\partial \mathbf{u}_f/\partial t) = -\nabla p'_f + \tfrac{4}{3}\mu_f \nabla(\nabla \cdot \mathbf{u}_f) \tag{6.2.14}$$

$$p'_f = c^2_{Tf}\rho'_f \tag{6.2.15}$$

where c_{Tf}, the isothermal sound speed, is evaluated at ambient density. The equations can be combined in several ways to produce a single equation, of higher order, for any of the unknowns appearing in them. But, since the fluid velocity is irrotational, we can simplify the procedure by putting $\mathbf{u}_f = \nabla\phi$, where ϕ, called the velocity potential, is a scalar function of time and position. Thus, using (6.2.15) to eliminate the density in favor of the pressure, (6.2.13) and (6.2.14) can be written as

$$\frac{\partial p'_f}{\partial t} + \rho_{f0}c^2_{T0}\nabla^2\phi = 0 \tag{6.2.16}$$

$$\nabla\left[p'_f + \rho_{f0}(\partial\phi/\partial t) - \tfrac{4}{3}\mu_f\nabla^2\phi\right] = 0$$

Integrating this between some point r and ∞, where all induced motions vanish, we obtain

$$p'_f = -\rho_{f0}\frac{\partial\phi}{\partial t} + \tfrac{4}{3}\mu_f\nabla^2\phi \tag{6.2.17}$$

Finally, taking the time derivative of this and substituting the result in (6.2.16) yields,

$$\frac{\partial^2\phi}{\partial t^2} = c^2_{Tf}\left[1 + \left(4\nu_f/3c^2_{Tf}\right)\frac{\partial}{\partial t}\right]\nabla^2\phi \tag{6.2.18}$$

When viscosity is neglected, this reduces to the wave equation, implying that the motion induced in the fluid will consist of sound waves. The existence of viscosity modifies these waves in various manners, making them dispersive. This means that the waves will propagate with a speed that depends on their frequency and that their amplitude will diminish faster than in the inviscid case.

We limit the discussion to pulsations that have a single frequency ω. For simplicity, we use the complex notation and express the radial displacement of points on the surface of the sphere, ε, as the real part of $\varepsilon_0 \exp(-i\omega t)$. If transients are ignored, the solution to (6.2.18) must be of the form $\phi = \Phi(r)\exp(-i\omega t)$, where $\Phi(r)$ is a complex function of the distance from the sphere's center. Substituting in (6.2.18) we obtain

$$\nabla^2\Phi + K^2\Phi = 0 \tag{6.2.19}$$

where $K^2 = k^2(1 - i\omega\tau_v)^{-1}$, $k = \omega/c_{Tf}$, and where

$$\tau_v = 4v_f/3c_{Tf}^2$$

This quantity has dimensions of time and represents the time scale between molecular collisions in the fluid. In water at 20°C, this time scale is equal to 6×10^{-13} sec. Thus, unless the frequency is exceedingly high, the quantity $\omega\tau_v$ will be very small.

In terms of ϕ, the fluctuating pressure produced in the external fluid by the pulsations follows from (6.2.17) and (6.2.18). Thus,

$$p'_f = \frac{i\omega\rho_{f0}\phi}{1 - i\omega\tau_v} \tag{6.2.20}$$

Now, the solution of (6.2.19) that vanishes at infinity can be expressed as $\Phi = Ch_0(Kr)$, so that

$$\phi = Ch_0(Kr)e^{-i\omega t} \tag{6.2.21}$$

where C is a constant whose value may obtained from the linearized boundary condition on the radial velocity: $u_f(a, t) = U_s$. Thus, if the velocity of the surface is expressed as $U_s = U_{s0}\exp(-i\omega t)$, and using $u_f = \partial\phi/\partial r$, we obtain $C = U_{s0}/Kh'_0(Ka)$. Hence,

$$\phi = U_s h_0(Kr)/Kh'_0(Ka) \tag{6.2.22}$$

$$u_f = U_s h'_0(Kr)/h'_0(Ka) \tag{6.2.23}$$

Because K is complex, these results are algebraically involved and not easily understood. To see their meaning, we take the viscosity to be zero, in which case K is real and equal to $k_0 = \omega/c_{Tf}$. We then have, on using $zh'_0(z) = -(1 - iz)h_0(z)$ and $h_0(z) = -(i/z)\exp(iz)$,

$$u_f = U_s \left(\frac{a}{r}\right)^2 \frac{1 - ikr}{1 - ib}e^{ik(r-a)} \tag{6.2.24}$$

where $b = k_0a$. This result shows that different points in the fluid experience oscillatory motions having different amplitudes and phases. The incompressible result comparable with this is, from (6.2.8) given by $U_s(a/r)^2$. Unlike the compressible case, this shows no phase differences between the oscillations induced at different points in the fluid; the sphere is simply displacing the external fluid as a whole. Furthermore, the amplitude of the compressible velocity contains two terms: one decaying as r^{-2}, corresponding to the incompressible velocity, and another decaying as r^{-1}. This second term is not important near the particle, where the r^{-2} term dominates, but takes over at large distances from it. It is this slowly decaying term that accounts for the sound waves emitted by the pulsating particle.

Similarly, when the fluid is inviscid, the pressure produced by the motion of the sphere follows from (6.2.20) and may be expressed as

$$p'_f = \rho_{f0} a \dot{U}_s \left(\frac{a}{r}\right) \frac{1}{1 - ib} e^{ik(r-a)} \tag{6.2.25}$$

where $\dot{U}_s = -i\omega U_s$. Comparison with the corresponding incompressible result, $p_f = p_\infty + \rho_{f0} a^2 \dot{U}_s/r$, obtained by linearizing (6.2.9), shows the effects of compressibility. These are revealed in the phase of the compressible result, which shows that a given amplitude is felt at different places at different times, unlike the incompressible case where no phase differences occur.

6.3 Force on Pulsating Sphere

We now calculate the fluid force on the pulsating sphere that results from its own motion. This acts normal to the surface and is due to two effects. The first derives from the viscous stresses in the fluid, and the second is due to reaction of the external fluid on the sphere. This reaction, in turn, may be divided into purely inertial effects that are due to the fluid mass that must be displaced by the sphere and into forces that owe their origin to the compressibility of the fluid. These include an acoustic component that is of interest in wider contexts. Since the stresses are uniform over the surface of the sphere, the radial force produced by the motion is given by $F_r = 4\pi a^2 \sigma'_{rr}(a)$, where σ'_{rr} is the normal component of the stress tensor in the fluid and is given by

$$\sigma'_{rr} = -p'_f + \frac{4}{3}\mu_f \left(\frac{\partial u_f}{\partial r} - \frac{u_f}{r}\right) \tag{6.3.1}$$

The fluctuation pressure is given, in terms of ϕ, by (6.2.20). At the mean position of surface of the sphere, that result becomes

$$\phi(a) = -\frac{a}{1 - iKa} U_s \tag{6.3.2}$$

where we have again used $zh'_0(z) = -(1 - iz)h_0(z)$. Similarly, using $u_r = \partial\phi/\partial r$, we find that

$$\left(\frac{\partial u_f}{\partial r} - \frac{u_f}{r}\right)_a = [3(1 - iKa) - (Ka)^2]\phi(a)/a^2 \tag{6.3.3}$$

where we have used the defining equation for $h_0(z)$ to eliminate $h_0''(Ka)$. Thus, using these expressions in (6.3.1), yields

$$\sigma_{rr}'(a) = i\rho_{f0}\omega U_s[(1 - iKa)^{-1} + 4\mu_f/\rho_{f0}\omega a^2] \tag{6.3.4}$$

Finally, making use of the smallness of $\omega\tau_v$, we have $(1 - iKa)^{-1} \approx (1 + ib)/(1 + b^2)$ so that using $\dot{U}_s = -i\omega U_s$ we can write this in the real form

$$\sigma_{rr}'(a) = -\frac{a\rho_{f0}}{1 + b^2}\dot{U}_s - \frac{\rho_{f0}\omega b}{1 + b^2}U_s - \frac{4\mu_f U_s}{a} \tag{6.3.5}$$

Multiplying this by the mean surface area of the sphere, we obtain the desired result:

$$F_r = -[16\pi a\mu_f + M_0 b\omega]U_s - M_0\dot{U}_s \tag{6.3.6}$$

where

$$M_0 = \frac{4\pi a^3\rho_{f0}}{1 + b^2} \tag{6.3.7}$$

This quantity is a mass of external fluid having a volume nearly equal to three times the volume of the pulsating sphere, the difference being due to compressibility, as implied by the term b^2 in the denominator. We note that this mass multiplies, in (6.3.6), the acceleration of points on the surface separating fluid from particle. If that surface had a mass of its own, the net effect of the term $M_0\dot{U}_s$ would be to add an amount M_0 to that mass in its equation of motion. This is the reason we call M_0 an *added* mass. But since the surface separating the two materials has no mass of its own, we may regard M_0 as *the* mass of the surface, This identification will be discussed later in more detail.

Now, in addition to $M_0\dot{U}_s$, the force on the sphere contains a term proportional to the velocity of the surface. This represents a dissipative force, and is due to viscous and compressible effects. This may be seen by computing the power that must be spent against to maintain the motion, which, so far as the oscillation is concerned, is lost. Thus, since in a unit time the surface moves a distance U_s, and the motion is harmonic in time, that power is given by $-\frac{1}{2}\Re\langle FU_s^*\rangle$. The term $M_0\dot{U}_s$, in the complex force, does not contribute to the power, since the time average of the product of the velocity with the acceleration vanishes for a harmonic motion. Thus, the average energy dissipation rate is

$$\langle\dot{e}\rangle_{loss} = \left(8\pi a\mu_f + \tfrac{1}{2}M_0 b\omega\right)U_{s0}^2 \tag{6.3.8}$$

This shows that the dissipation rate is the sum of two terms. The first is clearly due to viscous effects, and the second is due to acoustic radiation into the external fluid. It should be noted that, at very low frequencies, the viscous term is dominant.

6.4 Internal Fields

Let us now consider the fluid inside the sphere. As the sphere pulsates, a radial motion is induced in it which is accompanied by density and pressure changes. These fields are of interest, and are needed to determine the motion of the surface of the particle and to assess the assumption that the pressure and density are either uniform or are well represented by their spatial averages.

Uniform Density

To obtain the velocity field inside the particle, we first assume that the density in the particle is either uniform or is well represented by its average value. Then, the particle velocity may then be easily obtained from the continuity equation for the fluid in the particle. Thus, since the motion has central symmetry and the particle density is given by its average value, that equation gives

$$\nabla \cdot \mathbf{u}_p = -\frac{1}{\rho_p}\dot{\rho}_p \tag{6.4.1}$$

Since the motion has central symmetry, $\mathbf{u}_p = \{u_p(r, t), 0, 0\}$, we have $\nabla \cdot \mathbf{u}_p = r^{-2}[\partial(r^2 u_p)/\partial r]$. Therefore, the solution of (6.4.1), which is finite at the origin, is $u_p = -(\dot{\rho}_p/3\rho_p)r$. At the instantaneous position R of the surface, this velocity must equal the radial velocity $\dot{R}$ of the surface. Hence,

$$u_p = \dot{R}(r/R) \tag{6.4.2}$$

Thus, the velocity in the particle is zero at the origin and increases linearly with distance from it. Now consider the pressure. Since by assumption the density in the particle is uniform, the momentum equation gives, on using (6.4.2),

$$\frac{\partial}{\partial r}\left(\frac{p_p}{\rho_p} + \tfrac{1}{2}u_p^2\right) = -\frac{\partial u_p}{\partial t} \tag{6.4.3}$$

Using (6.4.2) and integrating between the particle surface, where $p_p = p_{ps}$, and some point r, yields

$$p_p(r, t) = p_{ps} + \tfrac{1}{2}\rho_p R\ddot{R} - \tfrac{1}{2}\rho_p \ddot{R}r^2/R \tag{6.4.4}$$

Integrating over the volume of the particle and dividing the result by it, yields

$$\overline{p}_p(t) = p_{ps} + \frac{1}{5}\rho_p R\ddot{R} \tag{6.4.5}$$

Since, according to this $\overline{p}_p(t) \neq p_{ps}$, we conclude that the pressure in the particle is not uniform and that the degree of nonuniformity is proportional to the acceleration of the particle surface. Furthermore, since (6.4.5) was obtained on the assumption that the density was uniform, we must conclude that the spatial variations of the pressure and the density can be disregarded only for small accelerations of the surface. A more precise condition will be given in the next section for small-amplitude motions. In what follows, we will assume that the surface accelerations are always small so that the particle pressure and density are uniform.

For small surface accelerations, the pressure is uniform and may be obtained from the density using the isothermal equation of state for the particle material, if this is known. This is of the form $p_p = p_p(\rho_p)$ and is generally nonlinear. A notable exception occurs in the case of perfect gas bubbles, for which $p_p = C\rho_p$, where C is a constant. For other substances, we may expand $p_p = p_p(\rho_p)$ in Taylor series about the ambient value. Thus,

$$p_p = p_{p0} + c_{Tp}^2(\rho_p - \rho_{p0}) + \tfrac{1}{2}\left(\partial^2 p_p/\partial\rho_p^2\right)_{\rho_{p0}}(\rho_p - \rho_{p0})^2 + \dots \quad (6.4.6)$$

with $\rho_p - \rho_{p0} = \rho_{p0}[(a/R)^3 - 1]$. A similar relation could be written for the volume averages. When the departures from ambient conditions are small, $R = a + \varepsilon$, with ε being a very small displacement, the pressure becomes

$$p_p - p_{p0} = -3\rho_{p0}c_{Tp}^2(\varepsilon/a) \quad (6.4.7)$$

The corresponding values for the density is obtained from (6.1.3)

$$\rho_p - \rho_{p0} = -3\rho_{p0}(\varepsilon/a) \quad (6.4.8)$$

whereas the radial velocity follows from (6.4.2) and is $u_p = (r/a)\dot{U}_s$.

Nonuniform Fields

To assess the validity of the uniform density assumption, it is useful to consider the internal motions in the linear approximation, this time allowing the internal pressure and density to depend on position. Thus, we consider small deviations from the ambient conditions and write $p_p = p_{p0} + p_p'(r, t)$, with $p_p' \ll p_{p0}$. The pressure fluctuation and its corresponding velocity field can then be obtained from the wave equation for the velocity potential ϕ_i through the usual relations, namely $\mathbf{u}_p = \nabla\phi_i$ and $p_p' = -\rho_{p0}(\partial\phi_i/\partial t)$. For harmonic motions, $\phi_i = \Phi_i(r)\exp(-i\omega t)$, that equation gives

$$\nabla^2\Phi_i + k_i^2\Phi_i = 0 \quad (6.4.9)$$

where $k_i = \omega/c_{Tp}$ is the internal wavenumber. The solution to this equation, which is finite at the origin, is $\Phi_i = Cj_0(k_i r)$, where C is a constant and $j_0(k_i r)$ is the spherical Bessel function of zero order. To evaluate the constant, we make use of the linearized boundary condition on the velocity, namely $u_p(a, t) = U_{s0}\exp(-i\omega t)$. Since $u_p = \partial\phi_i/\partial r$, we obtain $C = U_{s0}/k_i j_0'(b_i)$, where $b_i = k_i a$. Thus,

$$\phi_i = U_s a \frac{j_0(k_i r)}{b_i j_0'(b_i)} \quad (6.4.10)$$

The pressure fluctuation is obtained from this by means of $p_p' = -\rho_{p0}(\partial\phi/\partial t)$, which gives

$$p_p'(r, t) = -\rho_{p0}\dot{U}_s a \frac{j_0(k_i r)}{b_i j_0'(b_i)} \quad (6.4.11)$$

This obviously varies with position. Furthermore, since $j_0(b_i) = \sin b_i/b_i$, we see that p'_p becomes infinite for values of b_i that satisfy $b_i \cos b_i = \tan b_i$. These correspond to the acoustic natural frequencies of the spherical particle. The lowest value of b_i satisfying this transcendental equation is close to 4.49. For a 10-μm radius air bubble, this occurs at a frequency equal to about 5×10^7 Hz. But, for values of b_i smaller than this, those indefinite amplitudes are absent. In addition, if $z \ll 1$, we have $j_0(z) \approx 1 - z^2/3! + z^4/5!$, so that to leading order $zj'_0(z) \approx -z^2/3$. We then have, on writing $\ddot{U}_s = -\omega^2\varepsilon$,

$$p'_p \approx -3\rho_{p0}c_{Tp}^2(\varepsilon/a)\left[1 - \tfrac{1}{6}(k_i r)^2\right] \qquad (6.4.12)$$

On the other hand, the average pressure in the particle can be obtained from (6.4.11). This requires $\overline{j_0(k_i r)}$, where

$$\overline{j_0(k_i r)} = \frac{4\pi}{v_{p0}} \int_0^{v_{p0}} j_0(k_i r) r^2 dr$$

Using the definition of j_0, we obtain $\overline{j_0(k_i r)} = 3j'_0(b_i)/b_i$ so that

$$\overline{p}'_p = -3\rho_{p0}c_{Tp}^2(\varepsilon/a) \qquad (6.4.13)$$

This is equal to (6.4.7), an equation that was obtained by assuming that the density was uniform. Comparison with (6.4.12) shows that the pressure may be considered uniform and equal to $\overline{p}'_p$, provided $b_i^2 \ll 1$.

The representation of the particle pressure in terms of $\overline{p}'_p$, as given by (6.4.13) would seem applicable to larger frequencies, but it should be remembered that, in deriving (6.4.11), we assumed that the external pressure field was uniform around the particle. That assumption requires, in turn, that the external wavelength be much larger than the particle radius, or that $b^2 \ll 1$. Since $b = b_i(c_{Tp}/c_{Tf})$ and since the sound speeds are generally of the same order of magnitude, we see that the uniformity assumption also requires that $b^2 \ll 1$. When these conditions are satisfied, we have $p'_p = \overline{p}'_p$. In what follows, we shall use p'_p to denote either quantity.

6.5 Surface Motion

We have, so far, examined the internal and external fluid motions produced by a sphere pulsating at a frequency ω and with a small amplitude ε. We have not, however, considered the origin of these pulsations, nor have we determined their amplitude. Normally, a fluid particle immersed in another fluid pulsates only if the external conditions change. Of course, transient disturbances may excite those pulsations, but in the absence of continued external forcing, a pulsating particle will eventually come to rest, as a result of dissipation. In some instances, however, stationary radial particle pulsations are produced, for example, when a continuous sound wave – that is, a wave having no front or tail – propagates in the external fluid. We wish to study these stationary pulsations in the simplest situation. This occurs when the external pressure variations are harmonic and have very small amplitude. Here, the surface is displaced from the equilibrium position $R = a$, only by a small amount ε – that is, the instantaneous radius of the particle is $R = a(1 + \varepsilon/a)$, with $\varepsilon \ll 1$. Furthermore, in steady-state conditions, the surface displacement, ε, is also a harmonic function of time having the same frequency as that of the external pressure variations, but whose amplitude is not yet known. To obtain it, we consider the dynamic equilibrium between all the forces acting on the particle.

Consider first the forces due to the particle's own motion in a fluid that, far from the particle, is at rest. The first part of that force was obtained in Section 6.3 for a harmonic time dependence. If the radial velocity of the surface is U_s, that self-induced force is given by (6.3.6). In addition, there is the surface tension force, which, as we know, produce an excess pressure $2\sigma/a$ on a particle in equilibrium. When the surface of the particle is displaced to position $a + \varepsilon$, the excess pressure results in a force on the particle equal to $-8\pi a^2 \sigma/(a + \varepsilon)$.

Now consider the external forces responsible for the motion of the particle. These are due to external fields in the bulk of the fluid that might exist even in the absence of the particle. We take these fields to have length scales that are much larger than the diameter of the particle, and furthermore assume that all exterior boundaries are very far from the particle. In such idealized conditions, the force driving the motion is uniform around the particle and is due to the pressure difference between the fluid, far from the particle, and that in the particle itself, multiplied by the surface area of the particle. Since the displacements are small, this part of the force is thus equal to $-4\pi a^2(p_{f\infty} - p_p)$. In dynamic equilibrium, the sum of these forces must vanish. Hence,

$$M_0 \ddot{\varepsilon} + (4\mu_f/\rho_{f0}a^2 + b\omega)M_0 \dot{\varepsilon} = -4\pi a^2 [p_{f\infty} - p_p + 2\sigma/(a + \varepsilon)] \quad (6.5.1)$$

Let us now consider the pressure difference appearing on the right-hand side of (6.5.1). We first note that the appearance of p_p implicitly takes the particle pressure to be uniform. This requires, as we have seen, that $b^2 \ll 1$, so that the value of M_0 that should be used here is $M_0 = 4\pi a^3 \rho_{f0}$, instead of the value given by (6.3.7). Second, the appearance in (6.5.1) of $p_{f\infty}$ requires some explanation because the pressure in the fluid includes the pressure disturbance produced by the particle. However, the effects of that disturbance have already been taken into account in (6.5.1). This may be seen as follows: the pressure disturbance can be separated into compressible and incompressible components. The first of these is an acoustic wave moving away from the particle. Its effects on the motion of the particle's surface are accounted by the radiation term $b\omega M_0 \dot{\varepsilon}$ appearing on the left-hand side of (6.5.1). The incompressible component of the pressure disturbance decays far more rapidly and can be thought of as being confined to a finite region around the sphere's surface; its effects have also been incorporated into (6.5.1) through $M_0 \ddot{\varepsilon}$, as may be seen from the calculation of the self-induced force (6.3.6). This force effectively endows the surface of the particle with a mass M_0. While this mass is due to dynamic effects taking place in the external

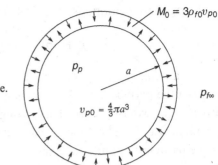

Figure 6.5.1. Particle-fluid interface.

fluid, we may regard it as belonging to the surface, as sketched in Fig. 6.5.1. Here, a spherical particle of radius a and mass m_p has a "skin" of mass M_0 whose volume is equal to $3v_{p0}$. The particle-shell system is embedded in a fluid whose pressure is $p_{f\infty}$ everywhere, thus explaining the term on the right-hand side of (6.5.1).

With these preliminaries in mind, we go back to (6.5.1) and put $p_p = p_{p0} + p'_p$ and $p_{f\infty} = p_0 + P'_f$, where P'_f is the small pressure fluctuation far from the particle, and where the mean pressure in the particle, p_{p0}, is related to that outside through $p_{p0} = p_0 + 2\sigma/a$. Thus, using (6.4.7) for p_p, we obtain

$$M_0\ddot{\varepsilon} + (4\mu_f/\rho_{f0}a^2 + b\omega)M_0\dot{\varepsilon} - 8\pi\sigma\varepsilon = -4\pi a^2(P'_f - p'_p) \qquad (6.5.2)$$

Let us now consider the pressure fluctuation in the particle, p'_p. If we take it to be equal to P'_f, as it would be the case in strict equilibrium, the right-hand side would vanish, and the solution of the remaining equation for ε would represent decaying, nonoscillatory behavior. Of course, the pressure in the particle is generally different from that in the fluid, and, for isothermal conditions, it varies with ε as (6.4.7) shows. Using that equation in (6.5.2), we obtain

$$M_0\ddot{\varepsilon} + (4\mu_f/\rho_{f0}a^2 + b\omega)M_0\dot{\varepsilon} + 4\pi a\left(3\rho_{p0}c_{Tp}^2 - 2\sigma/a\right)\varepsilon = -4\pi a^2 P'_f \qquad (6.5.3)$$

We now simplify this equation by introducing some abbreviations. First we note that the coefficient of ε is a positive quantity for all values of the equilibrium radius a. This may be seen by considering perfect gas bubbles. For them, $\rho_{p0}c_{Tp}^2 \equiv p_{p0}$, but the equilibrium pressure in the particle is given by $p_{p0} = p_0 + 2\sigma/a$, from which it follows that the quantity in question is positive. The same is true for other substances. We may therefore express $(3\rho_{p0}c_{Tp}^2 - 2\sigma/a)$ as the square of some positive quantity, and write it as $M_0\omega_{T0}^2$, where

$$\omega_{T0}^2 = \frac{1}{\rho_{f0}a^2}\left(3\rho_{p0}c_{Tp}^2 - 2\sigma/a\right) \qquad (6.5.4)$$

It is also convenient to put

$$\beta = 2\mu_f/\rho_{f0}a^2 + \tfrac{1}{2}\,b\omega \qquad (6.5.5)$$

noting that both terms on the right-hand side are positive.

Using these definitions in (6.5.3) and dividing the result by M_0, we obtain

$$\ddot{\varepsilon} + 2\beta\dot{\varepsilon} + \omega_{T0}^2\varepsilon = -(4\pi a^2/M_0)P_f' \qquad (6.5.6)$$

This is a remarkable result. It shows that, when the motions are linear, the displacement of the particle surface satisfies the equation for a damped, harmonic oscillator driven by an external force. This analogy does not mean that the particle behaves as a harmonic oscillator because the oscillator described by (6.5.6) is the system composed of the particle-fluid interface and the added fluid mass M_0. That is, every portion of the surface of the particle is a harmonic oscillator whose mass is provided by the reaction of the external fluid, whose restoring force is largely due to the compressibility of the material inside the surface, and whose damping is due to the viscosity and compressibility in the external fluid.

Seen in this manner, we may identify the quantity β defined by (6.5.5) as the damping coefficient for the pulsations. Also, we may express β as the sum of two damping coefficients: β_μ, due to viscous effects, and β_{ac}, due to acoustic radiation. Thus,

$$\beta_\mu = 2\mu_f/\rho_{f0}a^2 \quad \text{and} \quad \beta_{ac} = \tfrac{1}{2}b\omega \qquad (6.5.7\text{a,b})$$

This identification is derived by inspection of (6.5.3) and will be shown to be correct later, when we consider the energy dissipated by viscosity and compressibility.

Equation (6.5.6) also shows that the quantity ω_{T0} plays the role of the natural frequency of the system when the temperature is constant everywhere. The value of ω_{T0}, given by (6.5.4), includes the effects of surface tension, but these can be neglected when $(2\sigma/a)/3\rho_{p0}c_{Tp}^2 \ll 1$. For a 100-$\mu$m diameter air bubble in water at 1 atm, for example, this ratio is less than 0.03. As the size is decreased, surface tension effects increase, as can be seen in Fig. 6.5.2, which displays the natural frequency for air

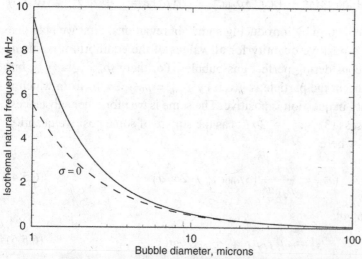

Figure 6.5.2. Surface tension effects on the isothermal natural frequency for air bubbles in water.

Table 6.5.1. *Value of* $\omega_{T0}/2\pi$, *in Hz, for air bubbles and toluene droplets in water*

Diameter (μm)	Air bubbles	Toluene droplets
10	6.02×10^5	6.62×10^7
50	1.13×10^5	1.24×10^7
75	7.49×10^4	8.20×10^6
100	5.59×10^4	6.12×10^6

bubbles at 1 atm, as a function of bubble diameter. Although the size-independent value is reached only asymptotically, the graph shows that the effects of surface tension may be ignored for particles whose radii are larger than about 10 μm. Similar conclusions apply to droplets in liquids, although the natural frequencies of droplets are considerably larger than those of bubbles, because of the decreased compressibility and increased mass.

These isothermal results also apply to solid particles in liquids or gases, or droplets in gases, provided they are regarded as perfectly elastic. The natural frequencies of these particles are considerably larger that those of gas bubbles and droplets in liquids because their compressibilities are very small. Table 6.5.1 shows some values for the isothermal natural frequencies of some particle-fluid combinations at 1 atm and 15°C. Given the large values of these frequencies, it is important that we examine the validity of (6.5.6). As we know, that equation is based on the assumption that the pressure in the particle is uniform at all frequencies. This requires that $b_i = \omega a/c_{Tp}$ be small. Let us evaluate this parameter at $\omega = \omega_{T0}$ and denote that value by $b_{i,0}$. In the absence of surface tension effects, this is given by

$$b_{i,0} = \sqrt{3/\delta} \qquad (6.5.8)$$

Thus, $b_{i,0}$ depends only on the density ratio δ. For air bubbles in water, this ratio is of the order of 10^3, making $b_{i,0}$ equal to about 0.05. Hence, the pressure in isothermal bubbles pulsating at their natural frequency may be taken to be uniform. The same is not true for the other cases because, for them, this equation shows that b_i can be larger than 1 at the frequency prescribed by (6.5.5), meaning that the results are only approximate for these particles.

One additional comment regarding the equation of motion for the surface, (6.5.6): its similarity with that of a harmonic oscillator proves to be very valuable in the study of radial motions of spherical particles. However, it is good to keep in mind that differences exist. Thus, the equation, as written, hides the fact that it is a third-order differential equation, owing to the frequency dependence of the acoustic damping coefficient. Thus, since $b = \omega a/c_{Tf}$, $\beta_{ac} = b\omega/2$, and $\omega^2 \varepsilon \equiv \ddot{\varepsilon}$, we can write that equation as $(a/c_{Tf})\dddot{\varepsilon} + \ddot{\varepsilon} + 2\beta_\mu \dot{\varepsilon} + \omega_{T0}^2 \varepsilon = -4\pi a^2 P'_f/M_0$. Only when the external fluid is incompressible does this reduce to the constant-coefficient harmonic-oscillator equation.

Damping Coefficients

Let us now reconsider the damping coefficients, β_μ and β_{ac}, identified from the equation of motion. Physically, these coefficients originate in the energy dissipation caused by viscosity and acoustic radiation, and it is instructive to evaluate them using the prescription given in Section 2.5. That prescription states that, if the energy dissipation rates are small, the corresponding damping coefficients can be obtained from

$$\beta_x = \frac{\langle \dot{e}_x \rangle_{loss}}{2e_0} \tag{6.5.9}$$

where $\langle \dot{e}_x \rangle_{loss}$ is the average energy dissipation rate associated with mechanism x, say, and e_0, the reference energy, is the amplitude of the energy of oscillation in steady-state conditions. Both quantities must be separately known. Consider first e_0. This is given by the sum of kinetic and potential energies; but, for harmonic oscillations, these two components are equal so that e_0 is equal to twice either one. The average kinetic energy is easier to calculate, but first we must define our oscillator. As pointed out previously, the oscillator is not the particle, but the massless surface of the particle and the added mass, M_0. Hence, the average kinetic energy is equal to $\frac{1}{4}M_0 U_{so}^2$, so that

$$e_0 = \tfrac{1}{2} M_0 U_{s0}^2 \tag{6.5.10}$$

Next come the energy loss rates. These were calculated previously, from the work done to overcome the resistance provided by the fluid, and are given by (6.3.8). Using that result and (6.5.10) in (6.5.9), we obtain

$$\beta = 2\mu_f(1 + b^2)/\rho_{f0}a^2 + \tfrac{1}{2}b\omega \tag{6.5.11}$$

The first term contains both the effects of viscosity and compressibility, but since b^2 is, by assumption, small, we may express β as the sum of physically different terms: one viscous and one acoustic (i.e., $\beta = \beta_\mu + \beta_{ac}$), so that we recover the values obtained from the equation of motion.

Driven Pulsations

Having examined the various terms in the equation of motion for the surface of the particle, we now consider that motion under a harmonic driving pressure fluctuation in the fluid. The starting point is (6.5.6). Thus, if $P_f' = P_{f0}' \exp(-i\omega t)$, where P_{f0}' is the amplitude of the pressure fluctuations and ω its frequency, the steady oscillations will also be harmonic in time and have the same frequency as the pressure – that is, $\varepsilon = \varepsilon_0 \exp(-i\omega t)$. Substitution of this into (6.5.6) yields

$$\left[\omega_{T0}^2 - \omega^2 - 2i(\beta_{ac} + \beta_\mu)\omega\right]\varepsilon = -(4\pi a^2/M_0)P_f' \tag{6.5.12}$$

Dividing through by ω_{T0}^2 and using $M_0\omega_{T0}^2 = 4\pi a(3\rho_{p0}c_{Tp}^2 - 2\sigma/a)$, we obtain

$$\frac{\varepsilon}{a} = -\chi_T \frac{P_f'}{3\rho_{p0}c_{Tp}^2 - 2\sigma/a} \qquad (6.5.13)$$

where

$$\chi_T = [1 - (\omega/\omega_{T0})^2 - i\hat{d}_T]^{-1} \qquad (6.5.14)$$

and where we have introduced a nondimensional, isothermal damping coefficient, $\hat{d}_T$, by means of

$$\hat{d}_T = 2\beta\omega/\omega_{T0}^2 \qquad (6.5.15)$$

We can now obtain the pressure fluctuation in the particle. Using (6.4.13) and (6.5.13), we have

$$p_p' = \chi_T \frac{P_f'}{1 - 2\sigma/3a\rho_{p0}c_{Tp}^2} \qquad (6.5.16)$$

The surface tension correction appearing in the denominator of this result is thus seen to increase the magnitude of the applied pressure fluctuation when the particles are very small. But, for larger particle sizes, we may neglect the correction and write (6.5.16) as

$$p_p' = \chi_T P_f' \qquad (6.5.17)$$

This shows clearly that the pressure fluctuation in the particle differs from that in the fluid in both magnitude and phase. The phase difference implies that the oscillations dissipate some energy because of viscosity and acoustic radiation.

Let us now consider the function χ_T. As (6.5.13) and (6.5.16) show, that function describes the response of the particle surface and pressure to an applied harmonic pressure fluctuation P_f'. Because of this, χ_T is usually referred to as a *response function*. It, and its nonisothermal counterpart, described later in this chapter, plays a very important role in describing linear pulsational motions.

In the present case, both $|\varepsilon/a|$ and $|p_p'/P_f'|$ are determined by $|\chi_T|$. Examination of (6.5.14) shows that at very low frequencies, the pressure fluctuation in the particle equals that in the fluid, whereas at very large frequencies it vanishes. In that high-frequency limit, $|\chi_T| \to 0$, meaning that the particle's surface essentially remains fixed, or frozen, while the external pressure field varies rapidly. In between those extreme frequency values, the pressure change in the particle can be larger than the driving pressure in the fluid outside. In Fig. 6.5.3, we show $|p_p'/P_f'|$ as a function of the frequency for an air bubble and a toluene droplet, both 100 μm in radius. We have not shown the equivalent result applicable to solid particles or to droplets in gases, because for them χ_T is very nearly equal to 1 for all frequencies to which the above theory applies. But for gas bubbles and droplets in liquids, we see that $|p_p'/P_f'|$ is larger than 1 in certain frequency ranges. In the case of the air bubble, the increase is

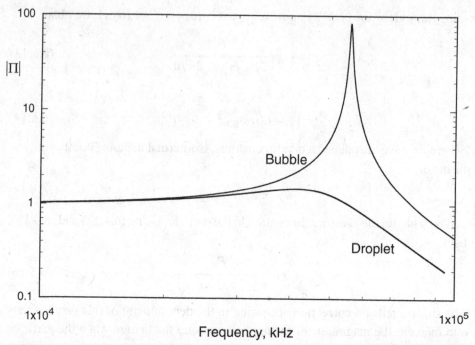

Figure 6.5.3. Magnitude of p'_p/P'_f for an air bubble and a toluene droplet in water. $D = 100 \, \mu m$.

rather large and occurs at a frequency that closely corresponds to its natural frequency. For droplets, on the other hand, the maximum value is, by comparison, rather small, and, furthermore, does not occur at the droplets natural frequency. The reason for the amplitude difference is, of course, the much larger compressibility of the gas bubbles. The shift in the location of the maximum pressure is due to the larger value of $b = \omega a/c_{Tf}$ applicable to droplets in liquids at their resonance frequency. Thus, if that value is denoted by b_0, we find that, in the absence of viscous effects, the maximum pressure occurs at a value of ω/ω_{T0} given by

$$(\omega/\omega_{T0})_{\chi T,\text{max}} = \left(\sqrt{1 + 12b_0^2} - 1\right)/6b_0 \qquad (6.5.18)$$

As pointed out previously, b_0^2 is negligible for gas bubbles in liquids, showing that, for them, $(\omega/\omega_{T0})_{\chi T,\text{max}} \approx 1 - \frac{3}{2}b_0$. But, for droplets in liquids, b_0^2 is not as small. Hence, the maximum peak for the latter occurs at smaller values of ω/ω_{T0}.

The maximum value of $|p'_p/P'_f|$ is obtained by substitution of (6.5.18) into the equation for χ_T and decreases, as expected, as the damping is increased. Another important mechanism that further limits the amplitude of the oscillations is heat transfer to and from the particle. This is considered in the next section.

6.6 Thermal Effects

We have so far ignored any temperature changes that might take place when a particle pulsates. This has enabled us to study the main features of the motion in a simple manner that has also elucidated two of the mechanisms that come into play to limit the amplitude of the pulsations, namely acoustic radiation and viscous damping. However, when a particle pulsates, its pressure and density vary, which generally means that its temperature also changes. Of course, these changes can also be induced by temperature variations in the external medium, as demonstrated for rigid particles in Chapter 3. Regardless of their origin, temperature changes are important in determining the overall response of a compressible particle to applied forces. In the linear approximation, this importance derives from the fact that such changes are normally accompanied by thermal dissipation, which, in some frequency ranges, is more significant than the dissipation produced by other mechanisms.

Before we consider the equations that govern the pulsational motion when temperature changes occur, it is useful to recall that temperature differences that may exist within a particle, tend to be eliminated by the action of thermal waves whose speed, $\sqrt{2\omega\kappa}$, is normally low. Thus, temperature changes induced in one part of a particle will not be felt in another until a certain delay has taken place. Consider, for example, a particle immersed in a fluid whose temperature is changing. Since the surface separating the particle from the fluid is not usually covered with adiabatic walls, it follows that heat exchange between particles and fluid will normally take place and that it will produce thermal waves traveling toward both sides of the interface. These waves tend to equalize temperature differences by conducting energy from a high-temperature region to another having a lower temperature. In the process, energy losses occur.

The losses arising from thermal dissipation are generally affected by viscosity and acoustic radiation; but, in the linear approximation, we may disregard those coupling effects because they are small. A similar assumption was already used in Section 6.5 to calculate the losses induced by acoustic radiation. There, we ignored the effects of heat conduction. However, to put the approximation in its proper context, we first consider the equations that would be required to study all effects simultaneously.

In the linear approximation, the equations of motion in both internal and external materials, both regarded as fluids, are

$$\frac{\partial \rho'}{\partial t} + \rho_0 \nabla \cdot \mathbf{u} = 0 \tag{6.6.1}$$

$$\rho_0 \frac{\partial \mathbf{u}}{\partial t} + \nabla p' = \frac{4}{3}\mu \nabla(\nabla \cdot \mathbf{u}) - \mu \nabla \times (\nabla \times \mathbf{u}) \tag{6.6.2}$$

$$\rho_0 c_p \frac{\partial T'}{\partial t} - \beta T_0 \frac{\partial p'}{\partial t} = k \nabla^2 T' \tag{6.6.3}$$

We also require an equation of state, say $p = p(\rho, T)$. In linearized form, this can be written as

$$p' = c_T^2 \rho' + \frac{\rho_0 c_p}{\beta T_0} \frac{\gamma - 1}{\gamma} T' \tag{6.6.4}$$

The equations are to be solved subject to the same boundary conditions for the pressure and the velocity as used earlier, together with the conditions that the temperatures

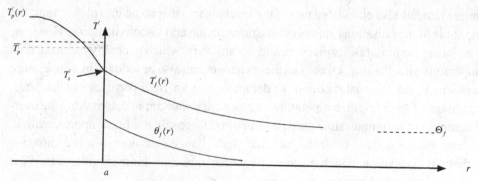

Figure 6.6.1. Schematic temperature distribution.

and heat fluxes at the interface be equal. That is,

$$T'_p(a, t) = T'_f(a, t), \quad \text{and} \quad k_f \left(\frac{\partial T'_f}{\partial r} \right)_{r=a} = k_p \left(\frac{\partial T'_p}{\partial r} \right)_{r=a} \quad (6.6.5\text{a,b})$$

Figure 6.6.1 shows, schematically, an instantaneous temperature distribution along a radial direction that satisfies those conditions. In the figure, we have denoted the common surface temperature by $T'_s(t)$. On the other hand, the temperature derivatives are generally different because of the thermal conductivities being generally different. As we move away from the surface, the magnitude of the temperature fluctuation changes. On the fluid side, it eventually becomes equal to the temperature that the fluid would have in the absence of the particle. For the particle, we can, at this stage, only say the following. If the particle were rigid, thermal waves would be generated at the surface that would penetrate less and less into the particle as the frequency increases. Furthermore, since the only manner in which their temperature can change is by heat transfer from the outside, we conclude that the temperature swing at points away from the surface, produce, in the higher frequency range, a temperature fluctuation that has a smaller magnitude than at the surface. This trend was shown in Fig. 3.5.2. For pulsating particles, the trend can be different because their motion is accompanied with temperature changes that exist in the particle, whether or not heat addition takes place at the boundary. It is thus possible that the average temperature inside pulsating particles be larger than that at the surface, as depicted in Fig. 6.6.1.

Particle Temperature

We now proceed to determine the temperature changes that exist when a sound wave is propagating in the external fluid. When no particle is present, the wave is plane and is accompanied by pressure, density, and temperature fluctuations. We denote these by P'_f, Δ'_f and Θ'_f, respectively, noting that these may be regarded as known. Furthermore, the effects of viscosity and heat conductivity may be disregarded when no particles are present, in which case these quantities are related to one another through the isentropic relations, applicable to acoustic motions:

$$P'_f = (\rho_{f0}c_{pf}/\beta_f T_0)\Theta'_f = c^2_{sf}\Delta'_f \tag{6.6.6}$$

If the frequency of the acoustic wave is not very high, its wavelength can be taken to be long in comparison with the diameter of the particles we are interested in. Thus, over distances that are of the order of 1 particle diameter, the wave nature of the fluctuations can be ignored. This means that the previous quantities may be regarded as depending only on time and not on distance.

Now, a pulsating particle produces disturbances that change the pressure, density, and temperature in the external fluid. However, for the purpose of studying the thermal field in the external fluid, we may neglect the pressure and density disturbances produced there by the particle. On the other hand, the temperature disturbance may be important in some cases and must, therefore, be retained. If we denote this disturbance by $\theta'_f(r, t)$, the fluid temperature may be expressed as $T'_f = \Theta'_f(t) + \theta'_f(r, t)$. Furthermore, since P'_f is known, we see that in order to obtain θ'_f we only need (6.6.3), which because of the first of the relations in (6.6.6), can be written as

$$\frac{\partial \theta'_f}{\partial t} = \kappa_f \nabla^2 \theta'_f \tag{6.6.7}$$

Thus, the temperature disturbance in the external fluid satisfies the diffusion equation. This is a direct consequence of our neglect of compressibility there and means that the field can be regarded as a thermal wave emanating from the particle and extending to infinity with speed $\sqrt{2\omega\kappa_f}$. For a harmonic time dependence, this equation gives

$$\nabla^2\tau_f + K^2\tau_f = 0 \tag{6.6.8}$$

where $\tau_f = \theta'_f/\Theta'_f$.

Consider now the pulsating particle. The temperature and pressure fluctuations in it, T'_p and p'_p, are unknown at this stage. To obtain them, we first note that *if* the pressure in the particle can be taken as uniform, we get from (6.6.3)

$$\frac{\partial}{\partial t}\left(T'_p - \frac{\beta_p T_0}{\rho_{p0}c_{pp}}p'_p\right) = \kappa_p\nabla^2\left(T'_p - \frac{\beta_p T_0}{\rho_{p0}c_{pp}}p'_p\right)$$

That is, $T'_p - (\beta_p T_0/\rho_{f0}c_{pp})p'_p$ also satisfies the diffusion equation. This quantity may be regarded as a modified particle temperature. The modification is due to thermal expansion and, since the particle pressure is uniform, it is the same everywhere in the particle. This modified temperature field travels in the particle as a wave whose phase velocity is $\sqrt{2\omega\kappa_p}$. This is generally different from the corresponding speed in the fluid, which means that the particle and the fluid are almost always at different temperatures – with the exception occurring at zero frequency.

Table 6.6.1. *Values of* $\xi = \frac{\beta_p/\rho_{p0}c_{pp}}{\beta_f/\rho_{f0}c_{pf}}$

Particle/fluid	ξ
Air bubbles in water	7.83×10^4
Oil droplets in air	1.38×10^{-4}
Toluene droplets in water	2.01×10^1

We now express the above equation in nondimensional form by putting $\tau_p(r, t) = T_p'/\Theta_f'$ and $\Pi = p_p'/P_f'$. This gives

$$\nabla^2(\tau_p - \xi\Pi) + K_i^2(\tau_p - \xi\Pi) = 0 \tag{6.6.9}$$

where

$$\xi = \frac{\beta_p/\rho_{p0}c_{pp}}{\beta_f/\rho_{f0}c_{pf}} \tag{6.6.10}$$

This property ratio provides a measure of the thermal expansion in the particle, relative to that in the fluid, and as Table 6.6.1 shows, varies widely from one fluid-particle combination to another, indicating the relative importance of the pressure term in (6.6.9).

Let us now consider (6.6.8) and (6.6.9). Since the temperature fields in and outside the droplets have central symmetry, their solutions are, respectively, $\tau_f = Dh_0(Kr)$ and $\tau_p - \xi\Pi = Ej_0(K_ir)$. To evaluate the constants D and E, it is useful to use the surface temperature fluctuation T_s'. This is also unknown but may be expressed as $T_s' = \Theta_f' + \theta_f'(a, t)$, or as $T_p'(a, t)$. Thus, putting $T_s = T_s'/\Theta_f'$, we obtain $D = (T_s - 1)/h_0(q)$, where $q = Ka$. Hence

$$\tau_f = (T_s - 1)\frac{h_0(Kr)}{h_0(q)}$$

Similarly, the inner solution yields, $E = (\tau_s - \xi\Pi)/j_0(q_i)$, where $q_i = K_ia$. Hence,

$$\tau_p = \xi\Pi + (T_s - \xi\Pi)\frac{j_0(K_ir)}{j_0(q_i)}$$

This gives the temperature distribution within the particle. Another useful quantity is the average temperature fluctuation. This can be obtained from τ_p by integration over the volume of the particle. Thus, using $\overline{j_0(K_ir)} = -3j_0'(q_i)/q_i$, we obtain

$$T = \xi\Pi - 3\frac{(T_s - \xi\Pi)}{q_i^2 G(q_i)} \tag{6.6.11}$$

where the function $G(q_i)$, equal to $j_0(q_i)/q_i j_0'(q_i)$, was discussed in Chapter 3, and

$$T = \overline{T}_p'/\Theta_f' = \overline{\tau}_p$$

This ratio, like the pressure ratio, Π, introduced previously in this section, plays an important role in subsequent sections.

We now return to T_s. To evaluate it, we make use of the boundary condition on the heat flux, (6.6.5b). First we evaluate the derivatives of the temperature fields at the surface of the particle. These are

$$\left(\frac{\partial \tau_p}{\partial r}\right)_a = (T_s - \xi \Pi)/a G(q_i) \qquad \left(\frac{\partial \tau_f}{\partial r}\right)_a = (1 - T_s)(1 - iq)/a \qquad (6.6.12a,b)$$

where we have used the identity $q h_0'(q)/h_0(q) = -(1 - iq)$. Putting these two results in (6.6.5b), and solving for T_s, we obtain

$$T_s = \xi \Pi + \frac{k_f}{k_p}(1 - \xi \Pi)\frac{G(q_i)}{F(q, q_i)} \qquad (6.6.13)$$

where

$$F(q, q_i) = \frac{1}{1 - iq} + \frac{k_f}{k_p} G(q_i)$$

This function also appears in the temperature distribution in a rigid particle (Section 3.5). Although the real and imaginary parts of $F(q, q_i)$ vary from one-particle fluid combinations to another, their values are independent on whether a particle is compressible or not. Also, reference to Fig. 3.5.1 shows that both real and imaginary parts of $F(q, q_i)$ are positive at all frequencies.

Let us now return to the temperature ratio. Substituting the value of T_s in (6.6.11), we get (Temkin, 1999)

$$T = \xi \Pi - 3\frac{k_f/k_p}{q_i^2 F(q, q_i)}(1 - \xi \Pi) \qquad (6.6.14)$$

The first terms on the right-hand side of (6.6.13) and (6.6.14) give the corresponding temperature ratios in the adiabatic limit. The second terms therefore account for heat conduction between particles and fluid. They also include the rigid-particle results, derived previously, and obtained from these by putting $\xi = 0$ in them.

Regarding (6.6.14), we see that particle compressibility contributes an amount equal to $\xi \Pi \left[1 + 3(k_f/k_p)/q_i^2 F(q, q_i)\right]$ to the temperature changes, with the quantity $\xi \Pi$ providing a measure of the particles' ability to expand and contract. This ability derives from two mechanisms: thermal expansion in both media and acoustic emission into the external medium. However, in the present discussion, which is limited to determining the effects of temperature changes, we have disregarded the pressure disturbance produced by the particle. Consistency requires that we do the same when evaluating T. To do this, we express the factor $(1 - \xi \Pi)$ appearing in (6.6.14) as $(1 - \xi) \Pi - (\Pi - 1)$. The difference $\Pi - 1$ appearing here represents the pressure disturbance, evaluated at the surface of the particle. Neglecting this disturbance is warranted provided the thermal properties of the internal and external media are different, so that ξ is not too close to 1. In all other cases, the approximation is a valid one so far as the temperature fields is concerned. We can therefore write (6.6.13) and (6.6.14) as

$$\frac{T}{\Pi} = \xi + \frac{3i}{2} \frac{k_f/k_p}{z_p^2} \frac{1-\xi}{F(q,q_i)} \tag{6.6.15}$$

$$\frac{T_s}{\Pi} = \xi - (\xi - 1)\frac{k_f}{k_p} \frac{G(q_i)}{F(q,q_i)} \tag{6.6.16}$$

We will also need the heat transfer rate, $\dot{Q}_p$. This is obtained by substituting (6.6.13) in (6.6.12a) and multiplying the result by $4\pi a^2 k_f$. Thus,

$$\dot{Q}_p = 4\pi a k_f \frac{(1-\xi\Pi)}{F(q,q_i)}\Theta_f' \approx 4\pi a k_f \frac{(1-\xi)\Pi}{F(q,q_i)}\Theta_f' \tag{6.6.17}$$

The first of these forms shows that the rigid-particle result, (3.5.19), is obtained when $\xi = 0$. The second provides an approximation to the first that is consistent with the previous results for T/Π.

It should be noted that, while the ratio T/Π is completely determined by (6.6.15), that equation is insufficient to determine T because the particle pressure ratio, Π, is still unknown. For isothermal pulsations, the pressure ratio was easily obtained from (6.4.13) and from the equation of motion for the surface of the particle. But the appearance in the formulation of temperature changes renders that approach inapplicable. Nevertheless, a variation of that procedure that takes advantage of our knowledge of T/Π is possible, as shown below.

Particle Pressure

Let us consider the pressure fluctuations in the particle. For isothermal pulsations, p_p' is proportional to the surface displacement ε. Use of this $p_p' - \varepsilon$ relation in the Rayleigh-Plesset equation suffices to complete the solution. But, when the temperature varies, that proportionality ceases to be valid. Instead, the thermal equation of state for the particle material gives, when linearized and averaged over the volume of the particle,

$$3\varepsilon/a = -p_p'/\rho_{p0}c_{Tp}^2 + \beta_p \overline{T}_p' \tag{6.6.18}$$

where we have used $v_p'/v_{p0} = -3\varepsilon/a$ because of the sphericity of the particles. The first term on the right-hand side gives the isothermal displacement, whereas the second is the displacement arising from thermal expansions. It should be noted that the particle may expand *or* contract, depending on the relative magnitude of these two contributions. For gas bubbles in liquids, for example, the pressure term dominates, so that a pressure increase always results in a contraction. However, for droplets in gases, thermal expansion can take over.

Keeping these possibilities in mind, we write (6.6.18) in a manner reminiscent of the isothermal result, limiting the discussion to harmonic pulsations. This allows us to use complex quantities to represent the fluctuations appearing in that equation. Then, we can write

$$p'_p = -3\rho_{p0}c^2_{Tp}\kappa(\varepsilon/a) \qquad (6.6.19)$$

where (Temkin, 2001b)

$$\kappa = \kappa_r + i\kappa_i = \left(1 - \frac{\gamma_p - 1}{\gamma_p\xi}\frac{\mathrm{T}}{\Pi}\right)^{-1} \qquad (6.6.20)$$

In writing this equation, we have used $\overline{T}'_p = (\beta_p T_0/\rho_{p0}c_{pp}) P'_f\mathrm{T}$, as well as $p'_p = \Pi P'_f$. Sometimes it is useful to write (6.6.19) in terms of the equilibrium particle pressure, p_{p0}. This can be done for all particle materials, by introducing the quantity $\zeta = \rho_{p0}c^2_{sp}/\gamma_p p_{p0}$, whose value is unity for perfect gas bubbles. Thus, we can write

$$p'_p = -3p_{p0}\kappa'(\varepsilon/a),$$

where $\kappa' = \kappa\zeta$.

Now, although (6.6.19) resembles the isothermal result, (6.4.13), and would reduce to it if κ were real and equal to 1, the complex nature of κ shows that a lag generally exists between the pressure and the displacement. This is, of course due to thermal effects.

The quantity κ defined previously plays an important role in the theory of bubble pulsations, where it is known as the *polytropic index*, for reasons that will become clear below. We shall carry that designation to other fluid-particle combinations, noting that it derives directly from the linearized equation of state and not from an assumed particle-volume relationship.

The Polytropic Index

Let us consider the real and imaginary parts of the polytropic index. These quantities play important, but different, roles in the pulsation theory, and since T/Π is known, they can be obtained from (6.6.20). Inspection of that equation shows that κ_r and κ_i depend on the frequency and on the properties of *both* particle and external fluids. They are therefore different for various particle-fluid combinations. Figures 6.6.2 and 6.6.3 show κ_r and κ_i for some particle-fluid combinations. The figures show several interesting features. First, κ_r is positive at all frequencies for air bubbles and droplets in water; but, that for droplets in air, it changes sign. A similar trend is observed for alumina particles in both gases and liquids. We also note that, in all cases, κ_r approaches a constant value as $\omega \to \infty$. This value is equal to the specific heat ratio of the particle material. This follows from (6.6.15), which shows that, in that high frequency limit, $\mathrm{T}/\Pi \to \xi$, so that (6.6.20) gives $\kappa_r \to \gamma_p$ and $\kappa_i \to 0$. Thus, at high frequencies the

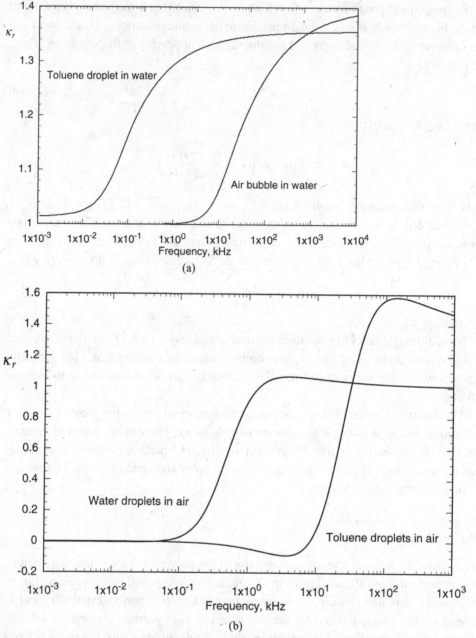

Figure 6.6.2. (a) Real part of the polytropic index for an air bubble and a toluene droplet, both in water. $D = 100\,\mu$m. (b) Real part of the polytropic index for a water droplet and a toluene droplet, both in air. $D = 100\,\mu$m.

polytropic index is real and equal to γ_p, showing that $p'_p \rightarrow -3\rho_{p0}c^2_{sp}(\varepsilon/a)$. This is the isentropic value for the pressure in the particle. The result obviously implies that no heat transfer takes then place and is due to the very small penetration depth of thermal waves into both particle and external fluids.

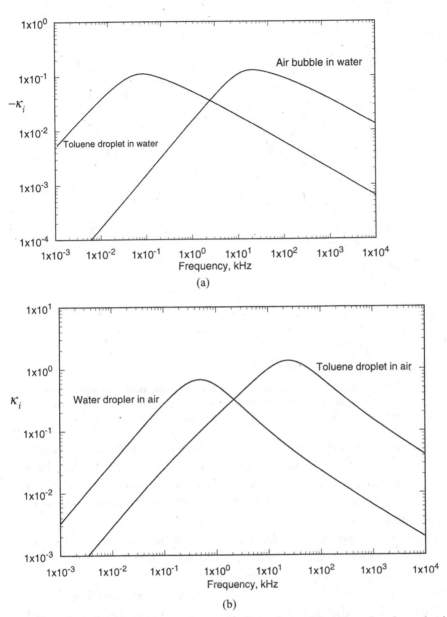

Figure 6.6.3. (a) Imaginary part of the polytropic index for an air bubble and a toluene droplet, both in water. $D = 100\,\mu$m. (b) Imaginary part of the polytropic index for a water droplet and a toluene droplet both in air. $D = 100\,\mu$m.

At the other end of the frequency spectrum, when $\omega \to 0$, we also find that κ_r becomes constant and that $\kappa_i \to 0$; but, this time, the limiting value of κ_r is obtained by putting $T/\Pi \to 1$. Thus, $\kappa_r \to [1 - (\gamma_p - 1)/\gamma_p \xi]^{-1}$. For gas bubbles in liquids, $\xi \gg 1$, giving $\kappa_r \approx 1$. That is, the bubble pulsates at nearly constant temperature. Thus, the variations of κ for gas bubbles in liquids are seen to result in isothermal behavior at low frequencies and isentropic behavior at large ones. In this sense, the bubble behaves

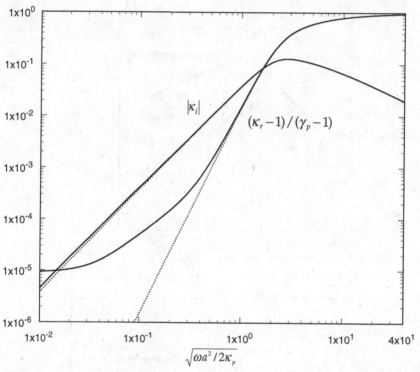

Figure 6.6.4. Variations of $|\kappa_i|$ and $(\kappa_r - 1)/(\gamma_p - 1)$ for a 100-μm air bubble in water. Solid lines were obtained from (6.6.20). Dotted line results exclude thermal effects in the water (Prosperetti, 1977). [Reprinted with permission from Temkin (2001). ©2001, Cambridge University Press.]

as a simple thermodynamic system undergoing different thermodynamic processes at different frequencies, with its behavior controlled by κ. We also see that when the gas in the bubble is perfect, (6.6.19) may be regarded as the first-order expansion of a polytropic equation of state, $p_p v_p^\kappa = C$, and this gives rise to the designation of κ as the polytropic index.

These results for the polytropic index for bubbles include the effects of thermal oscillations in the liquid outside the bubble. These are generally small as may be gathered by comparing solid and dotted lines in Fig. 6.6.4.

The behavior of κ_r predicted by (6.6.20) and by earlier works that did not include thermal effects in the liquid outside the bubble, has been confirmed experimentally by Crum (1983), who performed measurements with levitated gas bubbles, including air bubbles. His air bubble results are shown in Fig. 6.6.5, together with the theoretical results (Devin, 1959; Prosperetti, 1977; Temkin, 2001). In the figure, the real part of the polytropic index is shown as a function of $\sqrt{\omega/\omega_0}$, where ω_0 is the resonance frequency for a gas bubble when thermal effects are included. This frequency is given later in this section. It is seen in the figure that, although the data show some scatter, the overall trend predicted by the theory is confirmed.

Let us now consider κ_r for other fluid particle-fluid combinations. Here, the low-frequency behavior may differ from that of gas bubbles in liquids. For example, for

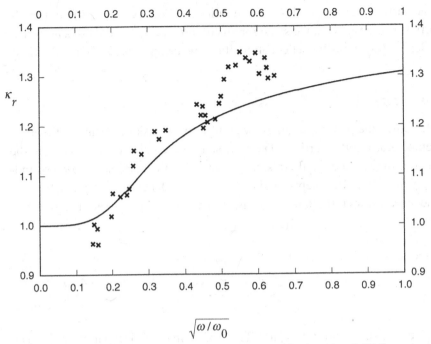

Figure 6.6.5. Real part of the polytropic index for air bubbles in water. Solid line was obtained from (6.6.20). Crosses represent the experimental results of Crum (1983).

liquid droplets in gases, $\xi \ll 1$, so that when the frequency is very small, we obtain $\kappa_r \to -\gamma_p \xi / (\gamma_p - 1)$. This is generally different from 1. Furthermore, since both $(\gamma_p - 1)$ and ξ are positive, we see from (6.6.19) and (6.6.20) that, contrary to the bubble case, the particle pressure increases as the particle expands. As explained earlier and below this behavior is due to thermal expansion effects.

For the imaginary part of κ, we note from Fig. 6.6.3 that it vanishes at both ends of the frequency spectrum for all particle-fluid combinations, as already pointed out. However, at finite frequencies, we observe that κ_i is negative for gas bubbles and droplets in liquids, and positive for the other cases. To see the origin of this behavior, we use (6.6.20) to write κ_i as

$$\kappa_i = -|\kappa|^2 \frac{\gamma_p - 1}{\gamma_p \xi} \Im \left\{ \frac{T}{\Pi} \right\}^*$$
$$= \frac{3}{2} |\kappa|^2 (1 - \xi) \frac{\gamma_p - 1}{\gamma_p \xi} \Re \left\{ \frac{1}{F} \right\}$$

(6.6.21)

where (6.6.15) was used to obtain the second form. Now, as pointed out previously, the real part of $F(q, q_i)$ is positive at all frequencies. It then follows that the sign of κ_i is determined by the value of ξ. For gas bubbles and droplets in liquids, $\xi > 1$, making κ_i negative. For other particle-fluid combinations, $\xi \ll 1$, giving $\kappa_i > 0$. Thus, the different behavior of κ_i stems from the different magnitudes of the thermal expansion effects in the particle, relative to those in the fluid [see Equation (6.6.10)] and imply that the particle response is not the same for all combinations. To study this response, we reconsider the Rayleigh-Plesset equation, this time using (6.6.19) for the pressure.

Surface Displacement

Having determined the polytropic index, we return to the task of obtaining the pressure and temperature in the particle. The next step in the derivation is to obtain the displacement of the surface. To do this, we assume that the surface equation of motion is still given by (6.5.2), except that the pressure now depends on the temperature. That dependence follows from (6.6.19). Thus, substitution of that equation into (6.5.2) gives

$$M_0\ddot{\varepsilon} + (4\mu_f/\rho_{f0}a^2 + b\omega)M_0\dot{\varepsilon} + 12\pi a\rho_{p0}c_{Tp}^2 i\kappa_i\varepsilon$$
$$+ 4\pi a(3\kappa_r\rho_{p0}c_{Tp}^2 - 2\sigma/a)\varepsilon = -4\pi a^2 P_f'$$

(6.6.22)

Although this is similar to (6.5.2), it differs from it in several important ways. The first is that since the real part of the polytropic index is not always positive, the quantity multiplying ε on the left-hand side is not always positive, as it was in the isothermal case. Therefore, the term does not always represent a restoring force, as it does when it is positive. One consequence of this is that a natural frequency does not always exist. However, when $\kappa_r > 0$, we may use (6.6.22) to define a natural frequency by means of

$$\omega_0^2 = (3\kappa_r\rho_{p0}c_{Tp}^2 - 2\sigma/a)/\rho_{f0}a^2$$

(6.6.23)

This important result is discussed later.

Returning to the general case, we note that the imaginary part of κ can be either negative or positive, depending the fluid-particle combination. This alters the nature of the equation, relative to the equation satisfied by an harmonic oscillator. Of course, these differences do not make the approach invalid. They simply imply that the particle response to an applied external pressure is not always analogous to that of a harmonic oscillator.

The solution to (6.6.22) can be obtained in the same manner as (6.5.2); but, to simplify the writing, we ignore surface tension effects. These can be easily retained, but their inclusion here increases the length of the equations without adding new physics. Thus, if the imposed pressure varies sinusoidally with circular frequency ω, the particle response will also be sinusoidal with the same frequency. Hence, since $i\varepsilon/\omega = -\dot{\varepsilon}$, we have

$$\ddot{\varepsilon} + (4\mu_f/\rho_{f0}a^2 + b\omega - \omega_{T0}^2\kappa_i/\omega)\dot{\varepsilon} + \omega_{T0}^2\kappa_r\varepsilon = -(4\pi a^2/M_0)P_f' \quad (6.6.24)$$

where the isothermal frequency, ω_{T0}, is now calculated from (6.5.4) with $\sigma = 0$. Regardless of the values of κ_r and κ_i, we find that the radial displacement is now given by

$$\frac{\varepsilon}{a} = -\chi \frac{P'_f}{3\rho_{p0}c_{Tp}^2} \tag{6.6.25}$$

where the response function, χ, is given by

$$\chi = \frac{1}{\kappa_r - (\omega/\omega_{T0})^2 - i\left[\omega(4\mu_f/\rho_{f0}a^2 + b\omega)/\omega_{T0}^2 - \kappa_i\right]} \tag{6.6.26}$$

It should be noted that, in writing this equation, we have not identified the quantity inside the bracket in the imaginary term in the denominator as a damping coefficient, as was done in the isothermal case. That identification can only be made when the equation of motion is analogous to that of a harmonic oscillator, which, as we have seen, is not always the case. Nevertheless, the above equations are sufficient to obtain the pressure and the temperature. Thus, the pressure is given by (6.6.19), or, in terms of $\Pi = p'_p/P'_f$, as

$$\Pi = \chi\kappa \tag{6.6.27}$$

while the temperature ratio, $T = \overline{T}'_p/\Theta'_f$, follows from (6.6.15). However, given that the real part of the polytropic index is not always positive for some particles, the above solution implies physically-different behavior for them. It is therefore useful to divide the discussion of the results, depending on the behavior of κ_r and κ_i.

Bubbles and Droplets in Liquids

For these two particle-fluid combinations, κ_r is positive at all frequencies, and κ_i is always negative so that the quantity inside the square bracket in (6.6.26) is positive. Hence, the corresponding particle-fluid systems are analogous to harmonic oscillators. Their resonance frequency and damping follow from (6.6.26) by division through κ_r. Thus, putting $\kappa_i = -|\kappa_i|$, we obtain

$$\chi = \frac{1/\kappa_r}{1 - (\omega/\omega_0)^2 - i\hat{d}/\kappa_r} \tag{6.6.28}$$

where

$$\omega_0^2 = \omega_{T0}^2\kappa_r \tag{6.6.29}$$

is the resonance frequency, and

$$\hat{d} = (\hat{d}_\mu + \hat{d}_{ac} + |\kappa_i|) \tag{6.6.30}$$

plays the role of the total, nondimensional damping coefficient.

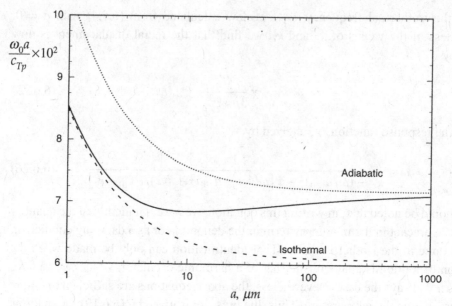

Figure 6.6.6. Resonance frequency for air bubbles as a function of bubble radius.

Consider first ω_0. Contrary to ω_{T0}, which is frequency-independent, ω_0 varies with the imposed frequency because of κ_r. As we have seen, this quantity varies between the isothermal and isentropic limits as the imposed frequency varies from small to large values. Corresponding to these variations, (6.6.29) states that ω_0 varies between the corresponding isothermal and isentropic frequencies, or $\omega_{T0} \le \omega_0 \le \omega_{S0}$, where $\omega_{S0} = \sqrt{\gamma_p}\,\omega_{T0}$ is the isentropic value. Thus, for a fixed particle size, (6.6.29) shows that the variations of ω_0 with the imposed frequency are those of κ_r. From the physical point of view, all of these effects are due to the variable penetration of the thermal waves into the particle.

It is also of interest to determine ω_0 as a function of the particle radius. Since κ_r may be regarded as a function of $z_p = \sqrt{\omega a^2/2\kappa_p}$, and since, at resonance $\omega = \omega_0$, we see that $\omega_0 = \omega_{T0}\sqrt{\kappa_r(z_{p0})}$, or $\omega_0 = f(\omega_0, a)$ – which is a transcendental equation for ω_0 – the solution may be determined by trial and error. Figure 6.6.6 shows the resulting frequency values, in nondimensional form, for an air bubble in water, as a function of the bubble radius. The curves shown in this figure were calculated using (6.6.23), and include the corresponding isothermal and adiabatic values. It is noted that the resonance frequency is a single-valued function of the radius for the three curves shown in the figure. But the converse is not true; that is, the radius corresponding to a given resonance frequency is not unique within a range of frequencies. This means that a determination of the size of a bubble through a measurement of its resonance frequency could result in some error.

Let us now consider the total, nondimensional damping coefficient, as given by (6.6.30). The appearance of the imaginary part of the polytropic index in that

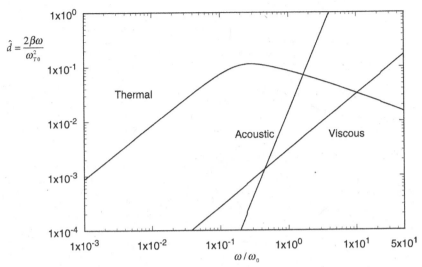

Figure 6.6.7. Nondimensional damping coefficients for a 100-μm air bubble in water.

equation implies that $|\kappa_i|$ is equal to the nondimensional thermal damping coefficient, that is

$$\hat{d}_{th} = |\kappa_i| \qquad (6.6.31)$$

so that we can also write (6.6.30) as $\hat{d} = (\hat{d}_\mu + \hat{d}_{ac} + \hat{d}_{th})$. Incidentally, each of these nondimensional damping coefficients are defined in terms of ω_{T0} via $\hat{d} = 2\beta\omega/\omega_{T0}^2$. The appearance of κ_r in the damping term in (6.6.28) indicates that the effective coefficients are those defined earlier, divided by κ_r. This would give, for example, $\hat{d}_{th}/\kappa_r = 2\beta_{th}\omega/\omega_0^2$. To avoid defining yet another quantity, we shall, however, continue to use the same definition for the $\hat{d}$s as before.

To get an idea of the importance of the thermal damping, relative to the viscous and acoustic counterparts, we show the three in Fig. 6.6.7 for a 100-μm air bubble in water, as a function of the nondimensional frequency ω/ω_0. It is seen that, for this bubble size, the thermal damping coefficient has the largest magnitude for frequencies that range from the lowest to slightly above resonance, where the acoustic component takes over. These relative magnitudes depend on the frequency. Thus, for frequencies below resonance, viscous effects become larger as the size is decreased. However, for all sizes, acoustic damping is the dominant mechanism beyond resonance. The situation is somewhat similar for droplets in liquids, although the magnitudes of all damping coefficients are smaller.

Let us now compute the pressure and temperature fluctuations in the particle. The pressure is given by (6.6.27) and may also be written as

$$\Pi = \frac{1 - i\hat{d}_{th}/\kappa_r}{1 - (\omega/\omega_0)^2 - i\hat{d}/\kappa_r} \qquad (6.6.32)$$

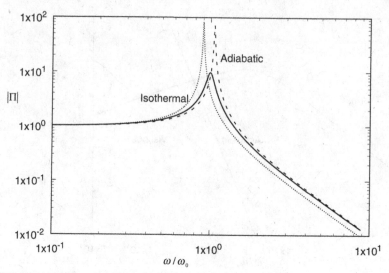

Figure 6.6.8. Pressure ratio amplitude for a 100-μm air bubble in water.

The magnitude of this quantity is shown in Fig. 6.6.8 for an air bubble in water. Also shown in the figure are the corresponding results applicable to isothermal and isentropic pulsations. These are obtained by neglecting thermal damping, and using the corresponding resonance frequencies, instead of ω_0. As expected, we see that thermal dissipation decreases the magnitude of the pressure ratio. The figure also demonstrates that the resonance frequency falls between the isothermal and the isentropic values. The exact relative position depends on the radius of the particle, and may be inferred from Fig. 6.6.6.

Figure 6.6.9 shows the corresponding pressure amplitudes for a toluene droplet pulsating in water under the influence of an external harmonic pressure. Here, we see that the isentropic pressure is nearly equal to the pressure obtained by retaining thermal effects. The reason is the small value of the thermal penetration depth at those values of ω/ω_0 emphasized in the figure.

Having obtained the pressure, we can now obtain the average and surface temperatures. These are given by (6.6.15) and (6.6.16), respectively, with Π given by (6.6.32). Figure 6.6.10 shows the magnitudes of T and T_s for an air bubble in water. It is seen that the average temperature near resonance is considerably higher than the surface temperature. Similar, but not as drastic, differences are obtained for droplets in liquids. The reasons for this were discussed at the beginning of this section and are due to the pulsational motion, which increases the magnitude of the temperature fluctuation. In the case of air bubbles, the temperature fluctuation amplitude is several orders of magnitude larger than the fluctuation in the surrounding liquid. Thus, considerable heating can take place in a bubble. These remarks apply, of course, to linear oscillations.

We should add that, because the compressibility ratio, N_s, applicable to droplets in liquids is not far from 1, their response may, for a wide range of frequencies, be

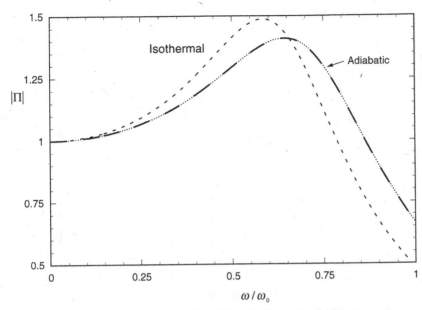

Figure 6.6.9. Pressure ratio amplitude for a 100-μm toluene droplet in water.

obtained by simply taking the variations of the pressure in the particle to be equal to those of the fluid outside (i.e., by putting $\Pi = 1$ in the results). This approximation, which ignores the resonant nature of the droplets, is useful in those situations where the frequency is not too large, as discussed below.

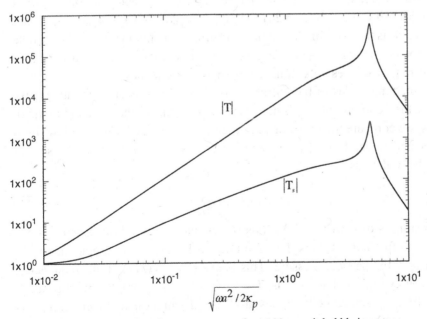

Figure 6.6.10. Average and surface temperatures for a 100-μm air bubble in water.

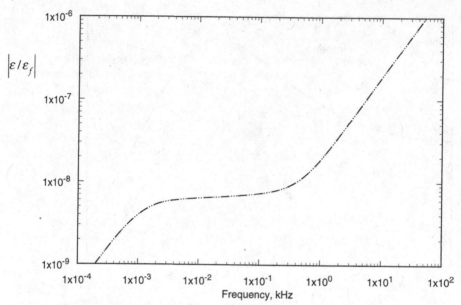

Figure 6.6.11. Radial displacement ratio for a 100-μm water droplet in air. Dotted line corresponds to $\Pi = 1$.

Aerosol and Hydrosol Particles

Here, we consider liquid and solid particles in gases, and solid particles in liquids; but, to shorten the description, we will use the above designation. For these particles, κ_r is negative in the lower frequency range, which means that the response is not resonant. Furthermore, $\kappa_i > 0$ at all frequencies, so that the factor multiplying $\dot{\varepsilon}$ in the surface equation of motion, (6.6.24) [i.e., $(4\mu_f/\rho_{f0}a^2 + b\omega - \omega_{T0}^2\kappa_i/\omega)$] is not always positive, as is the case for gas bubbles and droplets in liquids. Thus, the particle response to forced, harmonic forces is generally not similar to that of a harmonic oscillator, even if it executes pulsations at the imposed frequency.

The absolute amplitudes of the pulsations of these particles are, of course, very small because their compressibility is much smaller than that of the surrounding fluid. To show this, we compare the radial displacement, ε, to the displacement of the fluid in a plane, sound wave. That displacement is given by $\varepsilon_f = iP'_f/\rho_{f0}c_{sf}\omega$, so that (6.6.25) can be written as

$$\varepsilon/\varepsilon_f = {}^1\!/_3 i b \gamma_p N_s \chi \tag{6.6.33}$$

Given the smallness of both b and N_s, this shows that the radial displacements are very small, at all frequencies, as is shown in Fig. 6.6.11 for a water droplet in air.

Consider now the particle pressure. This is given by (6.6.27) at all frequencies. The low-frequency limit of that result is $\Pi = 1$. This follows from the limiting values of κ and χ. But, the same result applies at all but the highest frequencies. This is shown in Fig. 6.6.12, where $|\Pi - 1|$ is plotted against the frequency. That quantity

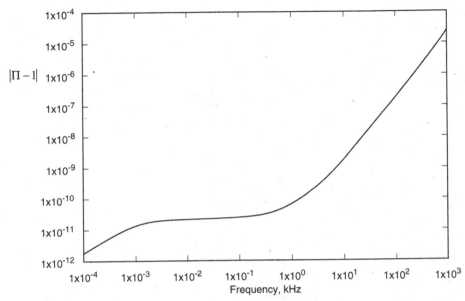

Figure 6.6.12. Magnitude of the pressure disturbance for water drops in air. $D = 100$-μm.

also represents the magnitude of the pressure disturbance produced by the particle. Given the very small value of $|\Pi - 1|$ at even the highest frequencies shown in the figure, we can safely put $\Pi = 1$ in the remaining equations. Thus, to a very good approximation, we find that the pressure variations in aerosol and hydrosol particles are equal to those in the fluid outside them. This was, of course, to be expected. In addition, it makes the analysis considerably simpler. Thus, since $\Pi = 1$, we find from (6.6.17) that the temperature is

$$T = \xi + \frac{3i}{2} \frac{k_f/k_p}{z_p^2} \frac{(1-\xi)}{F(q,q_i)} \tag{6.6.34}$$

The property ratio ξ appearing here, given by (6.6.10), is now very small and may be dropped, in which case we find that, as anticipated, the temperature of these particles is well described by the rigid limit result discussed in Chapter 3. But, retaining ξ in the equation provides a useful approximation for emulsion particles when the frequency is not too high. To show this, we consider the particle displacement again. Since, as (6.6.27) shows, $\chi = 1/\kappa$ when $\Pi = 1$, we obtain

$$\varepsilon/\varepsilon_f = {}^1/_3 \, i b \gamma_p N_s \left[1 - \frac{\gamma_p - 1}{\gamma_p \xi} T \right] \tag{6.6.35}$$

where the temperature ratio is given by (6.6.34). Figure 6.6.13 shows the variations of $\varepsilon/\varepsilon_f$ with frequency for a toluene droplet in water, using both (6.6.34) and (6.6.35). It is seen that, except for the deviations at high frequencies, the responses are nearly the same.

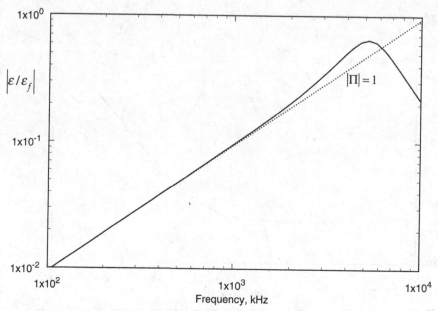

Figure 6.6.13. Radial displacement ratio for a 100-μm toluene droplet in air.

Similar results are noted for the temperature of these droplets, but here the differences are more noticeable, as shown in Fig. 6.6.14 for toluene droplets in water. The slight peak seen in the curve is, of course, produced by resonance effects. In any event, it is clear that, so long as the imposed frequency is smaller than ω_0, we may also put $\Pi = 1$ for emulsion droplets. The figure also shows the magnitude of the surface temperature. It is seen that, as with gas bubbles in liquids, this is smaller than

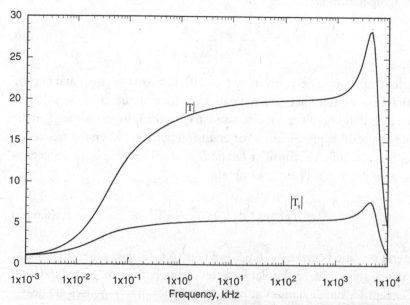

Figure 6.6.14. Average and surface temperatures for a 100-μm toluene droplet in water.

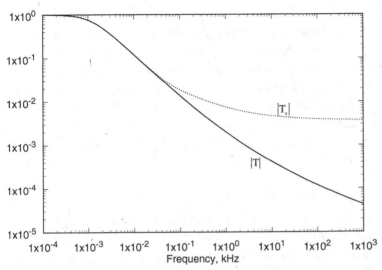

Figure 6.6.15. Average and surface temperatures for a 100-μm water droplet in air.

the average temperature inside. On the other hand, the surface temperature of aerosol particles is larger than the average temperature when the frequency is large. This is shown in Fig. 6.6.15 for a water droplet in air.

6.7 Energy Considerations

Let us now consider the energy losses that occur when a particle pulsates. These losses may be expressed in a variety of ways (e.g., through the damping coefficients). However, as we saw in the preceding section, the damping coefficients corresponding to thermal losses could not always be identified by inspection of the surface equation of motion. For these reasons, it is preferable to represent the losses in terms of the corresponding time-averaged dissipation rates. The rates corresponding to viscous effects and to acoustic radiation were already calculated in Section 6.5. The thermal component is considered in the next section.

Thermal Dissipation Rate

We would like to determine the thermal dissipation rate for all particle-fluid combinations of interest. For gas bubbles and droplets in liquids, the derivation is simple because the equations of motion applicable to them allowed us to identify the non-dimensional damping coefficient, $\hat{d}_{th} = 2\beta_{th}\omega/\omega_{T0}^2$, as being equal to $-\kappa_i$. On the other hand, β_{th} is related to the thermal energy dissipation rate, $\langle \dot{e}_{th} \rangle$, via $\beta_{th} = \langle \dot{e}_{th} \rangle /2e_0$, where the reference energy, e_0, is (see Section 6.3) equal to $\tfrac{1}{2}M_0\omega\omega_{T0}|\varepsilon|^2$. Thus, the thermal dissipation rate for gas bubbles and droplets in liquids is

$$\langle \dot{e}_{th} \rangle = \frac{1}{2}M_0\omega\omega_{T0}^2 |\varepsilon|^2 |\kappa_i| \qquad (6.7.1)$$

But the derivation does not work for aerosol and hydrosol particles because these do not pulsate as harmonic oscillators. Thus, a different approach is needed. This would also enable us to verify the above result for gas bubbles and droplets in liquids.

To obtain the thermal dissipation rate for all cases, we recall from thermodynamics that the energy losses that take place during any irreversible process, such as heat conduction, are equal to the total entropy increase during the process, multiplied by the lowest temperature available as a reservoir. In the present discussion, that temperature is T_0, the mean temperature, and the loss is due to thermal effects. Also, we need the time-averaged loss rate. Hence,

$$\langle \dot{e}_{th} \rangle = T_0 \langle \dot{S}_{total} \rangle \tag{6.7.2}$$

where $\dot{S}_{total} = \dot{S}_p + \dot{S}_f$. Now, each of these rates may be calculated by integration of the temperature fields in the particle and in the fluid, but a much simpler approach is to use the spatial average of the temperature in the particle. Then, since the particle receives an amount of heat $\dot{Q}_p$ per unit time at a temperature $\overline{T}_p$, and the fluid releases the same amount of heat at a temperature T_f, we have

$$\langle \dot{e}_{th} \rangle = T_0 \left\langle \dot{Q}_p \left(\frac{1}{\overline{T}_p} - \frac{1}{T_f} \right) \right\rangle \tag{6.7.3}$$

We now put $\overline{T}_p = T_0 + \overline{T}'_p$ and $T_f = T_0 + \Theta'_f$, and retain only leading terms to obtain

$$\langle \dot{e}_{th} \rangle = \frac{1}{2T_0} \Re \left\{ \dot{Q}_p (\Theta'_f - \overline{T}'_p)^* \right\} \tag{6.7.4}$$

The heat transfer rate is given by (6.6.17). For simplicity we use the first form given there. Thus,

$$\dot{Q}_p (\Theta'_f - \overline{T}'_p)^* = 4\pi k_f a \left| \Theta'_f \right|^2 (1 - \xi \Pi)(1 - T)^* / F \tag{6.7.5}$$

But from (6.6.14), we see that $(1 - T)^* = (1 - \xi \Pi)^* [1 + \frac{3}{2} i(k_f / k_p) / z_p^2 F^*]$. The second term here drops out when the real part of $\dot{Q}_p (\Theta'_f - \overline{T}'_p)^*$ is taken, as indicated by (6.7.4), thus giving

$$\langle \dot{e}_{th} \rangle = \frac{2\pi k_f a}{T_0} \left| \Theta'_f \right|^2 |1 - \xi \Pi|^2 \Re \left\{ \frac{1}{F(q, q_i)} \right\} \tag{6.7.6}$$

This is the desired result, but it is useful to express it in terms of the pressure fluctuation in the fluid and to remember that $1 - \xi \Pi \approx (1 - \xi) \Pi$. Thus, using $\Theta'_f = (\beta_f T_0 / \rho_{f0} c_{pf}) P'_f$ and $\Pi = \chi \kappa$, we obtain the final result

$$\langle \dot{e}_{th} \rangle = 2\pi \kappa_f a (\gamma_f - 1)(1 - \xi)^2 |\kappa|^2 |\chi|^2 \Re \left\{ \frac{1}{F} \right\} \frac{\left| P'_f \right|^2}{\rho_{f0} c_{sf}^2} \tag{6.7.7}$$

This applies to all particle-fluid combinations.

We now use this result to verify the result, obtained previously, for gas bubbles and droplets in liquids, namely, $\hat{d}_{th} = -\kappa_i$. The derivation is based on the prescription for obtaining β_{th} from $\langle \dot{e}_{th} \rangle$, namely $\beta_{th} = \langle \dot{e}_{th} \rangle / 2e_0$, or with $e_0 = {}^1/_2 M_0 \omega^2 |\varepsilon|^2$. To simplify the algebra, we make use of (6.6.20) for κ_i, and substitute these expressions in $d_{th} = 2\beta_{th}\omega / \omega_{T0}^2$, to obtain

$$\hat{d}_{th} = \frac{1 - \xi}{\xi} \kappa_i \tag{6.7.8}$$

The quantity ξ defined by (6.6.10) is, as pointed out previously, a measure of thermal expansion of the material in the particle, relative to that of the external fluid. It is very large for gas bubbles in liquids, so that for them (6.7.8) gives $\hat{d}_{th} = -\kappa_i$, as deduced from the equation of motion. But, for other particle-fluid combinations, discrepancies begin to appear. Thus, for toluene droplets in water, $\xi \approx 20$, and this shows that the previous result is, for them, not exactly the same, although the differences are small.

These differences originate in the Rayleigh-Plesset equation, which, as its derivation shows, does not take into account thermal expansion in the external fluid. The same conclusion can be arrived at if we ignore the thermal energy losses in the fluid. These are accounted for by the second term in (6.7.3); without them, we would find

$$\langle \dot{e}_{th} \rangle_p = 2\pi \kappa_f a (\gamma_f - 1) \xi (\xi - 1) |\kappa|^2 |\chi|^2 \, \Re \left\{ \frac{1}{F} \right\} \frac{|P_f'|^2}{\rho_{f0} c_{sf}^2} \tag{6.7.9}$$

where the subindex p in $\langle \dot{e}_{th} \rangle_p$ is used to remind us that these are the thermal losses associated with the particle. If we compute $\hat{d}_{th}$ on the basis of $\langle \dot{e}_{th} \rangle_p$ alone, we would find $\hat{d}_{th} = -\kappa_i$, which is the result obtained through use of the Rayleigh-Plesset equation and which clearly shows that thermal effects in the external fluid are not included in (6.7.9). This is not significant for bubbles in liquids; but, for other combinations, this may not be the case. Thus, for droplets in gases, the situation is reversed entirely in the sense that the fluid contribution to $\langle \dot{e}_{th} \rangle$, namely,

$$\langle \dot{e}_{th} \rangle_f = 2\pi \kappa_f a (\gamma_f - 1)(1 - \xi) |\kappa|^2 |\chi|^2 \, \Re \left\{ \frac{1}{F} \right\} \frac{|P_f'|^2}{\rho_{f0} c_{sf}^2} \tag{6.7.10}$$

is the largest. For this case, $\hat{d}_{th}$ loses its meaning as a thermal damping coefficient, as discussed previously, but the discussion shows that, in general, the thermal losses associated with particle pulsations in a sound wave are due to both the fluid and particle components, and are given by (6.7.6).

Absorption Cross-Sections

It is also useful to compare the pulsational losses to those encountered in the other basic motions when a particle is exposed to a sound wave. As discussed in Chapter 2, those motions include uniform translation, uniform rotation, deformation, and uniform expansion/compression. We have not considered particle rotations, nor have we discussed the losses that may be associated with deformation. Furthermore, we have only considered spherical particles. Thus, our particles have been allowed to undergo only two of these motions, namely uniform translations and radial pulsations. The losses associated with these have been calculated for several particle-fluid combinations, and the corresponding results appear in various places in the book.

Because these losses play a rather important role in the following chapters, it is important to put each of them in perspective by comparing their magnitudes for the suspension types to be considered. To do this, we need to express the losses in a manner that easily allows that comparison. One way to do this is to present the various dissipation rates in terms of the corresponding absorption cross-sections. The cross-section concept is simple. A wave, coming into a particle, may lose some of its energy owing to scattering and absorption by the particle. By scattering, we mean that process that spreads part of the incident wave energy around the particle, whereas by absorption we mean that part of the energy is absorbed by the particle and then dissipated by irreversible mechanisms.

Both scattering and absorption can be expressed in terms of a cross-section, and these may be defined as follows. Let a plane, acoustic wave traveling in the fluid external to the particle approach a spherical particle. The intensity of the wave, defined as the flux of acoustic energy per unit area, is $I_{inc} = \frac{1}{2}|P_f'|^2/\rho_{f0}c_{sf}$. Now, let σ_a be an area such that the energy flux through it equals the average energy dissipation rate. Then $\sigma_a I_{inc} = \langle \dot{e}_{loss} \rangle$. This area is called the absorption cross-section. A similar definition applies to the scattering cross-section, σ_s, but these are important only at frequencies that lie well beyond the validity of the pulsational theory given here.

The absorption cross-section produced by the pulsational motions is given by the sum of acoustic, thermal, and viscous components. Using the dissipation rates corresponding to each, we obtain

$$\frac{\sigma_{a,ac}}{\pi a^2} = 4\,|\chi|^2\,(\omega/\omega_{T0})^2 \tag{6.7.11}$$

$$\frac{\sigma_{a,th}}{\pi a^2} = \frac{3}{2}\,(1-\xi)^2\,(\gamma_f - 1)^2\frac{b}{z^2}\,|\Pi|^2\,\Re\left\{\frac{1}{F}\right\} \tag{6.7.12}$$

$$\frac{\sigma_{a,\mu}}{\pi a^2} = 8\,|\chi|^2\,(\omega/\omega_{T0})^2\,/by^2 \tag{6.7.13}$$

where $y = \sqrt{\omega a^2/2\nu_f}$.

The translational motion was considered in Section 4.11, where it was shown that the corresponding absorption cross-section is given by (4.11.19), or

$$\frac{\sigma_{a,tr}}{\pi a^2} = \frac{4}{3}\,(b/\delta)\,|\Im\{V\}| \tag{6.7.14}$$

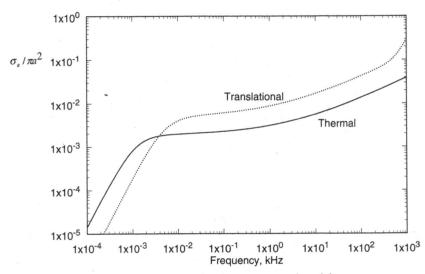

Figure 6.7.1. Absorption cross-sections for a 100-μm aerosol particle.

where, $V = u_p/u_f$, is given by (4.11.13). This result includes the sound radiated by the translating particle, usually referred to as self-scattering.

We are now ready to compare these results for the four types of particle-fluid combinations to which they apply. In alphabetical order, they are: *aerosols* (liquid or solid particles in gases); *bubbly liquids* leave; *emulsions* (liquid droplets in liquids); and *hydrosols* (solid particles in liquids). To provide a comparative basis for the four types, the four absorption cross-sections are shown for a 100-μm particle. The effects of varying size may be inferred from the previous equations, or from the discussion and figures given earlier. For the same reason, we have chosen the absolute frequency, in kHz, as the independent variable, with the highest value corresponding to $b = 1$.

Aerosols. Included in this type, we have both liquid and solid elastic particles in gases. We select a water droplet in air, as we have done before, but remind the reader that evaporation and condensation are not included in the formulation. Also, the complex pressure ratio, Π, for these particles has been taken to be equal to 1. Hence, the acoustic cross-section vanishes. Similarly, the viscous component of the pulsational cross-section is negligible at all frequencies. Thus, the only cross-sections that are meaningful to aerosols are the translational and the thermal, and are shown in Fig. 6.7.1. We see that both components produce cross sections that are generally small, and that at low frequencies the thermal component is dominant, whereas the translational one takes over at larger frequencies.

Bubbly Liquids. Here, the resonant nature of the response of a bubble results in much larger absorption cross-sections, particularly in the vicinity of resonance, as shown in Fig. 6.7.2. The figure shows that the thermal cross-section alone results in a cross-section of more than a thousand mean bubble cross-sections. We also see in the figure

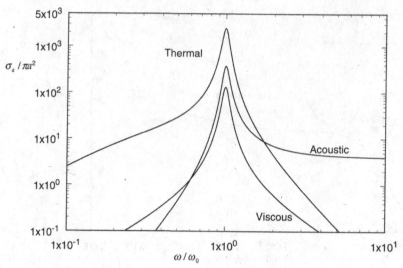

Figure 6.7.2. Absorption cross-sections for a 100-μm bubble in water.

that the viscous and acoustic components are also large. As we decrease the size of the bubbles, the viscous component increases, eventually becoming the largest one for the smallest particles. The figure does not show the translational component. This is several orders of magnitude smaller than those shown in the figure.

Emulsions. Figure 6.7.3 shows the absorption cross-sections for a toluene droplet in water. Although the thermal component is the largest, we see that, in general, all of the cross-sections are small and are of a comparable order of magnitude. This happens because the properties of the internal and external material have similar values.

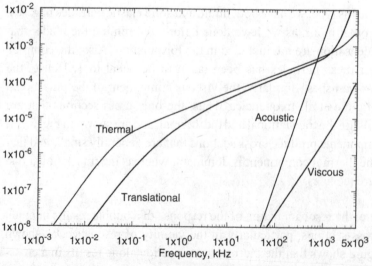

Figure 6.7.3. Absorption cross-sections for a 100-μm emulsion particle.

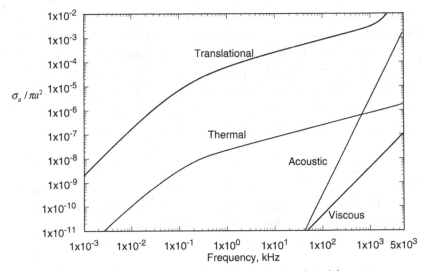

Figure 6.7.4. Absorption cross-sections for a 100-μm hydrosol particle.

Hydrosols. The cross-sections for hydrosols are also quite small, as shown in Fig. 6.7.4, but here, the translational cross-sections are larger at all frequencies below resonance, where the acoustic component takes over.

Thermodynamics of a Bubble Cycle

We close this section with a brief discussion of the thermodynamics of a bubble executing radial pulsations in water as a result of the external variations of pressure. When these variations are harmonic in time, the bubble's pulsations are cyclical, so that all of the internal variables will undergo cyclic variations. Now, since the particle pressure is uniform and since the particle temperature can be represented by its spatial average, $\overline{T}_p$, we may regard the particle as a homogeneous thermodynamic system, exchanging energy with the surroundings, as shown, schematically, in Fig. 6.7.5. The exchanges must satisfy the first law of thermodynamics, that we write as

$$\frac{dE_p}{dt} = \dot{Q}_p + \dot{W}_{in} - \dot{W}_{out} \qquad (6.7.15)$$

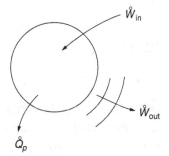

Figure 6.7.5. Energy balance for a radially driven, compressible sphere.

Here, $\dot{Q}_p$ is the instantaneous heat rate added to the particle, and $\dot{W}_{in} - \dot{W}_{out}$ is the instantaneous *net* work rate done on it. Contrary to the work rates, which are defined as positive quantities, the sign of the heat transfer rate is left undefined for now.

We are interested in the amounts of energy that are exchanged during any one cycle. These are given by averaging (6.7.15) over one period of the pulsations. Since the exchange is cyclical, the internal energy of the particle returns each cycle to its previous value, giving $\langle dE_p/dt \rangle = 0$. Thus, .

$$\langle \dot{Q}_p \rangle = \langle \dot{W}_{out} \rangle - \langle \dot{W}_{in} \rangle \tag{6.7.16}$$

It should be noted that, if (6.6.17) were used to calculate $\langle \dot{Q}_p \rangle$, we would obtain a zero value because that heat transfer rate is a first-order quantity. On the other hand, the above equation can be used to obtain $\langle \dot{Q}_p \rangle$ to second order. Anticipating that $\langle \dot{W}_{out} \rangle \le \langle \dot{W}_{in} \rangle$, (6.7.16) states that, during each cycle, an amount of heat $\langle -\dot{Q}_p \rangle$ leaves the particle.

Let us now calculate the work rates. The work rate done on the system is the work of compression, the so-called pdV work. It is important, however, to remember that the processes are irreversible. Hence, the "p" here is the *external* pressure, which is $p_\infty = p_0 + P_f'$, and not the particle pressure, as it would be for reversible transformations. Thus, since $\langle p_0 \dot{v}_p \rangle = 0$, we have $\langle \dot{W}_{in} \rangle = -\frac{1}{2}\Re\{P_f'(\dot{v}_p')^*\}$. Now, $\dot{v}_p = -4\pi a^2 i\omega\varepsilon$ and by (6.6.33), $P_f' = -(M_0\omega_{T0}^2/4\pi a^2)\varepsilon/\chi$, we obtain

$$\langle \dot{W}_{in} \rangle = \frac{1}{2}M_0\omega_0^2\omega |\varepsilon|^2 \, \hat{d} \tag{6.7.17}$$

where we used the fact that $\Re\{i/\chi\} = \hat{d}$.

Consider now the work produced by the pulsating particle on the external medium. This is composed of two parts. The first is the work done against the viscous stresses in the external fluid and was calculated in Section 6.3; it is given by the first term in (6.3.8) and may be expressed as $\langle \dot{W}_{out} \rangle_\mu = \frac{1}{2}M_0\omega_{T0}^2\omega |\varepsilon|^2 \, \hat{d}_\mu$. This is simply equal to the viscous energy dissipation rate, $\langle \dot{e}_\mu \rangle$. The second part of the work done is the rate associated with acoustic radiation; it is given by $\langle \dot{W}_{out} \rangle_{ac} = \frac{1}{2}M_0\omega_{T0}^2\omega |\varepsilon|^2 \, \hat{d}_{ac}$. Adding these two work rates, we find,

$$\langle \dot{W}_{out} \rangle = \frac{1}{2}M_0\omega_{T0}^2\omega |\varepsilon|^2 \left(\hat{d}_{ac} + \hat{d}_\mu \right) \tag{6.7.18}$$

Substituting (6.7.17) and (6.7.18) into (6.7.16), and using $\hat{d} = \hat{d}_{ac} + \hat{d}_\mu + \hat{d}_{th}$, we obtain

$$\langle \dot{Q}_p \rangle = -\frac{1}{2}M_0\omega_{T0}^2\omega |\varepsilon|^2 \, \hat{d}_{th} \tag{6.7.19}$$

This is the average heat flow rate. As anticipated, it is negative, meaning that it flows out of the particle. Comparison with (6.7.1) shows that this is the heat that is generated owing to thermal dissipation in the particle, also an expected result.

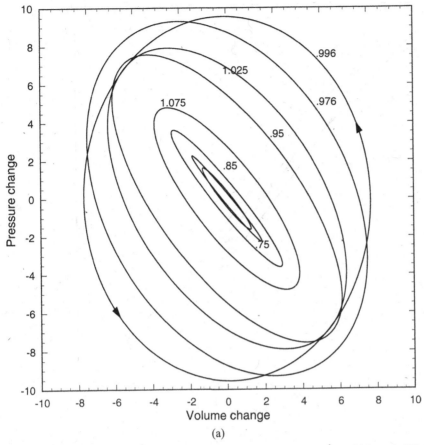

Figure 6.7.6. Thermodynamic cycle for an air bubble. (a) Near resonance. (b) Low and high frequencies.

The cycle can be visualized by plotting the variations of pressure against those of the volume. To do this, we express those quantities in real form and scale them with P'_{f0}/p_0. Thus,

$$\frac{p'_p/p_0}{P'_{f0}/p_0} = |\Pi| \cos(\omega t - \tan^{-1} \Pi) \tag{6.7.20}$$

$$\frac{v'_p/v_{p0}}{P'_{f0}/p_0} = -|\chi| \cos(\omega t - \tan^{-1} \chi) \tag{6.7.21}$$

In writing the second equation, we have treated the gas in the bubble as perfect. Figures 6.7.6a and 6.7.6b show a few cycles, corresponding to several values of the frequency ratio ω/ω_0. The arrows indicate increasing values of time in a cycle. The shapes of the closed curves are seen to be ellipses having different eccentricities, depending on the value of the frequency ratio. Before we discuss these, consider the area enclosed by each of the closed curves in the figure. These are given by $-\oint p_p dv_p = -\int_0^{2\pi/\omega} p_p \dot{v}_p dt \equiv -(2\pi/\omega)\langle p_p \dot{v}_p \rangle$. That is, the integral around the

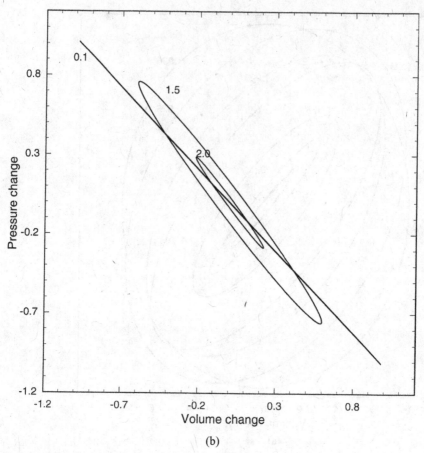

(b)

Figure 6.7.6 *Continued.*

closed curve is simply equal to the average value of the integrand, multiplied by the duration of the cycle – a physically obvious result. The remaining part of the derivation is straightforward and gives

$$-\oint p_p dv_p = -(2\pi/\omega)\frac{1}{2} M_0 \omega \omega_{T0}^2 |\varepsilon|^2 \hat{d}_{th} \qquad (6.7.22)$$

Comparison with (6.7.19) shows that the area enclosed by each cycle is equal to average heat loss rate multiplied by the duration of the cycle. This area varies with the frequency, being small at both small and large frequencies. At very small frequencies, Fig. 6.7.6b, the ellipses degenerate into straight lines, making a 45-degree angle with the vertical, implying that the pressure and the volume are in phase. Here, the pressure and temperature fluctuations in the bubble are equal to those in the fluid outside. Furthermore, since the enclosed area is now negligible, we see that there are no losses, even if the pressure and volume change. As the frequency increases, Fig. 6.7.6a, eccentric ellipses are formed whose area increases while the slope of their major

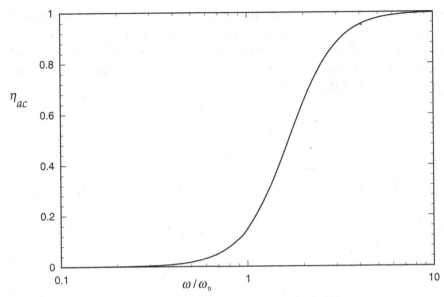

Figure 6.7.7. Thermal efficiency of a 100-μm radially pulsating bubble.

axis rotates slowly clockwise. At resonance, the ellipses' major axis points along the vertical. Beyond that point, the trends are reversed, with the areas decreasing while the main axis rotates counterclockwise, Fig. 6.7.6a.

Although the areas are small at small and large frequencies, the physical reasons for this are different. Thus, at low frequencies, almost all of the energy fed into the bubble is dissipated in the form of heat, whereas at high frequencies, almost all of that energy is sent back into the fluid in the form of acoustic waves.

Seen in this manner, the bubble is a thermodynamic engine, working on a cycle, with acoustic radiation playing the role of the "useful" work. Thus, this engine receives an amount of work $\langle \dot{W}_{in} \rangle$ from a work "source," performs some useful work $\langle \dot{W}_{ac} \rangle$, and releases an amount of heat $\langle -\dot{Q}_p \rangle$ to a "heat sink" and an amount of work $\langle \dot{W}_{\mu} \rangle$ to a "work sink." The efficiency of such an engine is $\eta_{ac} = \langle \dot{W}_{ac} \rangle / \langle \dot{W}_{in} \rangle$ and is shown in Figure 6.7.7.

This concludes our discussion of the thermal effects that come into play during the radial pulsations of a compressible particle. The discussion has been limited to linear motions. However, given the large amplification ratios that exist when a bubble is driven externally, it is very likely that nonlinear effects come into play. Some of those effects can be dealt with in the nonlinear form of the Rayleigh-Plesset equation, which was briefly discussed in Section 6.2.

6.8 Concluding Remarks

This chapter has examined volume pulsations of spherical particles in fluids. The discussion has included particles made of different materials, but has largely been limited to linear motions in both isothermal and nonisothermal conditions. Several important features have been studied, including the response of the particles to applied, monochromatic pressure fluctuations and the concomitant dissipation rates. The material is of some importance on its own and is central to the acoustic propagation theories discussed in the following chapters.

7

Thermodynamics of Suspensions

7.1 Introduction

Having examined some of the essential particle motions in fluids, we can begin our study of suspensions. The main interest is the motion of suspensions subject to applied forces. As the last four chapters demonstrate, forces applied to a suspension by the motion of an external wall are transmitted to the particles by the fluid. But, while small particles respond quickly to the fluid, their velocities, pressures, and temperatures will normally lag behind the corresponding quantities in the fluid. Thus, the internal state of a suspension is not one of equilibrium between particles and fluid. This means that it is not generally possible to define such quantities as the temperature or density at a point, as it is in a fluid without particles. One consequence of this is that the physical description of suspensions in motion generally requires methods that differ from those used for homogeneous fluids, where such definitions are possible. Some of those methods will be introduced in the following chapter, with the specific goal of applying them to the study of sound propagation in suspensions.

But, some information may be obtained for suspensions that are either at rest or that are sustaining such slow motions that make it possible to treat them from a thermodynamic point of view. These suspensions are said to be in equilibrium. In them, it is possible to define basic thermodynamic variables, and this enables us to obtain certain properties that can be useful in other contexts (e.g., when we discuss suspensions not far from equilibrium, as is the case with suspensions sustaining sound waves).

In this chapter, we present some elements of the thermodynamics of suspensions. These apply to homogeneous suspensions that consist of particles having a constant mass and are made of the same material. The discussion focuses in essential thermodynamic properties and excludes issues concerning particle formation and mixing. These are important in some contexts, but are not essential to the goals of this book.

7.2 Equilibrium

From a practical point of view, it would be useful to define suspension properties that have the same or similar meaning as those for a homogeneous fluid. For example, we would like to obtain suspension compressibilities, specific heats, sound speeds, etc. At the outset, it may be stated that these thermodynamic properties can be always defined, provided the suspension is in equilibrium. However, before we proceed to obtain them, it is necessary to explain what is meant by a suspension in equilibrium. To do so, we mentally isolate a volume element in the suspension and imagine it enclosed by an adiabatic container fitted with a piston. Furthermore, we imagine each particle to be surrounded by a fictitious, massless wall that can be manipulated to allow energy, but not matter, to flow through it. Since the flow of energy can occur in the form of work or heat, it is useful to consider walls that restrict either form. So long as the walls are impervious to the flow of work or heat, nothing can be said about equilibrium between particles and fluid. A definite statement can only be made by observing changes that might take place after the restrictive conditions have been removed. In the next section, we consider hypothetical situations in which each of the two types of flows is allowed to take place alone. In each case, the walls are massless and occupy no volume. For simplicity, we ignore surface-tension effects that may act at the boundary dividing particle from fluid, and assume that body forces – such as gravity – are absent.

Mechanical Equilibrium

Suppose first that the massless walls separating each of the particles from the fluid outside are adiabatic and that the suspension is initially at rest. Then, there can be no heat exchanged between particles and fluid, and the surface forces on a fluid element are those resulting from a uniform pressure. Regardless of the orientation of the element, the force acts along a direction that is perpendicular to the surface of the element. Thus, if we consider an element of area δA at the boundary between any particle and the surrounding fluid, it will be seen that the fluid exerts *on* each particle a force of magnitude $p_f \delta A$ along the normal to the boundary.

The situation on the particles' side can be more complicated, and it is useful to consider first fluid particles (e.g., droplets and bubbles). The internal forces in them will also consist of a uniform, isotropic pressure; but, to allow a variation from one particle to the next, we denote the pressure in the ith particle as $p_p^{(i)}$. This pressure produces a normal force $p_p^{(i)} \delta A$ on an element of area δA. Since, generally, $p_p^{(i)} \neq p_f$, the release of the walls produces a displacement of the boundaries between particles and fluid that cease when the pressures are equalized. Thus, a necessary condition for mechanical equilibrium is that $p_p^{(1)} = p_p^{(2)} = \ldots = p_f$, or

$$p_p = p_f \qquad\qquad (7.2.1)$$

where p_p is the pressure in each of the particles. Furthermore, if we now impose an infinitesimal pressure change on the system by means of an infinitesimal displacement of the walls surrounding the suspension, the pressures will change to $p_f \rightarrow p_f + dp_f$, and, for each of the particles, $p_p^{(i)} \rightarrow p_p^{(i)} + dp_p^{(i)}$. But, in view of (7.2.1), it follows that, for equilibrium to exist, we must also have $dp_p^{(1)} = dp_p^{(2)} = \ldots = dp_f$, or simply

$$dp_p = dp_f \qquad (7.2.2)$$

Equations (7.2.1) and (7.2.2) are the conditions for mechanical equilibrium. It is to be noted that because the infinitesimal change was produced by a slight change of the total volume of the element, the concentration of particles by volume will in general also change. However, since no particles leave the container, we see that the concentration by mass does not change. Thus, the previous conditions are complemented by

$$\phi_m = \text{constant}. \qquad (7.2.3)$$

Now consider solid particles. Their mechanical state is determined by a stress system that is generally more complicated than the simple isotropic one for a fluid at rest. Fortunately, the stress system in solid particles plays only a small, passive role in suspension dynamics. We will therefore adopt, for solid particles, the simplest possible stress model, namely that of an isotropic, elastic solid under a hydrostatic load. Shearing stresses that might exist will be ignored, as will residual stresses and hysterises effects. Thus, the stress in a solid particle is therefore represented by an isotropic pressure so that the conditions for mechanical equilibrium in a suspension of solid particles in a fluid will be also given by (7.2.1)–(7.2.3).

Thermal Equilibrium

To obtain the conditions for thermal equilibrium, we proceed in a similar fashion. Let the temperature of the fluid in the element be T_f and that in the ith particle be $T_p^{(i)}$, with $T_p^{(i)} \neq T_f$. Now suppose we replace the rigid, adiabatic walls around each particle with rigid, diathermal walls. This will result in a flow of heat that will cease when the particles and the fluid have the same temperature. Thus, $T_p^{(1)} = T_p^{(2)} = \ldots = T_f$, or simply

$$T_p = T_f \qquad (7.2.4)$$

Again, we introduce an infinitesimal temperature change, for example, by allowing some heat to flow into the element so that $T_f \rightarrow T_f + dT_f$, and $T_p^{(i)} \rightarrow T_p^{(i)} + dT_p^{(i)}$. But, in view of (7.2.4), it follows that, for equilibrium to exist, we must also have $dT_p^{(1)} = dT_p^{(2)} = \ldots = dT_f$, or

$$dT_p = dT_f \qquad (7.2.5)$$

It should be noted that, because substances generally expand or contract due to temperature changes, the volume concentration is not generally constant during the infinitesimal change. But, as it was the case for mechanical equilibrium, the mass concentration remains constant. Thus, we again add the condition (7.2.3).

Thermodynamic Equilibrium

Finally, if both mechanical and thermal equilibrium exist, we say that the suspension is in *thermodynamic* equilibrium (i.e., equilibrium requires the two sets of conditions given previously). Because they imply that the temperature and the pressure are uniform, we write these conditions as

$$T_f = T_p = T; \qquad\qquad dT_f = dT_p = dT$$
$$p_f = p_p = p; \qquad\qquad dp_f = dp_p = dp$$
$$\phi_m = \text{Constant}. \tag{7.2.6}$$

Quantitative criteria that determine whether these conditions apply during an actual change cannot be obtained from equilibrium considerations alone; they must be determined from dynamic considerations. But, on physical grounds, we may expect that if the changes occur very slowly, equilibrium will approximately exist because small particles can adjust rapidly to them.

To fix ideas, we consider two examples of isothermal motions in equilibrium. In the first, we consider a translational motion imposed on a dilute suspension by the action of a moving wall. For such low concentrations, only a very small fraction of all the particles in the container will be in contact with the wall. Therefore, the wall's motion primarily affects the fluid. Thus, whatever velocity is acquired by the particles, it must be acquired by the transfer of momentum from the fluid. Now suppose that the fluid velocity is changed suddenly. The particles, as we have seen earlier, cannot adjust immediately to the new conditions. However, if the external fluid changes occur over time scales that are very long, compared with the times required by the particles to react, it is clear that, except for a very short time, the particles may be regarded as being in equilibrium with the fluid.

As a second example, we consider a small, cylindrical container filled with a bubbly liquid and fitted with a massless piston at one end. The piston is initially held in position by means of a pin so that the internal pressure is maintained at some uniform value, p, that is only slightly different from p_0, the pressure outside. If we now release the piston slowly until the pressure in the container becomes p_0, the bubbles will expand or contract slowly, but their pressure will equal that of the fluid outside, so that equilibrium exists. It is clear that if the applied excess pressure were to be released suddenly, the piston would oscillate about a mean position eventually coming to a rest, and that the only equilibrium state that might exist would occur at the end of the oscillations.

7.3 Thermodynamic Properties

As pointed out previously, suspensions in equilibrium are relatively simple thermo-dynamic systems. They can therefore be described in terms of a small number of macroscopic properties by the usual thermodynamic equilibrium arguments. The most direct means of doing this would be to resort to a fundamental equation of state (i.e., an equation for the entropy, internal energy, or some other derived potential, such as the free energy of the system). But, some of the basic properties only require the density for their definition, and it is useful to consider them first.

Isothermal Compressibility

An easily measured mechanical property of a suspension in equilibrium is its com-pressibility. This measures the fractional change of volume due to the application of a small pressure. This pressure can be applied in various manners, but the simplest occurs when the temperature is held constant (e.g., by immersion in a large thermal bath). Thus, the isothermal compressibility is defined by

$$K_T(0) = -\frac{1}{\delta V}\left[\frac{\partial(\delta V)}{\partial p}\right]_T \qquad (7.3.1)$$

where δV is the volume of the element and the subscript T indicates that the temper-ature is held constant. The notation used in (7.31) needs some clarification. First, the argument in K_T is meant to imply that the equation applies to very small frequencies (i.e., to very small motions). A more appropriate notation would be $K_T(\omega \to 0)$; but, for simplicity, we shall use instead $K_T(0)$. Later on, we shall use the same notation for the isentropic compressibility and for the equilibrium sound speeds. Second, the negative sign in (7.3.1) is used on the expectation that the volume of the suspension decreases when the pressure increases. If so, $K_T(0) \geq 0$.

Now, to ensure that equilibrium be maintained, we must take the derivative in (7.3.1) while holding the mass concentration constant. To do this, we use the definition (1.3.7) of the suspension density ρ to write $\delta V = \delta M/\rho$. Substituting this into (7.3.1), remembering that ϕ_m is constant, we have

$$K_T(0) = \frac{1}{\rho}\left(\frac{\partial\rho}{\partial p}\right)_{T,\phi_m} \qquad (7.3.2)$$

To simplify the writing, we will omit the subscript, ϕ_m, in this and subsequent deriva-tives. Now, the density of the suspension can be expressed in terms of the material densities of the particles and fluid, which, in turn, can be expressed in terms of equa-tions of state:

$$\rho_f = \rho_f(p_f, T_f) \quad \text{and} \quad \rho_p = \rho_p(p_p, T_p) \qquad (7.3.3a,b)$$

The existence of the fluid's equation of state is obvious. That one for the particle also exists follows from our initial assumption that the smallest particle to be considered would have a very large number of molecules. Thus, the material in the particle is a homogenous substance whose state is also described by an equation of state.

Now, since the temperature is constant and the pressure has a uniform value, which we denote by p, the previous equations of state reduce to $\rho_f = \rho_f(p)$ and $\rho_p = \rho_p(p)$. Furthermore, since the mass concentration is constant, it is advantageous to use (1.3.9) to obtain $d\rho$. Thus,

$$d\rho = \rho^2 \left[(1 - \phi_m)\frac{d\rho_f}{\rho_f^2} + \phi_m \frac{d\rho_p}{\rho_p^2} \right] \tag{7.3.4}$$

Hence, forming the derivative of ρ with respect to the pressure and eliminating the mass concentration in favor of the volume concentration we obtain

$$K_T(0) = (1 - \phi_v)K_{Tf} + \phi_v K_{Tp} \tag{7.3.5}$$

where $K_{Tf} = 1/\rho_f c_{Tf}^2$ and $K_{Tp} = 1/\rho_f c_{Tp}^2$ are the isothermal compressibilities of the fluid and particle material, respectively, with $c_{Tf}^2 = (\partial p/\partial \rho_f)_T$ and $c_{Tp}^2 = (\partial p/\partial \rho_p)_T$ being, respectively, the isothermal sound speeds of the fluid and particle materials. Thus, the isothermal compressibility of the suspension is given by the sum of the fluid and particle compressibilities, each multiplied by the respective volume fraction. In other words, $K_T(0)$ follows the same distributive law as the effective density, (1.3.8). As shown later, this does not hold when temperature changes occur.

An idea of the effects of the particles on the behavior of a fluid may be obtained by considering the important case of bubbly liquids. Here, the compressibility of the gas in the bubbles is considerably larger than that of the liquid around them so that even a small volume concentration of gas bubbles produces large effects. Thus, for example, a 1% volume concentration of air bubbles in water produces a compressibility that is about 200 times larger than that of the water alone. On the other hand, the density of the bubbly liquid is, for that low concentration, very nearly equal to that of the liquid alone. Thus, a low concentration bubbly liquid has the density of a liquid and the compressibility of a gas. As shown by the simple experiment described in Chapter 1, this combination produces some remarkable effects.

Isothermal Sound Speed

It is known from thermodynamics that stability of a system requires that $(\partial\rho/\partial p)_T \geq 0$. Thus, in analogy with the simple substance case, we define the isothermal sound speed for a suspension as the nonnegative quantity $c_T^2(0) = (\partial p/\partial \rho)_T$, in which case we can write

$$c_T^2(0) = 1/\rho K_T(0) \tag{7.3.6}$$

where the notation on c_T^2 has the same meaning as that used for K_T.

Equation (7.3.6) may be written in various forms, including

$$\frac{c_{Tf}^2}{c_T^2(0)} = (1 - \phi_v + \phi_v \rho_p/\rho_f)(1 - \phi_v + \phi_v N_T), \qquad (7.3.7)$$

where $N_T = K_{Tp}/K_{Tf}$ is the ratio of particle-to-fluid compressibilities,

$$N_T = \rho_f c_{Tf}^2 / \rho_p c_{Tp}^2 \qquad (7.3.8)$$

We postpone the discussion of the isothermal sound speed until the corresponding result for the adiabatic sound speed has been obtained.

Coefficient of Thermal Expansion

The effective density also suffices to obtain the coefficient of thermal expansion of a suspension in equilibrium. This coefficient measures the fractional change of volume with temperature in a suspension held at constant pressure and is defined by

$$\beta = -\frac{1}{\rho}\left(\frac{\partial \rho}{\partial T}\right)_p \qquad (7.3.9)$$

Since the mass concentration is constant, we obtain from (1.3.9)

$$\frac{\beta}{\rho} = (1 - \phi_m)\frac{\beta_f}{\rho_f} + \phi_m\frac{\beta_p}{\rho_p} \qquad (7.3.10)$$

where β_f and β_p are the coefficients of thermal expansion of the fluid and of the particles, respectively. This result may be expressed more succinctly by eliminating ϕ_m in favor of ϕ_v:

$$\beta = (1 - \phi_v)\beta_f + \phi_v\beta_p \qquad (7.3.11)$$

Thus, the suspension's coefficient of thermal expansion obeys the same distributive law as the suspension's density and isothermal compressibility. Figure 7.3.1 shows β/β_f as a function of the volume concentration for an aerosol, a bubbly liquid, and an emulsion for $0 \le \phi_v \le 0.1$. The equation applies to even larger values, provided the suspension can still be regarded as homogeneous. Since ϕ_v is small for all cases shown in the figure, we see that the differences between them are due to the different values of β_p/β_f applicable to each suspension type. This ratio is largest for bubbly liquids and smallest for aerosols (see Table 7.4.1 in the next section).

Specific Heats

As the two previous examples show, some physical properties of fairly general classes of suspensions may be obtained from their density. Other properties require a fundamental equation of state (e.g., equations for the internal energy or the entropy). In what follows, we shall assume that such equations exist for the particle and fluid materials.

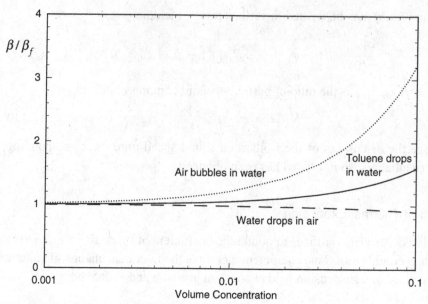

Figure 7.3.1. Value of β/β_f for an aerosol, a bubbly liquid, and an emulsion as a function of the volume concentration.

Knowledge of such an equation makes it possible to determine every thermodynamic property of a system. Particularly important are the specific heats at constant pressure, c_p, and at constant volume, c_v. These may be obtained from the entropy equation by means of the following identities:

$$c_p = T(\partial s/\partial T)_p \qquad (7.3.12)$$

$$c_v = T(\partial s/\partial T)_\rho \qquad (7.3.13)$$

As before, these derivatives are to be evaluated while holding the mass concentration constant. Also, these specific heats also apply in the limit of small frequencies, so that they ought to be written in the same manner as the compressibility. But, we shall omit that notation for them as well as for other equilibrium thermodynamic properties, with the exception of the compressibilities and the sound speeds, because these play a role in dynamic conditions.

Let us now determine the equilibrium changes of entropy and internal energy. To do this, we note that these quantities are extensive properties. This means that the internal energy (or the entropy) of a system that is composed of several parts, is equal to the sum of the energies (or the entropies) of its parts. Thus, the *internal energy*, δE, of a suspension is given by the sum of the separate internal energies δE_f and δE_p of the fluid and particles, respectively, or

$$\delta E = \delta E_f + \delta E_p \qquad (7.3.14)$$

This equation implies that the two contributions to the energy are separate, and that not other contributions are present. Thus, contributions that owe their origin to effects that depend on *both* fluid and particles would alter this result. One such contribution is surface energy. This gives rise to surface tension. Also implied in (7.3.14) is the idea that both energies can be defined unequivocally. This requirement will be examined herein.

Now, if e_f and e_p are the internal energies per unit mass of fluid and of particle materials, respectively, the fluid and particle's internal energies contained in δV are

$$\delta E_f = (1 - \phi_m)e_f\delta M \quad \text{and} \quad \delta E_p = \phi_m e_p\delta M \qquad (7.3.15\text{a,b})$$

where δM is the suspension mass in δV. Thus, $\delta E = [(1 - \phi_m)e_f + \phi_m e_p]\delta M$. Hence, we may define the suspension's internal energy per *unit mass* as

$$e = (1 - \phi_m)e_f + \phi_m e_p \qquad (7.3.16)$$

We can similarly define an entropy per unit mass for the suspension. Thus, starting with

$$\delta S = \delta S_f + \delta S_p \qquad (7.3.17)$$

and proceeding in the same manner, we obtain

$$s = (1 - \phi_m)s_f + \phi_m s_p \qquad (7.3.18)$$

For future reference, we note from (7.3.17) that, in those processes that are deemed to occur in isentropic conditions, the entropies of the particles and fluid may change provided the changes are equal but opposite (i.e., provided $\delta S_p = -\delta S_f$). This implies that, while there is no heat transfer to the suspension as a whole, such a transfer may exist between particles and fluid.

Let us now return to (7.3.16) and (7.3.18). Noting that in thermodynamic equilibrium, all quantities appearing in these equations have well-defined meanings, and that the mass concentration remain constant during equilibrium changes, we may compute with them infinitesimal changes of energy and entropy. Thus,

$$de = (1 - \phi_m)de_f + \phi_m de_p \quad \text{and} \quad ds = (1 - \phi_m)ds_f + \phi_m ds_p \quad (7.3.19\text{a,b})$$

We now express the differentials on their right-hand side of these expressions in terms of measurable quantities (e.g., the temperature and the pressure). To do this, we note that because the fluid and particle materials are separately homogeneous substances, the fluid and particle entropies per unit mass may be obtained from fundamental equations of state for *each* substance, i.e., from

$$s_f = s_f(p_f, T_f), \qquad s_p = s_p(p_p, T_p) \qquad (7.3.20\text{a,b})$$

Implicit in these equations is the assumption that the variables p_f, T_f and p_p, T_p have well-defined values. If that is the case, the differentials for s_f and s_p can be obtained from them. Thus,

$$ds_f = c_{pf}\frac{dT_f}{T_f} - \frac{\beta_f}{\rho_f}dp_f \quad \text{and} \quad ds_p = c_{pp}\frac{dT_p}{T_p} - \frac{\beta_p}{\rho_p}dp_p \qquad (7.3.21\text{a,b})$$

where c_{pf} and c_{pp} are the constant-pressure specific heats of the fluid and particle materials, respectively. We may also express these differentials in terms of the
temperature and density. Thus,

$$ds_f = c_{vf}\frac{dT_f}{T_f} - \frac{\beta_f c_{Tf}^2}{\rho_f}d\rho_f \quad \text{and} \quad ds_p = c_{vp}\frac{dT_p}{T_p} - \frac{\beta_p c_{Tp}^2}{\rho_p}d\rho_p \qquad (7.3.22a,b)$$

where c_{vf} and c_{vp} are the corresponding constant-volume specific heats.

For future reference, we note that if we don't take the pressures and temperatures
to be equal, we can write (7.3.19b) as

$$(1 + \eta_m)ds = c_{pf}\frac{dT_f}{T_f} - \frac{\beta_f}{\rho_f}dp_f + \eta_m\left(c_{pp}\frac{dT_p}{T_p} - \frac{\beta_p}{\rho_p}dp_p\right)$$

where we have used $\phi_m = \eta_m/(1 + \eta_m)$. Of course, the expression is only meaningful
if the temperatures and pressure variations appearing on the right-hand side can be
defined. The expression is used in Chapter 9.

The expressions for ds_f and ds_p apply to the particles and to the fluid when the
two media are separately undergoing equilibrium changes. We are here interested
in those changes for which the suspension as a whole may be considered to be in
equilibrium. As discussed previously, this requires that the temperature and pressures
satisfy (7.2.6). Thus, substituting (7.3.21) into (7.3.19b), we obtain

$$ds = [(1 - \phi_m)c_{pf} + \phi_m c_{pp}]\frac{dT}{T} - \frac{\beta}{\rho}dp \qquad (7.3.23)$$

Similarly, the pair (7.3.22) yields

$$ds = [(1 - \phi_m)c_{vf} + \phi_m c_{vp}]\frac{dT}{T} - \left((1 - \phi_m)\frac{\beta_f c_{Tf}^2}{\rho_f}d\rho_f + \phi_m\frac{\beta_p c_{Tp}^2}{\rho_p}d\rho_p\right)$$

$$(7.3.24)$$

Specific Heat at Constant Pressure

This is defined by (7.3.12) and is easily obtained from the change in entropy given by
(7.3.23). Thus,

$$c_p = (1 - \phi_m)c_{pf} + \phi_m c_{pp} \qquad (7.3.25)$$

The constant pressure specific heat per unit mass of suspension is given by the sum of
the respective fluid and particle contributions. Also, use of this in (7.3.23) gives

$$Tds = c_p dT - \frac{\beta T}{\rho}dp$$

This is the first law of thermodynamics, with TdS being equal to the reversible heat
transfer and is in complete analogy with the simple substance case.

For future reference, we note that (7.3.25) can be also expressed as

$$c_p = c_{pf} \frac{1 + \eta_m c_{pp}/c_{pf}}{1 + \eta_m} \tag{7.3.26}$$

Specific Heat at Constant Volume

To obtain c_v, it would be desirable to express the second term in (7.3.24) in terms of $d\rho$ alone, rather than in terms of the fluid and particle density changes. This is, however, not generally possible. However, when either one of the two terms inside the large bracket in that equation is small, compared with the other, that is, when *either* $(\beta_f c_{Tf}^2/\rho_f)d\rho_f \ll (\beta_p c_{Tp}^2/\rho_p)d\rho_p$ or vice-versa, the second term on the right-hand side of (7.3.24) will be proportional to $d\rho$. In that case, the definition, (7.3.13) of c_v gives,

$$\bar{c}_v = (1 - \phi_m)c_{vf} + \phi_m c_{vp} \tag{7.3.27}$$

where the overbar on $\bar{c}_v$ is used as a reminder that the result refers to cases where one of the above inequalities hold. Thus, (7.3.27) applies to both gas bubbles in liquids and to liquid/solid particles in gases.

For other cases (e.g., emulsions), neither inequality holds. To obtain c_v for these cases, we proceed as follows. First, since ϕ_m is constant, we have $p = p(\rho, T)$, so that ds can be expressed as

$$ds = \left[c_p - \frac{\beta T}{\rho} \left(\frac{\partial p}{\partial T} \right)_\rho \right] \frac{dT}{T} - \frac{\beta}{\rho} \left(\frac{\partial p}{\partial \rho} \right)_T d\rho \tag{7.3.28}$$

The coefficient of dT/T is, by definition, c_v, but it is useful to write it in a different form. Thus, remembering that

$$\left(\frac{\partial p}{\partial T} \right)_\rho \left(\frac{\partial T}{\partial \rho} \right)_p \left(\frac{\partial \rho}{\partial p} \right)_T = -1, \tag{7.3.29}$$

the derivative inside the square bracket on the right-hand side of (7.3.28) may be expressed as

$$\left(\frac{\partial p}{\partial T} \right)_\rho = \rho \beta c_T^2(0) \tag{7.3.30}$$

where we used the definition of the coefficient of thermal expansion and of the isothermal sound speed. Hence,

$$T ds = \left[c_p - T\beta^2 c_T^2(0) \right] dT - \frac{\beta T c_T^2(0)}{\rho} d\rho \tag{7.3.31}$$

so that

$$c_p - c_v = T\beta^2 c_T^2 \tag{7.3.32}$$

This expression is identical to that which applies for single substances. As is the case for them, this expression shows that the suspension's specific heat at constant pressure is larger than that at constant volume. Using (7.3.32), we have

$$T ds = c_v dT - \frac{\beta T c_T^2(0)}{\rho} d\rho$$

also in complete analogy with the single substance case.

Specific Heat Ratio

Let us now consider the specific heat ratio, $\gamma = c_p/c_v$. To obtain it, we use the identity $c_p - c_v = c_p(\gamma - 1)/\gamma$ in (7.3.32) and solve for γ. Thus,

$$\gamma = \frac{1}{1 - T\beta^2 c_T^2(0)/c_p} \qquad (7.3.33)$$

Since the quantities β, c_p, and $c_T(0)$ are known, this is sufficient to calculate the specific heat ratio. The result applies to general suspensions, and may be simplified considerably for aerosols if the particles are rigid and to bubbly liquids if the liquid is nearly incompressible. We consider each case separately.

Aerosols. We consider the general expression for γ in the limit when $N_T \to 0$ (i.e., when the particle's compressibility is very small, compared with that of the external fluid). In that case, the particles may be considered rigid. In this limit, the constant volume specific heat is given by the special value (7.3.27), so that

$$\bar{\gamma} = \gamma_f \frac{1 + \phi_m(c_{pp}/c_{pf} - 1)}{1 + \phi_m(\gamma_f c_{pp}/c_{pf} - 1)} \qquad (7.3.34)$$

where we have used the symbol c_{pp} instead of c_{vp} – there being no difference between the two for rigid materials (see Section 2.4). Equation (7.3.34) can be expressed more succinctly in terms of the mass loading, η_m. Thus (Marble, 1963; Rudinger, 1965),

$$\frac{\bar{\gamma}}{\gamma_f} = \frac{1 + \eta_m c_{pp}/c_{pf}}{1 + \gamma_f \eta_m c_{pp}/c_{pf}}, \qquad (7.3.35)$$

Since $\gamma_f \geq 1$, it follows that $\bar{\gamma} \leq \gamma_f$, with the equality sign taking place when $\eta_m \to 0$. At the other end of the range, when η_m is very large, $\bar{\gamma} = 1$. In this limit, the heat capacity of the suspension is dominated by the particles.

Bubbly Liquids. Here, the compressibility of the liquid is small, compared with that of the gas in the bubbles, so that the special value (7.3.27) for c_v also applies. However,

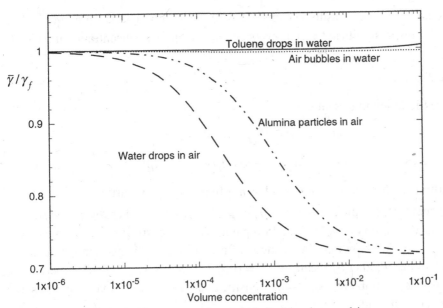

Figure 7.3.2. Value of $\overline{\gamma}/\gamma_f$ for two aerosols: a bubbly liquid and an emulsion.

since in bubbly liquids the density ratio, δ, is very large, it follows that for small values of the volume concentration, $\phi_m \approx \phi_v/\delta$ is always very small. We therefore have

$$\overline{\gamma} = \gamma_f \left[1 + \frac{\rho_p c_{pp}}{\rho_f c_{pf}} \frac{\gamma_f - 1}{\gamma_f} \phi_v \right] \qquad (7.3.36)$$

Furthermore, to a very good approximation, the second term inside the square bracket is very small, compared with the first, showing that the specific heat ratio in a dilute bubbly liquid differs little from that in the liquid alone. Since this itself is very nearly equal to 1, we see that $\overline{\gamma} \approx 1$. This means that, in very slow processes, both the entropy and the temperature remain nearly constant.

Emulsions. Here, the particle and fluid properties are comparable, so that (7.3.33) is needed to obtain γ.

Hydrosols. For solid particles in liquids, the second term in the denominator of that equation is normally small on account of the small compressibilities of both particles and fluid. Hence, we obtain, to leading order

$$\gamma \approx 1 + \frac{T\beta^2 c_T^2(0)}{c_p} \qquad (7.3.37)$$

However, the second term may be neglected, so that $\gamma = 1$.

Figure 7.3.2 displays the variations of γ/γ_f versus ϕ_v, for two aerosols: a bubbly liquid and an emulsion. We see that the ratio of specific heats for emulsions is also very nearly equal to 1, which was to be expected, given that both materials are liquids. For aerosols, however, significant variations occur as the volume concentration increases.

PROBLEM

1. Derive the following equations:

$$\rho de = c_v dT + \left[p/\rho - \beta T c_T^2(0)\right] d\rho,$$

$$\rho de = \rho c_p dT - \beta T dp + (p/\rho)d\rho$$

2. Determine c_p for a suspension that has n different types of particles.

3. Consider the isentropic change $p = p_0 + p'$, $T = T_0 + T'$, where the primed quantities have very small magnitudes, compared with the corresponding ambient values, p_0 and T_0. Show that the changes of pressure, density, and temperature are related to one another by means of $p' = (\partial p/\partial \rho)_s \rho' = (\rho_0 c_p/\beta T_0)T'$, where ρ_0 is the suspension ambient suspension density, and where all thermal properties are evaluated at p_0 and T_0.

7.4 Isentropic Sound Speed

In analogy with the isothermal sound speed defined earlier, we define an isentropic sound speed by means of

$$c_s^2(0) = (\partial p/\partial \rho)_s$$

Before evaluating the derivative, we note that the identification of $c_s(0)$, like that for $c_T(0)$, with a sound speed cannot be formally made until dynamic equations of motion for the suspension are introduced. This will be done in the next chapter, where it is shown that the above definition is obtained from a frequency-dependent sound speed by taking the frequency approach very small values. Because of this, we have denoted the derivative by $c_s^2(0)$.

To evaluate the derivative in (7.4.1), we first write it as

$$\left(\frac{\partial p}{\partial \rho}\right)_s = \frac{(\partial s/\partial \rho)_T}{(\partial s/\partial T)_\rho}\left(\frac{\partial p}{\partial \rho}\right)_T$$

The ratio of thermodynamic derivatives appearing on the right-hand side is, by definition, equal to the specific heat ratio. Hence, we can write

$$c_s^2(0) = \gamma c_T^2(0) \tag{7.4.1}$$

This relationship is identical to that applicable to any homogeneous material and can be directly derived from the definitions of the specific heats. This approach is simple because both quantities on the right-hand side of (7.4.1) are known. The result can be expressed in various forms; but, in order to assess the effects of the particles, we consider the ratio c_{sf}/c_s, where c_{sf} is the isentropic sound speed in the fluid alone. Since this is related to the isothermal sound speed c_{Tf} by means of $c_{sf}^2 = \gamma_f c_{Tf}^2$, we obtain, using (7.3.6) and (7.3.33) in (7.4.1)

$$\frac{c_{sf}^2}{c_s^2(0)} = \frac{c_{sf}^2}{c_T^2(0)} - (\gamma_f - 1)\left(\frac{\beta}{\beta_f}\right)^2 \left(\frac{c_{pf}}{c_p}\right) \tag{7.4.2}$$

Finally, using the explicit values of $c_T(0)$, β, and c_p, we obtain (Temkin, 1992)

$$\left(\frac{c_{sf}}{c_s(0)}\right)^2 = (1 - \phi_v + \phi_v \rho_p/\rho_f)$$
$$\times \left[\gamma_f(1 - \phi_v + \phi_v N_T) - (\gamma_f - 1)\frac{(1 - \phi_v + \phi_v \beta_p/\beta_f)^2}{(1 - \phi_v)(1 + \eta_m c_{pp}/c_{pf})}\right] \tag{7.4.3}$$

This result is discussed below; but, first, we give the corresponding result for the adiabatic compressibility, $K_s(0) = 1/\rho c_s^2(0)$. Since the front factor is equal to ρ/ρ_f, this gives

$$\frac{K_s(0)}{K_{sf}} = \gamma_f(1 - \phi_v + \phi_v N_T) - (\gamma_f - 1)\frac{(1 - \phi_v + \phi_v \beta_p/\beta_f)^2}{(1 - \phi_v)(1 + \eta_m c_{pp}/c_{pf})} \tag{7.4.4}$$

where $K_{sf} = 1/\rho_f c_{sf}^2$ is the isentropic compressibility of the fluid. Comparison of this to the corresponding isothermal result, (7.3.5), shows that, contrary to it, the adiabatic compressibility does not generally obey a simple distributive law.

Let us now return to the sound speed, given by (7.4.3). That expression was used to produce the solid line shown in Fig. 1.1.1 and will be useful when we discuss sound waves in the following chapters. Presently, we note that the expression indicates those properties that affect the equilibrium sound speed. These are more clearly distinguished by reducing the expression for suspensions that have very small volume concentrations. The reduction to such values has to be made carefully because the property ratios appearing in (7.4.3) can be large for some suspensions. Thus, for example, N_T is of the order of 10^4 for bubbly liquids. Hence, the product $\phi_v N_T$ can be finite, even if $\phi_v \ll 1$. Similarly, the density ratio, δ, is very small for aerosols. Therefore, the ratio ϕ_v/δ can also be finite for such small concentrations. On the other hand, the ratio of thermal expansion coefficients is, at most, of order 10. Hence, the product $\phi_v(\beta_p/\beta_f)$ is normally small when ϕ_v is small. Thus, we first approximate (7.4.3) by retaining $\phi_v N_T$ and ϕ_v/δ, and find

$$\frac{c_{sf}^2/c_s^2(0) - 1}{\phi_v} = \rho_p/\rho_f - 2 + \left(1 + \frac{\phi_v}{\delta}\right)\gamma_p N_s$$
$$+ \frac{(\gamma_f - 1)(1 + \phi_v/\delta)}{1 + (\phi_v/\delta)c_{pp}/c_{pf}}\left(\frac{\rho_p c_{pp}}{\rho_f c_{pf}} - 2\frac{\beta_p}{\beta_f}\right)$$

Table 7.4.1. *Property ratios in (7.4.4) for some suspensions at 1 atm and 15°C*

Suspension	ρ_f/ρ_p	$\rho_f c_{sf}^2/\rho_p c_{sp}^2$	$\rho_p c_{pp}/\rho_f c_{pf}$	β_p/β_f
Air bubbles in water	8.12×10^2	1.54×10^4	2.96×10^{-4}	23.2
Alumina particles in air	3.1×10^{-4}	1.0×10^{-7}	7.23×10^2	7.2×10^{-6}
Silica particles in water	0.45	5.95×10^{-2}	4.39×10^{-1}	2.10
Toluene drops in water	1.15	1.34	3.44×10^{-1}	6.9

where we have used $\gamma_p N_T = \gamma_f N_s$ and where

$$N_s = \rho_f c_{sf}^2/\rho_p c_{sp}^2 \tag{7.4.5}$$

is the ratio of the isentropic particle compressibility, $K_{sp} = 1/\rho_p c_{sp}^2$, to that of the fluid, and c_{sp} is the adiabatic sound speed for the particle material.

A further approximation applies for such small values of ϕ_v that ϕ_v/δ is also small, compared with 1. In aerosols, this usually requires very small volume concentrations. For these, we obtain

$$\frac{c_{sf}^2/c_s^2(0) - 1}{\phi_v} = \rho_p/\rho_f + \gamma_p N_s + (\gamma_f - 1)\left(\rho_p c_{pp}/\rho_f c_{pf} - 2\beta_p/\beta_f\right) - 2 \tag{7.4.6}$$

Let us examine (7.4.6), term by term. We see that the first term on the right-hand side is the contribution to the speed difference caused by the difference of densities; this is the only contribution that does not explicitly depend on thermal effects. The second term is due to the different compressibilities of the materials when no heat is added. The third term takes into account the different rates of expansion due to different heat capacities and thermal expansion coefficients. The first two contributions are positive, but the third can be negative for some fluid-particle combinations (e.g., bubbly liquids). In general, however, the right-hand side of (7.4.6) is positive so that, when ϕ_v is small, the speed of sound in a homogeneous suspension is smaller than in the fluid alone. The equation also shows that when the properties of the fluid and the particle have the same values, we have $c_s(0) = c_{sf}$, as expected. However, the various property ratios appearing in the equation generally differ from one. Furthermore, those ratios vary considerably from one suspension to another (as shown in Table 7.4.1). We may therefore anticipate different sound speeds applicable to each suspension type.

The previous equations for the sound speed and the compressibility apply to aerosols, bubbly liquids, and hydrosols, and show how the various mechanisms affect the speed and the compressibility in a suspension. Because of the difference in the values of the property ratios that different suspensions have, the general expressions reduce to simpler forms. We consider these separately.

Aerosols. Here, the compressibility of the suspension is not affected by the particles because of the very small value of the compressibility ratio, N_T, applicable to aerosols. However, the different thermal properties produced by the particles – notably the specific heats – can affect the speed significantly. Thus, consider solid or liquid particles in a gas. Here, we can set $\gamma_p = 1$, $\beta_p = 0$, and $N_T = 0$, so that without other approximations, (7.4.3) gives

$$\left(\frac{c_{sf}}{c_s(0)}\right)^2 = (1 - \phi_v)^2(1 + \eta_m)\frac{1 + \eta_m\gamma_f c_{pp}/c_{pf}}{1 + \eta_m c_{pp}/c_{pf}} \qquad (7.4.7)$$

This can be further simplified for dilute suspensions. Thus, for aerosols so dilute that both ϕ_v and η_m are very small, compared with unity, we obtain

$$\frac{c_{sf}^2}{c_s^2(0)} = 1 + \frac{\phi_v}{\delta}\left[1 + (\gamma_f - 1)c_{pp}/c_{pf}\right] \qquad (7.4.8)$$

Bubbly Liquids. The situation changes when the particles in the liquid are gas bubbles. Here, we can put $\gamma_f = 1$, but the particle-specific heat ratio is for most gases larger than one. Hence, γ_p has to be retained. Thus, (7.4.3) becomes

$$\frac{c_{sf}^2}{c_s^2(0)} = (1 - \phi_v + \phi_v/\delta)[1 + \phi_v(\gamma_p N_s - 1)] \qquad (7.4.9)$$

Because $N_s \gg 1$ for gas bubbles in liquids, the second term inside the second bracket can, even for small volume concentrations, be of the order of 1 or larger. Thus, when $\phi_v \ll 1$, the previous equation may be approximated by

$$c_{sf}^2/c_s^2(0) = 1 + \gamma_p\phi_v N_s \qquad (7.4.10)$$

showing that even a very low-volume concentration can produce a sound speed significantly smaller than that of the liquid, as was shown in Fig. 1.1.1.

Emulsions and Hydrosols. Here, when the volume concentration is small, the differences between $c_s(0)$ and c_{sf} are small, because N_s is of the order of 1 and $(\gamma_f - 1) \ll 1$. The differences increase at larger concentrations, depending on the materials involved.

Figure 7.4.1 shows the variations of $c_s(0)$ with volume concentration for two aerosols, a bubbly liquid, and an emulsion. As expected, the changes are far more drastic for bubbly liquids, but in fractional terms, those occurring in aerosols are also significant.

The above results are of considerable importance from the dynamical perspective, because the variations of sound speed in dynamic conditions are departures from the nearly static values given by (7.4.3). Although we do not yet know the magnitude of those departures, we expect, on the basis of this figure that, when ϕ_v is small, they will be large for bubbly liquids, moderate for aerosols, and small for emulsions and hydrosols.

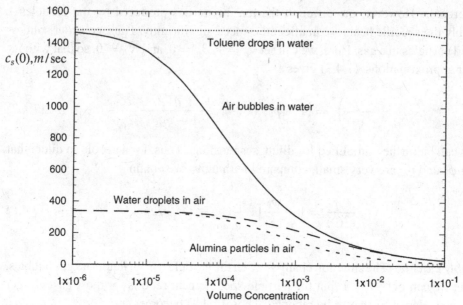

Figure 7.4.1. Value of $c_s(0)$ for two aerosols: a bubbly liquid and an emulsion.

PROBLEM

Derive (7.4.6).

7.5 Equations of State

It is also possible to obtain explicit equilibrium equations of state for some suspensions. Although of limited validity, these equations may be useful in some contexts. To obtain them, however, we require *explicit* equations of state for both fluid and particle materials, and these are not generally available. An important exception occurs in *dilute* aerosols, when the gas can be considered perfect. For this case, we note that the isentropic sound speed in an aerosol may be written as

$$c_s^2(0) = c_{sf}^2(\overline{\gamma}/\gamma_f)/(1 + \eta_m) \tag{7.5.1}$$

when use is made of (7.3.35). Furthermore, when the gas in the aerosol can be treated as perfect, the isentropic sound speed at temperature T is

$$c_{sf}^2 = \gamma_f \Re_f T \tag{7.5.2}$$

where $\Re_f$ is the gas constant. Hence, we may write the aerosol's isentropic sound speed as

$$c_s^2 = \overline{\gamma} \Re T \tag{7.5.3}$$

where

$$\Re = \frac{\Re_f}{1 + \eta_m} \tag{7.5.4}$$

is the aerosol's constant. But, since the gas in the aerosol is perfect, its equation of state is $p = \rho_f \Re_f T$. Thus, using $\Re_f = \Re(1 + \eta_m)$ and $\rho_f = \rho/(1 + \eta_m)$, we obtain (Marble, 1963)

$$p = \rho \Re T \qquad (7.5.5)$$

A dilute aerosol composed of nearly rigid particles in a perfect gas satisfies the perfect gas equation of state. Because of this, such an aerosol is also called a *dusty gas*.

Isentropic Equations of State

Although, as discussed previously, it is not possible to obtain equations of state for a general suspension, it is sometimes possible to obtain *isentropic* equations of state. These connect sequences of equilibrium states all having the same entropy. As the suspension passes from one such state to another, its total entropy, S, will remain constant. It is important to keep in mind that since the entropy of the suspension is the sum of the entropies of the fluid and particles, $S = S_f + S_p$, we would, of course have $dS = 0$. But, as pointed out previously, constancy of suspension entropy does not require that the entropies of the fluid and particles remain separately constant. Of course, if S_f and S_p are separately constant, we would, indeed, have $dS = 0$, but this would generally require the particles to be covered with adiabatic walls. Rather, the vanishing of dS for a general suspension only requires that the entropy increase of the fluid be balanced by a decrease of the particle's having the same magnitude, or $dS_f = -dS_p$. It is also clear that when this balance holds, heat transfer between particles and fluid may take place. But, if it does, it has to take place reversibly.

We now use (7.3.23) to derive an isentropic equation of state. Since s and ϕ_m are constant, the changes in temperature can be expressed as

$$dT = K_s \frac{\gamma - 1}{\beta} dp$$

where we have expressed $\beta/\rho c_p$ in terms of the specific heat ratio and the isentropic compressibility K_s. This equation applies to all three general types of suspensions. We consider only the following case:

Calorically Perfect Dusty Gases. Here, $\gamma = \bar{\gamma}$ and $\beta T = 1$, as may be shown from the dusty gas equation of state. Furthermore, in view of (7.5.2), (7.5.4), and (7.5.5), we have $\rho c_s^2 = \bar{\gamma} p$, so that

$$\frac{dT}{T} = \frac{(\bar{\gamma} - 1)}{\bar{\gamma}} dp \qquad (7.5.6)$$

Now, if the specific heats of the gas and particles are constant during the process, then $\bar{\gamma}$ is a constant for changes occurring in equilibrium, and we can integrate (7.4.6) to obtain

$$T p^{-(\bar{\gamma}-1)/\bar{\gamma}} = C \qquad (7.5.7)$$

where C is a constant. This is the familiar temperature-pressure relation for a perfect gas undergoing isentropic changes. Relations between p and ρ can be also obtained. Thus,

$$pp^{-\gamma} = C \tag{7.5.8}$$

where C is some other constant.

PROBLEMS

1. Show that the isentropic equation of state for a dilute bubbly liquid may be expressed as $Tp^{-(\gamma_f - 1)} = C$.
2. Show that, in a dusty gas, the isentropic pressure changes are also isothermal when the mass loading is large. Plot an isentrope for a dusty gas and compare it with the perfect gas result.

7.6 Concluding Remarks

This chapter has looked at suspensions in equilibrium. Although these are rarely found, the properties obtained for them provide a solid foundation on which dynamic theories can be constructed and to which all theories must reduce for very slow motions.

8

The Two-Phase Model

8.1 Introduction

Having determined some properties of suspensions in equilibrium, we turn our attention to suspensions that are subject to external forces and seek ways of describing their response. It is evident that the approach taken to study the response of a single particle to applied forces is not useful in suspensions. That approach would require the solution of the fluid equations of motion subject to the usual boundary conditions on the surface of each of the particles. The level of difficulty may be appreciated from the fact that the fluid force acting on either one of the particles in a freely moving pair is known only approximately.

It is also evident that a suspension in motion cannot be regarded as a homogeneous medium, as we did in the last chapter, where we took advantage of the fact that, in a homogeneous suspension in equilibrium, the particles have the same pressures, temperatures and velocities as the fluid around them. As Chapters 3–6 have shown, this enormous simplification seldom occurs. Thus, the equations that apply to homogeneous fluids in motion are not generally applicable to suspensions.

A method that has proven successful in the description of suspension motions is based on the idea that the particles, being large in number, may be described as a continuous medium, obeying its own set of conservation equations. In this context, we recall that we regard fluids as continuous media even though they are composed of molecules and molecular voids. This is possible because of the very large number of molecules that volume elements of microscopic dimensions normally have, and because the length scales in most situations of interest are considerably larger than those dimensions. It is thus possible to ascribe physical meaning to the continuum idea of a point as being a volume whose lateral dimensions are very small in comparison with all distances of interest, but that nevertheless contains a very large number of molecules.

Although very significant differences exist between the molecules in a fluid and the small particles in a suspension, the two cases are somewhat similar, at least for suspensions that contain large numbers of particles. One of the differences is that the space between particles is not a void, but it is occupied by a fluid. Another is that a "point" in the suspension must be considerably larger than a "point" in a fluid, because it must contain a sufficiently large number of particles so that such concepts as the particle density, pressure, and velocity at a "point" have a meaning. This is more stringent in dilute suspensions, where the distance between neighboring particles is large.

In what follows, we shall assume that owing to the large number of particles in small volume elements of many suspensions, average particle quantities can be defined in such a manner that we may regard the variables associated with the particles in those elements as being distributed continuously, rather than discretely as they in fact are. The precise specification of the conditions required to represent the particles in a suspension as a continuum phase is a delicate matter and will not be discussed here. The interested reader is referred to the research literature listed in the bibliography section, where the issue is discussed at length. However, the following remarks should be kept in mind. First, the size of the particles must be large, compared with all molecular dimensions in the external fluid, including mean molecular separations. This ensures that the fluid can be regarded as a continuous medium. Second, the elemental suspension volume must be sufficiently small so as to be regarded as a point from the macroscopic point of view, and yet it must contain such a large number of particles, that separate measurements of their number in the same macroscopic conditions will not detect significant departures from some value. Third, we require that all length scales of interest in a given problem be much larger than the lateral dimension of the elementary suspension element. For example, in the study of sound waves in suspensions, we require that the wavelength be large, compared with the lateral dimension. Fourth, it is assumed that, in force-free suspensions, particle collisions do not occur. Particle collisions usually result in a change of the number of particles in a suspension, making the analysis considerably more involved, as it will be demonstrated in a later chapter. Finally, we assume that the particles and the fluid in a suspension occupy distinct spaces – that is, the particles are assumed to have sizes that are much larger than the molecular voids in the fluid, as already assumed, and are impermeable to fluid molecules.

Seen in this manner, a suspension of nonevaporating particles may be regarded as a medium composed of two continuous phases: the fluid phase and the particulates phase. Within a suspension element, each phase is associated with a velocity, a pressure, and so on that represent average values of those quantities over the element. The values of these variables in one phase are, however, generally different from those in the other. For example, when external forces are applied to a suspension, the velocity acquired by one phase in a volume element will be different from that in the other. Stated differently, there will generally exist a relative, or *slip*, velocity between the two phases. The same is true of the pressures and the temperatures. But, although the pressure, velocity, and temperature in each phase obey their own system of equations, the two systems are coupled owing to the force and heat transfer between particles and fluid. Because of this coupling, the two-phase model is sometimes referred to as the *coupled* model.

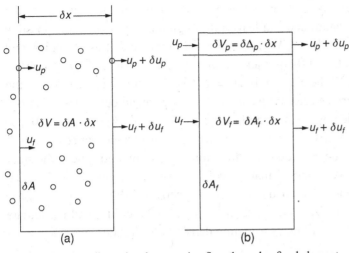

Figure 8.2.1. One-dimensional suspension flow through a fixed element.

But, aside from the name used, it is evident that the particle force and the heat transfer rate are of central importance in the model. One obvious reason for this is that those quantities are the driving equilibrating mechanisms, and that they start acting the moment velocity and/or temperature slips occurs between fluid and particles. Thus, we anticipate them to be active at nearly all instants in suspensions in motion. In the derivation below, we shall assume that both particle force and heat transfer rate are *known* functions of position and time. This is generally not true, except for dilute suspensions, where they are given by the single particle results obtained in earlier chapters.

In the following sections, we obtain the equations of motion for the fluid and particulate phases in dilute suspensions. These are based on the general conservation principles of mass, momentum, and energy. These are applied to each phase in the suspension using several additional simplifications. Most important among these is the assumption that viscosity and heat conductivity are unimportant, except at the surface of the particles. The assumption is based on the fact that those transport mechanisms are far more important for a fluid in contact with an external surface, such as a particle, than for an unbounded fluid. In addition, we shall also disregard surface tension effects, as well as particle rotations.

8.2 Conservation Equations

The derivation of the equations of motion for the fluid and particulate phase follows closely the derivation of the equations of motion for a single phase fluid. However, the presence of a second phase makes the derivation slightly more involved, and it is useful to first consider a one-dimensional motion through a *fixed* volume element, which we take to be of rectangular shape, as sketched in Fig. 8.2.1a,b. The motion is assumed to be along the x-axis of a coordinate system. Fluid and particles enter the element through the left-hand face and leave it through the right hand face, possibly having experienced some changes. We are interested in the velocities, densities, and concentrations of the fluid and the particulate phases. The densities may be defined as follows. If the volume of the element is $\delta V = \delta A \, \delta x$ – where δx is the width of the element and δA is the surface area of the element's face perpendicular to the x-axis – then the volumes occupied by the fluid and by the particles are $\delta V_f = (1 - \phi_v)\delta A \delta x$ and $\delta V_p = \phi_v \delta A \delta x$, respectively. Here, ϕ_v is the particle volume concentration, a quantity that should now be regarded as a function of position and time. In writing these expressions, we have assumed that no volume overlap occurs. Therefore, the masses of fluid and particles contained in δV are, respectively, $\delta M_f = \rho_f(1 - \phi_v)\delta V$ and $\delta M_p = \rho_p \phi_v \delta V$, where ρ_f and ρ_p are the densities of the fluid and particulate materials. These expressions allow us to define densities of the fluid and particulate *phases* by means of

$$\sigma_f(x, t) = \rho_f(1 - \phi_v) \qquad \sigma_p(x, t) = \rho_p \phi_v \qquad \text{(8.2.1a,b)}$$

The sum of these phase densities can also be used to define the density of the suspension or

$$\sigma = \sigma_f + \sigma_p \qquad \text{(8.2.2)}$$

It should be noted that σ is equal to the suspension density ρ introduced in Chapter 1. That density only required that the suspension be homogeneous as a whole, whereas σ is assumed to be a continuous function of position. Of course, when each elemental volume in a suspension is homogeneous, the two definitions are equal.

The fact that the volumes of the phases do not overlap permits us to visualize the suspension element differently, in a manner that may be helpful in the derivation of the equations of motion. For this purpose, we reconsider the above volume element, this time *imagining* that the particulate and fluid phases are completely separated – each occupying an imaginary volume equal in magnitude to the volumes occupied in the actual suspension element and containing the same instantaneous masses. The division is sketched in Fig. 8.2.1b. From the condition that the volumes of each phase be instantaneously the same as before, we find that the areas δA_p and δA_f are given by $\delta A_f = (1 - \phi_v)\delta A$ and $\delta A_p = \phi_v \delta A$, respectively, with $\phi_v = \phi_v(x, t)$.

A similar subdivision can conceptually be made for more general volumes. Because the dividing surface is, generally, in motion, it is sometimes useful to consider material volume elements (i.e., volumes that contain the same amount of matter). To avoid confusion, we shall denote the corresponding volumes as $\delta \tau$ for the suspension, $\delta \tau_f = (1 - \phi_v)\delta \tau$ for the volume of fluid in $\delta \tau$, and $\delta \tau_p = \phi_v \delta \tau$ for the corresponding volume of the particulate phase.

Conservation of Mass

Let us now obtain equations expressing the conservation of mass for the fluid and particulate phases. To fix ideas, we start with the particulate phase, using the separate-phase volume element of Fig. 8.2.1b, assuming that all motions are strictly one-dimensional. Thus, aligning the x-axis of a system of coordinates with the direction of the motion, we see that the particulate phase velocity is $\mathbf{u}_p = \{u_p(x, t), 0, 0\}$. Now, since the particulate phase volume, δV_p, has, by definition, only particulate matter whose density is ρ_p, we see that the particulate mass inside it is $\int_{\delta V_p} \rho_p dV$. Taking δV_p to be small, this integral is approximately given by $\rho_p \, \delta A_p \delta x$, or using $\delta A_p = \phi_v \delta A$, by $\sigma_p(x, t)\delta V$. This mass may change in time, and the rate of change is given by $(\partial \sigma_p/\partial t)\delta V$. Now, the only manner this can occur is if, as a result of the motion, a net amount of particulate mass enters or leaves the volume. Since the mass flow rate entering the element is $\rho_p u_p \delta A_p$, the net mass flow rate out is $\delta x \partial(\rho_p u_p \delta A_p)/\partial x$. Equating this to $(\partial \sigma_p/\partial t)\delta V$ and using $\delta A_p = \phi_v \delta A$ and $\sigma_p = \rho_p \phi_v$, we obtain the one-dimensional continuity equation for the particulate phase:

$$\frac{\partial \sigma_p}{\partial t} + \frac{\partial}{\partial x}(\sigma_p u_p) = 0 \tag{8.2.3}$$

The fluid's conservation mass equation can be derived in a similar fashion, with the result being

$$\frac{\partial \sigma_f}{\partial t} + \frac{\partial}{\partial x}(\sigma_f u_f) = 0 \tag{8.2.4}$$

Generalization to three-dimensional motions is simple, but it is useful to obtain the three-dimensional forms of these equations from the Reynolds transport theorem. This will enable us to obtain the remaining conservation equations in a simple manner. The theorem states that if $g(\mathbf{x}, t)$ is a continuous, scalar function of position and time – defined in a region $\tau(t)$, bounded by a closed area where the velocity component along the outward normal is $\mathbf{u} \cdot \mathbf{n}$ – the time derivative of $\int_{\tau(t)} g(\mathbf{x}, t)$ may be expressed as

$$\frac{d}{dt} \int_{\tau(t)} g(\mathbf{x}, t)d\tau = \int_{\tau(t)} \left[\frac{\partial g}{\partial t} + \nabla \cdot (g\mathbf{u})\right]d\tau \tag{8.2.5}$$

Since the volume integral of the second term may be expressed as $\int_{A(t)} g\mathbf{u} \cdot \mathbf{n}dA$, the theorem states that the rate of change of the integral on the left-hand side is equal to the integral of the local rate of change of g plus the flux of g out of the region.

We may apply this theorem to any physical quantity we wish. For example, suppose $g = \sigma_f(\mathbf{x}, t)$ and $\tau(t) = \delta \tau(t)$, then $\sigma_f \delta \tau$ represents the total amount of fluid in $\delta \tau$. Since, by definition, $\delta \tau(t)$ contains the same mass of fluid mass as it moves, we see that the integral on the left-hand side of (8.2.5) vanishes. It then follows that the integral on the right-hand side must vanish for all choices of $\delta\tau(t)$. Hence, putting $\mathbf{u} = \mathbf{u_f}$, we obtain

$$\frac{\partial \sigma_f}{\partial t} + \nabla \cdot (\sigma_f \mathbf{u}_f) = 0 \tag{8.2.6}$$

Similarly, we obtain for the particulate phase,

$$\frac{\partial \sigma_p}{\partial t} + \nabla \cdot (\sigma_p \mathbf{u}_p) = 0 \tag{8.2.7}$$

As we see, these are the usual continuity equations for fluids of densities σ_f and σ_p, respectively. It is also possible to express these equations using the substantial derivative D/Dt. However, because the phases move with different velocities, it is important to distinguish the two cases. We therefore write,

$$\frac{D_f}{Dt} = \frac{\partial}{\partial t} + \mathbf{u}_f \cdot \nabla \quad \text{and} \quad \frac{D_p}{Dt} = \frac{\partial}{\partial t} + \mathbf{u}_p \cdot \nabla \tag{8.2.8a,b}$$

where the subscripts in the substantial derivative indicate whether it gives the time derivative following the fluid or the particulate phase. With these definitions, we can write the continuity equations as

$$\frac{D_f \sigma_f}{Dt} = -\sigma_f \nabla \cdot \mathbf{u}_f \quad \text{and} \quad \frac{D_p \sigma_p}{Dt} = -\sigma_p \nabla \cdot \mathbf{u}_p \tag{8.2.9a,b}$$

Linear Momentum

Let us first consider the fluid phase. The relevant conservation theorem here is Newton's second law, which states that the rate of change of momentum of a material object is equal to the forces applied to it. Thus, we need to compute the rate of change of the momentum of a given mass of fluid, following its motion in space, as well as the forces that act on it. Since the fluid momentum per unit volume is $\sigma_f \mathbf{u}_f$, and the fluid occupies a volume $\delta\tau_f$ in the element, we have

$$\frac{d}{dt} \int_{\delta\tau_f} \sigma_f \mathbf{u}_f d\tau = \mathbf{f}_T \tag{8.2.10}$$

where $\mathbf{f}_T$ is the total force acting on the fluid in the element. Let us consider first the left-hand side of this equation. Using the Reynolds transport theorem and the equation of continuity for the fluid, it may be written as

$$\frac{d}{dt} \int_{\delta\tau_f} \sigma_f \mathbf{u}_f d\tau = \int_{\delta\tau_f} \sigma_f \frac{D_f \mathbf{u}_f}{Dt} d\tau \approx \sigma_f \frac{D_f \mathbf{u}_f}{Dt} \delta\tau \tag{8.2.11}$$

Consider now the total force, $\mathbf{f}_T$. As is customary, we may divide it into body and surface forces. Since the fluid is considered inviscid, the only surface force that is applied on the fluid in the element is the normal component of the stress tensor (i.e., the pressure). With reference to Fig. 8.2.2, we may regard the fluid in the element

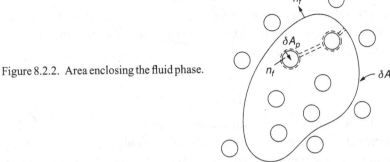

Figure 8.2.2. Area enclosing the fluid phase.

as being enclosed by the external surface δA, entirely drawn in the fluid, an internal surface area δA_p separating fluid from particles, and a number of two-way segments in the fluid. For simplicity, only two such segments are shown. Then, the surface area totally enclosing the fluid is $\delta A_f = \delta A + \delta A_p$. On δA, the external pressure acting on the fluid is p_f, whereas that acting on δA_p is p_p. Also, the integrals over the segments cancel out. Therefore, the surface force acting on the fluid is

$$- \int_{\delta A} p_f \mathbf{n}_f dA - \int_{\delta A_p} p_p \mathbf{n}_f dA$$

But, on δA_p, the boundary condition on the normal component of the pressure gives $p_p = p_f$. Thus, the two surface integrals combine to give $- \int_{\delta A_f} p_f \mathbf{n}_f dA$. We now change this surface integral into a integral over the volume occupied by the fluid in the element and obtain for the surface work

$$- \int_{\delta A_f} p_f \mathbf{n}_f dA = - \int_{\delta \tau_f} \nabla p_f d\tau$$
$$\approx -(1 - \phi_v) \nabla p_f \, \delta \tau \qquad (8.2.12)$$

Consider now the body force. We assume that it applies uniformly on the suspension volume; but, that, like gravity, its total effect on a given body is proportional to the mass of the body. Thus, if $\mathbf{f}_B(\mathbf{x}, t)$ is the body force per unit mass, acting on *both* fluid and particles, the total body force on the fluid will be

$$\int_{\delta \tau_f} \sigma_f \mathbf{f}_B(\mathbf{x}, t) d\tau \approx \sigma_f \mathbf{f}_B(\mathbf{x}, t) \, \delta \tau \qquad (8.2.13)$$

Next is the force exerted by the particles on the fluid. In a suspension, this force acts at discrete locations, namely the instantaneous locations of each particle; but, in the two-phase model, we must think of it as being distributed continuously within the suspension. Thus, if $\mathbf{f}_p(\mathbf{x}, t)$ is the particulate force acting on the fluid at $\mathbf{x}$ and t, per unit suspension volume, the total force in our element of volume is

$$\int_{\delta \tau} \mathbf{f}_p(\mathbf{x}, t) d\tau \approx \mathbf{f}_p(\mathbf{x}, t) \, \delta \tau \qquad (8.2.14)$$

We shall return to this force at a later time. For now, it is sufficient to assume that the actual particle force can be represented in this manner. Collecting the individual forces and substituting them in Newton's second law, we obtain the momentum equation for the fluid in the suspension, namely

$$\sigma_f \frac{D_f \mathbf{u}_f}{Dt} + (1 - \phi_v)\nabla p_f = \mathbf{f}_p + \sigma_f \mathbf{f}_B \tag{8.2.15}$$

Let us now write the momentum equation for the particulate phase. The derivation is entirely similar and need not be shown. We only note that the force on the particulate phase is due to a pressure term plus the distributed force with which the fluid acts on the particulate phase. The latter force is simply $-\mathbf{f}_p$. Thus,

$$\sigma_p \frac{D_p \mathbf{u}_p}{Dt} + \phi_v \nabla p_p = -\mathbf{f}_p + \sigma_p \mathbf{f}_B \tag{8.2.16}$$

The appearance in this equation of a particulate pressure requires further explanation. In fluids, the origin of pressure is molecular collisions. The "molecules" in our particulate phase are the particles, and these, by assumption, do not collide with one another. On the other hand, all materials of which particles are made are compressible to some degree. Thus, generally speaking, the particles pulsate in response to the fluid. The pulsation is accompanied, as we saw in detail in Chapter 6, by a fluctuation of the particle pressure. It is this pressure, assumed to be distributed in the particulate phase, that appears in (8.2.16). The matter is considered later.

Internal Energy

The last conservation principle we need is that for the energy. This includes both internal and kinetic energies; but, in order to take advantage of known thermodynamic principles, our derivation will focus on the internal energy. As before, we derive first the energy equation for the fluid phase. Our goal is to obtain an equation relating the temperature, pressure, and velocity distributions in the fluid phase to the exchanges of energy that occur between the two phases. The relation is provided by the first law of thermodynamics, which states that the rate of change of the internal energy of a system is equal to the rate at which heat is transferred to, and work is done on it. To apply the law, we note the following. First, a fraction of the net work done on a moving element increases the kinetic energy of the element and has no effect on the internal energy. That part of the work should therefore not appear in the equation for the internal energy. Second, since the element is not at rest, it is not normally in thermodynamic equilibrium. Hence, the equilibrium thermodynamics definition of the internal energy is not applicable. In what follows, the symbol e_f will represent the internal energy per unit mass of fluid, whether the fluid is at rest or not. In case the fluid is in motion, the value of e_f is that obtained if the element were isolated and allowed to come to rest. A similar meaning applies to e_p.

With these preliminaries in mind, consider the internal energy of the fluid in a suspension element having a volume $\delta\tau$. The total amount of that energy is $\int_{\delta\tau} \sigma_f e_f d\tau$, so that its rate of change becomes, on using the equation of continuity,

$$\frac{d}{dt}\int_{\delta\tau} \sigma_f e_f d\tau = \int_{\delta\tau} \sigma_f \frac{D_f e_f}{Dt} d\tau \qquad (8.2.17)$$

$$\approx \sigma_f \frac{D_f e_f}{Dt} \delta\tau$$

Let us now consider the energy addition rate. This is best done by calculating the separate components of that rate, according to their origin. First is the rate at which the particles do work on the fluid as they expand: $p_f(D_f\phi_v/Dt)\delta\tau$. Next is the rate at which the body force does work on the fluid inside the volume element. This is

$$\int_{\delta\tau} \sigma_f \mathbf{f}_B \cdot \mathbf{u}_f d\tau \approx \sigma_f \mathbf{f}_B \cdot \mathbf{u}_f \delta\tau \qquad (8.2.18)$$

Next is the rate at which the surface forces do compressive work. Since the surface force on δA_f is, as explained previously, the fluid pressure p_f, we have,

$$-\int_{\delta A_f} p_f \mathbf{u}_f \cdot \mathbf{n}_f dA = -\int_{\delta\tau_f} \nabla \cdot (p_f \mathbf{u}_f) d\tau \qquad (8.2.19)$$

$$\approx -(1 - \phi_v)\nabla \cdot (p_f \mathbf{u}_f)\delta\tau$$

because $\delta\tau_f = (1 - \phi_v)\delta\tau$. We also have the work done by the particulate force in displacing the fluid: $\mathbf{f}_p \cdot \mathbf{u}_f \delta\tau$. Adding all contributions, we obtain

$$[\sigma_f \mathbf{f}_B \cdot \mathbf{u}_f - (1 - \phi_v)\mathbf{u}_f \cdot \nabla p_f - (1 - \phi_v)p_f \nabla \cdot \mathbf{u}_f + \mathbf{f}_p \cdot \mathbf{u}_f + p_f(D_f\phi_v/Dt)]\delta\tau$$

On the other hand, by the momentum equation, (8.2.15), we see that the rate at which the kinetic energy of the fluid in the element increases by the action of the forces acting on it is

$$\sigma_f \mathbf{u}_f \cdot \frac{D_f \mathbf{u}_f}{Dt}\delta\tau = [\mathbf{f}_p \cdot \mathbf{u}_f + \sigma_f \mathbf{f}_B \cdot \mathbf{u}_f - (1 - \phi_v)\mathbf{u}_f \cdot \nabla p_f]\delta\tau$$

$$(8.2.20)$$

Subtracting this from the total work rate, we obtain that part which affects the internal energy of the element, namely

$$-(1 - \phi_v)p_f \nabla \cdot \mathbf{u}_f \delta\tau$$

There remain two other components. One is the energy dissipated in the fluid that is due to the translational motion of the particles. This is $\mathbf{f}_p \cdot (\mathbf{u}_p - \mathbf{u}_f)\delta\tau$. The other is the heat addition rate from the particles to the fluid. If we let $\dot{q}_p$ be the distributed particle heat transfer rate to the fluid, per unit volume of suspension, then the total heat transfer rate to the fluid in the element becomes

$$\int_{\delta\tau} \dot{q}_p(\mathbf{x}, t)d\tau \approx \dot{q}_p(\mathbf{x}, t)\delta\tau \qquad (8.2.21)$$

Collecting the energy addition terms, and rearranging, we obtain

$$\{\dot{q}_p(\mathbf{x}, t) - (1 - \phi_v)p_f \nabla \cdot \mathbf{u}_f + \mathbf{f}_p \cdot (\mathbf{u}_p - \mathbf{u}_f) + p_f(D_f\phi_v/Dt)\}\delta\tau$$

Equating this energy input rate to the rate of change of the internal energy yields the desired equation

$$\rho_f(1 - \phi_v)\frac{D_f e_f}{Dt} = \dot{q}_p + \mathbf{f}_p \cdot (\mathbf{u}_p - \mathbf{u}_f) - (1 - \phi_v)\frac{p_f}{\rho_f}\frac{D_f\rho_f}{Dt} \qquad (8.2.22)$$

where we have used (8.2.9a). This is, formally, the equation that expresses the conservation of internal energy for the fluid phase in a suspension. The first and third terms on the right-hand side require no further explanation. The middle term, $\mathbf{f}_p \cdot (\mathbf{u}_p - \mathbf{u}_f)$, represents the frictional heating that the relative motion produces. As we show later, this term is always positive.

Let us now express the derivative of internal energy in terms of measurable quantities. Of the several forms available, we chose (see the second equation in Problem 1, Section 7.3)

$$\rho_f\frac{D_f e_f}{Dt} = \rho_f c_{pf}\frac{D_f T_f}{Dt} - \beta_f T_f\frac{D_f p_f}{Dt} + \frac{p_f}{\rho_f}\frac{D_f\rho_f}{Dt} \qquad (8.2.23)$$

Thus, substitution in (8.2.22) yields the final result

$$(1 - \phi_v)\left(\rho_f c_{pf}\frac{D_f T_f}{Dt} - \beta_f T_f\frac{D_f p_f}{Dt}\right) = \dot{q}_p + \mathbf{f}_p \cdot (\mathbf{u}_p - \mathbf{u}_f) \qquad (8.2.24)$$

The right-hand side terms show the presence of the particulate phase, without which this reduces to the energy equation for a fluid undergoing isentropic changes.

The energy equation for the particulate phase may be derived in the same manner. It is

$$\phi_v\left(\rho_p c_{pp}\frac{D_p T_p}{Dt} - \beta_p T_p\frac{D_p p_p}{Dt}\right) = -\dot{q}_p \qquad (8.2.25)$$

where the heat addition rate term on the right-hand side is equal, but of opposite sign to that for the fluid.

Equations of State

The system of equations obtained above requires additional information to be complete. In particular, they require equations of state for each phases. Explicit analytical equations of state are available only in a very limited number of instances (e.g., perfect gases), but information for other situations exist in tabular or graphical forms. For our purposes, however, it is sufficient to assume that they are available for simple materials, and that the fluid and particle materials are "simple" in the sense that only two variables are required to determine the thermodynamic state of either phase. Which two variables depends on the situation; but, for now, we take the temperature and the density of either phase and write those equations as

$$p_f = p_f(\rho_f, T_f) \qquad p_p = p_p(\rho_p, T_p) \qquad (8.2.26a,b)$$

Particulate Pressure

As indicated previously, the particulate pressure appearing in the basic equations has its origin in the compressibility of the particles. If we disregard this compressibility and take the particles to be rigid – a step that is warranted for solid particles in gases, for example – then the particulate pressure vanishes. But, in general, particle compressibility effects exist that result in fluctuations of the pressure in a particle. In the two-phase model presented above, the particulate phase variables are regarded as continuous functions of position. But what values are we to assign to those variables, or, stated differently, how are the continuous variables related to those that apply to individual, finite-sized particles? A simple, though far from rigorous approach, is to assign to the continuous variables the values that an actual particle would have at that point in the same conditions. For the particulate pressure, then, we need the pressure in a compressible particle.

In Chapter 6, we calculated the average pressure in a particle immersed in a fluid whose pressure, far from the particle, is changing. Two calculations were in fact presented, one retaining nonlinear effects, but taking the external fluid to be incompressible. The second result applies to linear motions and retains external fluid compressibility. The first result, based on the Rayleigh-Plesset equation, is preferable in the present context because it permits the inclusion of several dissipative mechanisms through the damping coefficient. We will thus adopt it so far as the particulate pressure is concerned and emphasize that its use here does not make the external fluid incompressible. Thus, using (6.2.12) with $p_p \rightarrow p_p(\mathbf{x}, t)$, $p_{f\infty} \rightarrow p_f(\mathbf{x}, t)$, and $R \rightarrow R(\mathbf{x}, t)$, and disregarding surface tension, which we have not included in the model, we obtain

$$p_p = p_f + \rho_f R \frac{\partial^2 R}{\partial t^2} + \rho_f \left(\frac{4 v_f}{R} + \frac{3}{2} \frac{\partial R}{\partial t} \right) \frac{\partial R}{\partial t} \qquad (8.2.27)$$

It should be noted that $R(\mathbf{x}, t)$ generally depends on position, even if all the particles have the same equilibrium size, because their displacement varies from point to point in the suspension. Also, the appearance of the fluid viscosity here is consistent with the model, because the model assumes that viscosity effects are nonzero only at the surface of the particle, where (8.2.27) applies (see Section 6.2.1).

8.3 System of Equations

For future reference, we list below the two sets of equations that encompass the two-phase model. Within the limits stated earlier, the equations apply to the four basic types of suspensions considered in this book, namely aerosols, bubbly liquids, emulsions, and hydrosols. Among the limitations, we should repeat that the effects of viscosity and heat conductivity are taken into account only at the surface dividing particles from surrounding fluid. Thus, if other surfaces are present, such as the walls of a tube, such effects cannot be taken into account by these equations. To include these effects it would be necessary to add viscous stresses and heat conduction in the fluid's set. While this is not difficult for dilute suspensions, it would considerably increase the complexity of the system, which is already quite complex. We will therefore refer the interested reader to the bibliography, where some works dealing with this issue are listed.

Continuity

$$\frac{D_f \sigma_f}{Dt} = -\sigma_f \nabla \cdot \mathbf{u}_f \qquad \frac{D_p \sigma_p}{Dt} = -\sigma_p \nabla \cdot \mathbf{u}_p \qquad (8.3.1a,b)$$

Momentum

$$\sigma_f \left(\frac{D_f \mathbf{u}_f}{Dt} + \frac{1}{\rho_f} \nabla p_f \right) = \mathbf{f}_p + \sigma_f \mathbf{f}_B$$

$$\sigma_p \left(\frac{D_p \mathbf{u}_p}{Dt} + \frac{1}{\rho_p} \nabla p_p \right) = -\mathbf{f}_p + \sigma_p \mathbf{f}_B \qquad (8.3.2a,b)$$

Energy

$$\sigma_f \left(c_{pf} \frac{D_f T_f}{Dt} - \frac{\beta_f T_f}{\rho_f} \frac{D_f p_f}{Dt} \right) = \dot{q}_p + \mathbf{f}_p \cdot (\mathbf{u}_p - \mathbf{u}_f)$$

$$\sigma_p \left(c_{pp} \frac{D_p T_p}{Dt} - \frac{\beta_p T_p}{\rho_p} \frac{D_p p_p}{Dt} \right) = -\dot{q}_p \qquad (8.3.3a,b)$$

State

$$p_f = p_f(\rho_f, T_f) \qquad p_p = p_p(\rho_p, T_p) \qquad (8.3.4a,b)$$

Complementing the system, we have the definitions of the phase densities and the particulate pressure

$$\sigma_f = \rho_f(1 - \phi_v) \qquad \sigma_p = \rho_p \phi_v \qquad (8.3.5a,b)$$

$$p_p = p_f + \rho_f R \frac{\partial^2 R}{\partial t^2} + \rho_f \left(\frac{4\nu_f}{R} + \frac{3}{2} \frac{\partial R}{\partial t} \right) \frac{\partial R}{\partial t} \qquad (8.3.6)$$

As listed, the system consists of 11 equations, for 11 variables, viz: σ_f, σ_p, $\mathbf{u}_f$, $\mathbf{u}_p$, p_f, p_p, T_f, T_p, ρ_f, ρ_p and ϕ_v.

Also, for future reference, we add here some remarks concerning the volume concentration and the mass loading, which, like the volume concentration ϕ_v, is generally variable. Consider the volume concentration first. From the continuity equations we obtain

$$\frac{D_f \phi_v}{Dt} = (1 - \phi_v) \left[\frac{1}{\rho_f} \frac{D_f \rho_f}{Dt} + \nabla \cdot \mathbf{u}_f \right] \qquad (8.3.7)$$

$$\frac{D_p \phi_v}{Dt} = -\phi_v \left[\frac{1}{\rho_p} \frac{D_p \rho_p}{Dt} + \nabla \cdot \mathbf{u}_p \right] \qquad (8.3.8)$$

These are generally different because of the difference between the velocities. Now, it is seen that the quantities inside the square brackets are, for a one-phase system identically zero. But, for a two-phase medium, this is not the case because of the variations of the volume fraction. A simple example might be useful. Suppose we write the volume concentration in terms of the number of particles per unit volume, n, and of their individual volume, v_p, assuming them all to be equal. Then $\phi_v = n v_p$, so that

$$\frac{D_p \phi_v}{Dt} = n \frac{D_p v_p}{Dt} + v_p \frac{D_p n}{Dt} \qquad (8.3.9)$$

The last term is zero because the number of particles is necessarily constant when we follow the particle's motion, Thus, the changes in the volume concentration are due to changes in the individual particle volumes, a result that was to be expected.

Now consider a fixed surface element $\mathbf{n}\delta A$, across which some flow takes place. The mass flow rates of fluid and particles crossing the element are $\rho_f(1 - \phi_v)\mathbf{u}_f \cdot \mathbf{n}\delta A$ and $\rho_p \phi_v \mathbf{u}_p \cdot \mathbf{n}\delta A$, respectively. We now define the mass loading ratio as being equal to the ratio of particle mass flow rate to fluid mass flow rate. Thus,

$$\eta_m = \frac{\dot{m}_p}{\dot{m}_f} = \frac{\rho_p \phi_v}{\rho_f(1 - \phi_v)} \frac{\mathbf{u}_p \cdot \mathbf{n}}{\mathbf{u}_f \cdot \mathbf{n}} \qquad (8.3.10)$$

We have used the same symbol that in Chapter 1 was used for $\rho_p \phi_v / \rho_f(1 - \phi_v)$, because the dynamic definition above reduces to it when the two velocities are equal.

For the mass concentration, we proceed in the same manner, defining it as the ratio of particulate mass flow rate to total mass flow rate. Thus,

$$\phi_m = \eta_m/(1 + \eta_m) \qquad (8.3.11)$$

which is the definition given in Chapter 1.

8.4 Force and Heat Transfer Rate

In writing the two-phase equations of motion, we have assumed that the particle force and heat transfer are known. Therefore, before we can tackle specific dynamic problems, we need to specify those quantities in terms of the variables appearing in the system of equations and in terms of the results for single, finite-size particles obtained in Chapters 3 and 4. Let us first consider the force exerted by the particulate phase on the fluid. The total force on the fluid in the suspension element is the sum of the forces applied by each individual particle. Now, the force applied by each particle may be divided into a force acting along the direction of the particle translation, and a force acting along the normal to its surface, which is due to pulsations. This normal force, being uniform around the particle, vanishes for small particles because of cancellation of symmetrically opposed surface area elements. The remaining force is equal but opposite to the force, $\mathbf{F}_p$, acting on that particle. We thus have, adding all individual particle contributions, $\mathbf{f}_p = -(\mathbf{F}_{p1} + \mathbf{F}_{p2} + \cdots + \mathbf{F}_{pN})$, where N is the number of particles in the volume element. In general, these individual forces are different; but, if the particles are of identical material, shape, and size, and if they are distributed randomly within the element, then we may take them to be equal. Thus, if $\mathbf{F}_p$ is the force exerted on any one of the particles in the element, we simply have $\mathbf{f}_p = -N\mathbf{F}_p$, or dividing this by the volume of the element, we have $\mathbf{f}_p = -n\mathbf{F}_p$, where n is the number of particles per unit suspension volume.

The issue is, of course, the value of this individual force. Since $\mathbf{F}_p$ would have to be determined from the fluid equations of motion subject to the boundary conditions on each of the particles, it follows that it includes contributions due to the presence of other particles in the volume element. These contributions produce interactions between the particles that can be important in suspension that are not dilute. But for the dilute suspensions considered in this book, we may neglect those interactions, in which case $-\mathbf{F}_p$ is the force exerted on the fluid by an isolated particle.

But even then, the computation of this force is, in general, far from being simple. The reasons are that the particle's shapes are usually irregular, that the particle is moving near a boundary, or that the motions are such that nonlinear effects come into play. The first two difficulties are removed if the particles are spherical and if we ignore wall interactions on the basis that enclosing walls are far from most of the particles in a suspension. The nonlinearity of some motions is, of course, important in determining the response of fluids to applied forces; but, so far as the force on a single particle, nonlinear effects arise when the Reynolds number based on the relative velocity is not small. Because particles tend to adjust quickly to changes in the surrounding fluid, the relative Reynolds number is usually, but not always, small. But, when it is small, the particle force is then given by the Re $\ll 1$ results of Chapter 4. When the Reynolds number is not small, the particle force can be estimated by means of the empirical correlations mentioned in that chapter for the drag coefficient.

In most of the applications of the two-phase model that follow, we will limit the discussion to small-amplitude motions, and further assume that the particle force is equal to that calculated for rigid, spherical particles using the isothermal Stokes' equations. The fact that the particles are, for the purpose of computing the translational force, assumed to be rigid means that we neglect the effects of radial pulsations on the translational motion of the sphere. This is acceptable so long as the pulsations have a small amplitude. On the other hand, the fact that the isothermal equations were used to obtain that force means that thermal effects are included only in as much as the fluid velocity appearing in the particle force is affected by them. To fix ideas, consider the translational force on a solid aerosol particle moving slowly in a gas. As we showed in Chapter 4, the fluid force on it is well described by the Stokes drag, so that

$$\mathbf{f}_p(\mathbf{x}, t) = 6\pi\mu_f an(\mathbf{u}_f - \mathbf{u}_p) \tag{8.4.1}$$

Thus, thermal effects may influence the fluid velocity and, therefore $\mathbf{f}_p$, but the linear dependence of this force on the relative velocity is not affected by them.

Consider now the heat transfer rate to the fluid, $\dot{q}_p(x, t)$. We proceed as with the force, making the same restrictions, so that $\dot{q}_p = -n\dot{Q}_p$, where $\dot{Q}_p$ retains its meaning as the heat transfer rate to a single particle. Again, to fix ideas, we consider the rigid-particle heat transfer as described by (3.2.18). We then would have

$$\dot{q}_p(\mathbf{x}, t) = 4\pi k_f an(T_f - T_p) \tag{8.4.2}$$

More general expressions for the particle force and heat transfer rate than these were obtained in Chapters 3, 4, and 6, and will be used later to study some specific suspensions motions. But, before we do that, we consider situations where they play no role. These situations are of limited applicability, but illustrate some important features of suspensions in motion.

8.5 Near-Equilibrium Flow

The two-phase system of equations applies to suspensions in which the fluid and particulate phases are moving with the velocities, pressures, and temperatures that are prescribed by the conservation laws for given external conditions. In general, the values of those variables in one phase differ from the corresponding values in the other, and it is then necessary to use the complete system to determine them.

It is, however, instructive to consider an idealized situation in which the external body force vanishes, and all the variables have identical values, thus: $\mathbf{u}_p = \mathbf{u}_f = \mathbf{u}$, $p_p = p_f = p$ and $T_p = T_f = T$. Here, the particle force and heat transfer rate also vanish. Furthermore, noting that since the velocities are equal, the substantive derivatives, D_f/Dt and D_p/Dt are identical, and may thus be both written as D/Dt. Since $D/Dt = \partial/\partial t + \mathbf{u} \cdot \nabla$, the appearance of this derivative in the equations that follow would seem to imply that the conditions can be unsteady. However, unsteadiness normally induces a relative velocity, thus destroying the assumed equilibrium. Thus, in writing the following equations, we assume that whatever unsteadiness is present, it must be such that equilibrium conditions exist.

With these remarks in mind, we substitute the above variables in the two-phase system of equations. Thus, noting that $\sigma_f + \sigma_p = \sigma$, we obtain, on adding the two continuity equations,

$$\frac{D\sigma}{Dt} = -\sigma\nabla \cdot \mathbf{u} \tag{8.5.1}$$

Similarly, the two momentum equations give,

$$\sigma \frac{D\mathbf{u}}{Dt} + \nabla p = 0 \tag{8.5.2}$$

For the energy equation, we first note that $(1 - \phi_v)\rho_f c_{pf} + \rho_p c_{pp}\phi_v = \sigma c_p$ and that $(1 - \phi_v)\beta_f + \beta_p\phi_v = \beta$. Hence, adding the two energy equations we get

$$\sigma c_p \frac{DT}{Dt} - \beta T \frac{Dp}{Dt} = 0 \tag{8.5.3}$$

This equation can also be written as $T(DS/Dt) = 0$, meaning that the motions are isentropic. Thus, the assumption that the velocities, pressures, and temperatures are equal in the two phases, produces, as may have been anticipated, the equations of motion for an ideal fluid of density σ, constant pressure-specific heat c_p, and thermal expansion coefficient β. The equations are complemented by the corresponding equation of state. Since the entropy is constant, that equation is $p = p(\sigma)$.

Steady-Flow Conditions. Let us apply the equilibrium-flow equations to a steady, one-dimensional flow (e.g., flow in a constant cross-section tube). Here, $D/Dt = u(d/dx)$, if the flow is along the x-axis of a rectangular system of coordinates. Hence, the above system reduces to

$$\frac{d}{dx}(\sigma u) = 0 \tag{8.5.4}$$

$$\sigma u \frac{du}{dx} + \frac{dp}{dx} = 0 \tag{8.5.5}$$

$$\sigma u c_p \frac{dT}{dx} - \beta T u \frac{dp}{dx} = 0 \tag{8.5.6}$$

$$p = p(\sigma) \tag{8.5.7}$$

The equations apply in regions where the suspension is in equilibrium. So long as the equilibrium is maintained, the velocity, pressure, density, and temperature can vary, in which case the equations, when integrated, provide the relations that those quantities must satisfy. Suppose that at stations $x = x_1$ and $x = x_2$, equilibrium conditions hold. Then, under some restrictions, it is possible to integrate the equations so that if we are given the values of our variables at x_1, those at x_2 may be obtained – even if a region exists between the two stations where equilibrium does not exist. A specific example of this will be treated in Section 8.9.

Let us then proceed with the integration. The continuity equation shows that σu is a constant at every point in the flow where the particle and fluid velocities are, as assumed, equal. The quantity σu is, of course, the mass flow rate, $\dot{m}$, of particles and fluid passing through any fixed location in the tube. Hence, since equilibrium exists at

x_1 and at x_2, the suspension velocity at those two locations are u_1 and u_2, respectively, we have

$$\sigma_1 u_1 = \sigma_2 u_2 \qquad (8.5.8)$$

Similarly, since $\dot{m} = \sigma_1 u_1 = \sigma_2 u_2$ is constant, the momentum equation gives

$$\dot{m} u_1 + p_1 = \dot{m} u_2 + p_2 \qquad (8.5.9)$$

The energy equation cannot be integrated without making further assumptions. First, we use (8.5.5) to write $u(dp/dx) = -\frac{1}{2}\dot{m}d(u^2/2)$, so that (8.5.6) can be written as

$$\dot{m}[c_p dT/dx + \tfrac{1}{2}\beta T(du^2/dx)] = 0$$

Now, as (7.3.26) shows, $c_p = c_{pf}(1 + \eta_m c_{pp}/c_{pf})/(1 + \eta_m)$, where η_m is the mass loading ratio, obviously a constant. Thus, if the material-specific heats c_{pf} and c_{pp} can be regarded as constants, we may write the first term in the square bracket above as $d(c_p T)/dx$. For the second term, we take βT to be a constant, noting that the change in the kinetic energy due to variations of that product are of second order. To avoid confusion with the variable T appearing in the first term, we write that product as $\beta_0 T_0$. Under these assumptions, the energy equation can also be integrated between the two equilibrium stations, giving

$$c_p T_1 + \frac{1}{2}\beta_0 T_0 u_1^2 = c_p T_2 + \frac{1}{2}\beta_0 T_0 u_2^2 \qquad (8.5.10)$$

To complete the system, we may use the equation of state, which gives

$$p_1 = p_1(\sigma_1) \quad \text{and} \quad p_2 = p_2(\sigma_2) \qquad (8.5.11a,b)$$

Further progress requires an explicit equation of state, an example of which is provided by the dusty gas equation of state (Section 7.5). The matter is pursued further in Section 8.9, where we consider the steady flow of a perfect gas through a normal shock wave.

Low-Frequency Disturbance. As a second example of a near-equilibrium flow, we return to the equilibrium state described by (8.5.1)–(8.5.3) and ask what happens to our suspension if a very small-amplitude, time-dependent disturbance is introduced into it. To answer this, we first consider an undisturbed equilibrium flow. In it, the velocity, pressure, and temperature are constants, which we denote by U_0, P_0, and T_0, respectively. By definition, $u_p = u_f = u$, $p_p = p_f = p$, and $T_p = T_f = T$. Now suppose that the small disturbance introduced in the suspension varies so slowly in time that the particle pressure, temperature, and velocity remain equal to their fluid counterparts. Then, we can write

$$\mathbf{u} = \mathbf{U}_0 + \mathbf{u}'; \quad p = P_0 + p'; \quad T = T_0 + T' \quad \sigma = \sigma_0 + \sigma', \text{etc.}$$

where all primed quantities may be functions of position and time, but are very small, compared with the unprimed ones (e.g., $p' \ll P_0$). To simplify matters, we assume that the walls of the duct and the piston producing the motion are adiabatic. Then the motion remains isentropic. Hence, the pressure is only a function of the density, which means that the energy equation is unnecessary. Thus, we need consider only the continuity and momentum equations, (8.5.1) and (8.5.2), respectively. These give, after using the above expressions

$$\frac{\partial \sigma'}{\partial t} + \nabla \cdot [\sigma_0 \mathbf{u}' + \mathbf{U}_0 \sigma' + \sigma' \mathbf{u}'] = 0 \tag{8.5.12}$$

$$(\sigma_0 + \sigma')[\partial \mathbf{u}'/\partial t + (\mathbf{U}_0 + \mathbf{u}') \cdot \nabla \mathbf{u}'] + \nabla p' = 0 \tag{8.5.13}$$

$$p' = (dp/d\sigma)_{\sigma_0} \sigma' + \tfrac{1}{2}(d^2 p/d\sigma^2)_{\sigma_0}(\sigma')^2 + \dots \tag{8.5.14}$$

We now take advantage of the smallness of the disturbances and linearize these equations to obtain

$$\frac{\partial \sigma'}{\partial t} + \sigma_0 \nabla \cdot \mathbf{u}' + \mathbf{U}_0 \cdot \nabla \sigma' = 0 \tag{8.5.15}$$

$$\sigma_0 \left(\partial \mathbf{u}'/\partial t + \mathbf{U}_0 \cdot \nabla \mathbf{u}'\right) + \nabla p' = 0 \tag{8.5.16}$$

The pressure disturbance is then given by the first term in (8.5.14), but the derivative appearing in that term was calculated in Section 7.3, where it was denoted by the symbol $c_s^2(0)$. Thus, we write

$$p' = c_s^2(0)\sigma' \tag{8.5.17}$$

The last three equations may be combined to obtain a single equation for either σ' or p'. To do this, we first take the time derivative of (8.5.15), the spatial derivative of (8.5.16), and subtract the second from the first, obtaining

$$\frac{\partial^2 \sigma'}{\partial t^2} + \mathbf{U}_0 \cdot \nabla \frac{\partial \sigma'}{\partial t} - \sigma_0 \mathbf{U}_0 \cdot \nabla(\nabla \cdot \mathbf{u}') - \nabla^2 p' = 0 \tag{8.5.18}$$

To obtain an equation for one of the three variables appearing here, we first eliminate the velocity term by using the continuity equation to evaluate $\sigma_0 \nabla \cdot \mathbf{u}'$. We can then eliminate σ' in favor of p' by means of (8.5.17). These substitutions yield the desired result

$$\frac{\partial^2 p'}{\partial t^2} + 2\mathbf{U}_0 \cdot \nabla(\partial p'/\partial t) + (\mathbf{U}_0 \cdot \nabla)(\mathbf{U}_0 \cdot \nabla)p' = c_s^2(0)\nabla^2 p' \tag{8.5.19}$$

To simplify this, we note that the quantity $\mathbf{U}_0 \cdot \nabla$ represents a derivative in the direction of $\mathbf{U}_0$. Thus, aligning the positive x-axis of a fixed system of coordinates with $\mathbf{U}_0$, we obtain

$$\frac{\partial^2 p'}{\partial t^2} + 2U_0 \frac{\partial^2 p'}{\partial x \partial t} = c_s^2(0) \left(1 - M_0^2\right) \frac{\partial^2 p'}{\partial x^2} \tag{8.5.20}$$

where $M_0 = U_0/c_s(0)$. This reduces to the wave equation when U_0 is zero or very small, which means that the pressure disturbance travels as a wave, with speed $c_s(0)$. That is, $c_s(0)$ is the speed of sound for a suspension in equilibrium, an interpretation that was anticipated in the previous chapter. Thus, formally, that speed is, as anticipated, given by (7.4.3).

Let us now consider (8.5.20) with $U_0 \neq 0$. We note that the equation is, essentially, a wave equation, modified by the appearance of the second term on the left-hand side and by the constant factor $(1 - M_0^2)$ that changes the sound speed of propagation. To show this more clearly, we inquire whether wavelike solutions exist. For this purpose, we consider harmonic disturbances with circular frequency ω and assume a solution proportional to $\exp[i(kx - \omega t)]$, where k is to be determined. Substituting this into (8.5.20) shows that nontrivial solutions exist for values of k that satisfy

$$k^2(1 - M_0^2) + 2M_0 k k_0 - k_0^2 = 0 \tag{8.5.21}$$

where $k_0 = \omega/c_s(0)$. Since $c_s(0)$ has been identified with the sound speed in the medium in equilibrium, the quantity M_0 is seen to be the Mach number based on the base velocity U_0. This Mach number is not necessarily negligible, because $c_s(0)$ can, for some suspensions, be considerably smaller than the sound speed in fluids without particles. Thus, M_0^2 is not always negligible, compared with 1. Solving (8.5.21) for k we obtain

$$k_\pm = k_0 \frac{-M_0 \pm 1}{1 - M_0^2} \tag{8.5.22}$$

with the positive sign applying to waves moving toward larger values of x. For simplicity, we take the base flow velocity to be subsonic (i.e., $M_0 < 1$), in which case we obtain

$$k_+ = k_0/(1 + M_0) \quad \text{and} \quad k_- = -k_0/(1 - M_0) \tag{8.5.23a,b}$$

Thus, we find that there are two values of k for which the assumed solution exists. Hence, the pressure disturbance may be expressed as

$$p' = Ae^{i[k_0 x/(1+M_0) - \omega t]} + Be^{-i[k_0 x/(1-M_0) + \omega t]} \tag{8.5.24}$$

The first term on the right-hand side represents acoustic waves traveling in the direction of the positive x-axis, whereas the second represents a wave traveling in the opposite direction. The phase velocities of these waves are, respectively, $c_+ = c_s(0)(1 + M_0)$ and $c_- = -c_{s0}(1 - M_0)$. However, relative to an observer moving with the suspension, the disturbance waves travel with speed $c_s(0)$.

The following facts should also be noted. One is that the phase velocities we have obtained are independent of the frequency. This means that the profile of a disturbance (i.e., its spatial variations of pressure) remains constant as it travels. Second, the wavenumbers of the waves, k_+ and k_-, are real – which means that the amplitude of the waves are constant. A complex wavenumber would either mean an exponentially increasing or exponentially decreasing amplitude, as can be seen by adding to k_0 an imaginary part. Unless energy is being added to the wave, the first option is not feasible in actual systems. The second, a decay of the amplitude, is possible, but generally requires a dissipative mechanism. When the amplitude of the wave decays, we say that the waves are attenuated. Because the attenuation is associated with the imaginary part of the wavenumber, this part is therefore called an *attenuation coefficient*. One of the central problems in the acoustics of suspensions is to determine such coefficients for general suspensions. Also, a nonzero attenuation coefficient is usually accompanied by a frequency-dependent phase velocity. When these effects occur, the propagation is said to be *dispersive*, in which case the profile of the wave is distorted as it travels.

In the present example, none of these effects is present because we considered such low frequencies that the two phases remained in equilibrium at all times. But, as the frequency increases, the assumption of equilibrium necessarily breaks down because the particles are incapable of adjusting instantaneously to the changes imposed by a sound wave. This lack of adjustment, or lag, continues to exist so long as the frequency of the disturbance is finite. Of course, it is possible that some of the particle variables (e.g., the pressure), continues to be equal to the corresponding quantity in the fluid. If this happens, we would then have partial equilibrium between the two phases. Examples of suspensions in assumed partial equilibrium are given next to illustrate the use of the two-phase model to study sound propagation in suspensions. A more complete treatment of the problem is given in the following chapter.

8.6 Isothermal Sound Propagation in an Aerosol

As the first application of the two-phase equations to nonequilibrium motions, we consider acoustic wave propagation an aerosol composed of *equal-sized* solid particles in a gas ignoring thermal effects. These were included in an early application of the two-phase model (Temkin and Dobbins, 1966) and will be considered later. In this section, we merely want to show the two-phase system at work; for this purpose, we consider acoustic motions in isothermal conditions. For these, the energy equation is unnecessary, and the remaining equations forming the system – continuity and momentum – apply in linearized form. Furthermore, in the absence of the acoustic wave, we take the aerosol to be at rest, at uniform pressure p_0. Our interest is to determine what happens when a sound wave travels in the aerosol. To do this, we take the particles to behave as rigid spheres, which, of course, makes p'_p vanish. We also take the motion to be along the x-axis so that $\mathbf{u}_f = \{u_f(x, t), 0, 0\}$. Since the motion of the particulate phase occurs in response to that of the fluid, its velocity is along $\mathbf{u}_f$, or $\mathbf{u}_p = \{u_p(x, t), 0, 0\}$. The remaining variables will also depend only on x and t.

Hence

$$\frac{\partial \sigma'_f}{\partial t} + \sigma_{f0}\frac{\partial u_f}{\partial x} = 0 \qquad \frac{\partial \sigma'_p}{\partial t} + \sigma_{p0}\frac{\partial u_p}{\partial x} = 0 \qquad (8.6.1\text{a,b})$$

$$\sigma_{f0}\frac{\partial u_f}{\partial t} + (1 - \phi_v)\frac{\partial p'_f}{\partial x} = f'_p \qquad \sigma_{p0}\frac{\partial u_p}{\partial t} = -f'_p \qquad (8.6.2\text{a,b})$$

$$p'_f = c^2_{Tf}\rho'_f \qquad (8.6.3)$$

$$\sigma_f = \rho_f(1 - \phi_v) \qquad \sigma_p = \rho_p\phi_v \qquad (8.6.4\text{a,b})$$

The quantity f'_p appearing in the momentum equation is the linearized version of the particulate force. Since the particles are equal, $f_p = -nF_p$, where $n = n_0 + n'$, with $n' \ll n_0$. For aerosols, F_p is well described by Stokes' law. Thus,

$$f'_p = 6\pi\mu_f a n_0(u_p - u_f) \qquad (8.6.5)$$

Now, since the suspension is dilute, we have $1 - \phi_v \approx 1$. Then, evaluating the time derivatives of σ'_f and σ'_p using (8.6.4), we obtain

$$\frac{1}{\rho_{f0}c^2_{Tf}}\frac{\partial p'_f}{\partial t} - \frac{\partial \phi'_v}{\partial t} + \frac{\partial u_f}{\partial x} = 0 \qquad \frac{\partial \phi'_v}{\partial t} + \phi_{v0}\frac{\partial u_p}{\partial x} = 0 \qquad (8.6.6\text{a,b})$$

$$\rho_{f0}\frac{\partial u_f}{\partial t} + \frac{\partial p'_f}{\partial x} = 6\pi\mu_f a n_0(u_p - u_f)$$

$$\rho_{p0}\phi_{v0}\frac{\partial u_p}{\partial t} = -6\pi\mu_f a n_0(u_p - u_f) \qquad (8.6.7\text{a,b})$$

Next, eliminate $\partial\phi'_v/\partial t$ from the first of equations (8.6.6) by using the second. Thus,

$$\frac{\partial \pi'_f}{\partial t} + \frac{\partial u_f}{\partial x} + \phi_{v0}\frac{\partial u_p}{\partial x} = 0 \qquad (8.6.8)$$

where $\pi'_f = p'_f/\rho_{f0}c^2_{Tf}$ and where c_{Tf} is the isothermal sound speed in the fluid. For the momentum equations we note that

$$6\pi\mu_f a n_0 = \rho_{f0}\frac{\phi_{v0}/\delta}{\tau_d} \qquad (8.6.9)$$

where $v_f = \mu_f/\rho_{f0}$, $\delta = \rho_{f0}/\rho_{p0}$, and where τ_d is the dynamic relaxation of a particle. For aerosols, the density ratio δ is very small. Hence, the quantity ϕ_{v0}/δ is not necessarily small, even when $\phi_v \ll 1$. Furthermore, the quantity ϕ_{v0}/δ is equal to the mass loading, η_{m0}, at rest when $\phi_v \ll 1$. Using these definitions, we can write (8.6.7a,b) as

$$\frac{\partial u_f}{\partial t} + c^2_{Tf}\frac{\partial \pi'_f}{\partial x} - \frac{\eta_{m0}}{\tau_d}(u_p - u_f) = 0$$

$$\frac{\partial u_p}{\partial t} + \frac{1}{\tau_d}(u_p - u_f) = 0 \qquad (8.6.10\text{a,b})$$

We may combine the three remaining equations into a smaller number of equations of higher order, but since we are searching for waves traveling toward increasing values of x, we assume that all variables vary as $\exp(ikx - i\omega t)$, with ω taken to be real and positive. On the other hand, we expect that, because of dissipation resulting from the relative motion, k will be complex, with positive real and imaginary parts. The assumed dependence on x and t reduces the system of differential equations into an algebraic system, which we write in matrix form as

$$
\begin{pmatrix}
(1 - \phi_{v0})\omega & -k & -k\phi_{v0} \\
ikc_{Tf}^2 & -i\omega + \eta_{m0}/\tau_d & -\eta_{m0}/\tau_d \\
0 & -1/\tau_d & -i\omega + 1/\tau_d
\end{pmatrix}
\begin{pmatrix}
\pi_f' \\
u_f \\
u_p
\end{pmatrix}
= 0 \qquad (8.6.11)
$$

This has a nontrivial solution provided the determinant of the coefficient matrix vanishes, or

$$
\begin{vmatrix}
(1 - \phi_{v0})\omega & -k & -k\phi_{v0} \\
ikc_{Tf}^2 & -i\omega + \eta_{m0}/\tau_d & -\eta_{m0}/\tau_d \\
0 & -1/\tau_d & -i\omega + 1/\tau_d
\end{vmatrix}
= 0 \qquad (8.6.12)
$$

Expanding this, we obtain, after rearrangement

$$
k^2 c_{Tf}^2 (1 + \phi_{v0} - i\omega\tau_d) = (1 - \phi_{v0})\,\omega^2 (1 + \eta_{m0} - i\omega\tau_d) \qquad (8.6.13)
$$

This is called a dispersion relation; it specifies the values of the wavenumber, $k = k(\omega)$, that are allowable within the model. For the present case, it states that acoustic waves can propagate in an isothermal aerosol provided solutions for $k = k(\omega)$ with positive real and imaginary parts exist. The final result for k may be expressed in various manners, the simplest one being

$$
k^2 = \frac{\omega^2}{c_{Tf}^2} \left(1 + \frac{\eta_{m0}}{1 - i\omega\tau_d} \right) \qquad (8.6.14)
$$

In obtaining this from (8.6.13), we have dropped ϕ_{v0} compared with unity while retaining $\eta_{m0} = \phi_{v0}/\delta$, because of the smallness of the density ratio δ in aerosols.

Equation (8.6.14) gives the wavenumber for the waves in terms of the mass loading, the frequency ω, and the dynamical relaxation for the particles. Since the result depends on the frequency, we anticipate that the phase velocity is frequency-dependent. Furthermore, since k is complex, the waves are attenuated as they travel. To see this, we first write $k = k_r + i\alpha(\omega)$, which indicates that all variables depend on distance and time as $\exp[-\alpha x + i(k_r x - \omega t)]$. This shows that the waves' amplitude decays to a value $1/e$ in a distance equal to $1/\alpha$. It also shows that the phase velocity of the waves is equal to ω/k_r. We denote this velocity as $c_T(\omega)$ to remind ourselves that the result applies to isothermal conditions. Now, to obtain $c_T(\omega)$ and $\alpha(\omega)$ we put

$$
k = \omega/c_T(\omega) + i\alpha(\omega) \qquad (8.6.15)
$$

into (8.6.14) and equate the real an imaginary parts on both sides of the equation, getting

$$\frac{c_T^2(0)}{c_T^2(\omega)} - \frac{\alpha^2 c_T^2(0)}{\omega^2} = \frac{c_T^2(0)}{c_{Tf}^2}\left(1 + \eta_{m0}\frac{1}{1+\omega^2\tau_d^2}\right) \tag{8.6.16}$$

$$2\frac{\alpha c_{T0}}{\omega}\frac{c_T(0)}{c_T(\omega)} = \eta_{m0}\frac{c_T^2(0)}{c_{Tf}^2}\frac{\omega\tau_d}{1+\omega^2\tau_d^2} \tag{8.6.17}$$

where $c_T(0)$ is the low-frequency limit of $c_T(\omega)$, given by (7.3.7). Since $N_T \ll 1$, that equation gives $c_T^2(0)/c_{Tf}^2 = 1/(1+\eta_{m0})$. The choice of $c_T(0)$ to scale $c_T(\omega)$ is a natural one when the mass loading is finite, but when the aerosol is so dilute that the mass loading is also very small, it is advantageous to use c_{Tf} to scale the phase velocity and the attenuation. In those cases, we shall use the nondimensional ratio $\hat{\alpha} = \alpha c_{Tf}/\omega$. Since this is equal to $\alpha/2\pi\lambda$, where λ is the wavelength that a sound wave would have in a fluid without particles, $\hat{\alpha}$ is commonly referred as the attenuation per wave length.

Let us now return to (8.6.16) and (8.6.17). These are two equations for the phase velocity and the attenuation, and may be easily solved for arbitrary values of the mass loading compatible with $\phi_{v0} \ll 1$. But when the mass loading is also small, the solution can be obtained quickly by inspection. We first note that, if $\eta_{m0} \ll 1$, the equations imply that $\hat{\alpha} = O(\eta_{m0})$ so that we may drop its square in (8.6.16), set $c_T(0)/c_{Tf}$ equal to 1 on the right-hand side of both equations, and also set $c_T(0)/c_T(\omega)$ equal to 1 on the left-hand side of (8.6.17). Thus, when $\eta_{m0} \ll 1$, we obtain

$$\frac{c_{Tf}^2}{c_T^2(\omega)} \approx 1 + \eta_{m0}\frac{1}{1+\omega^2\tau_d^2} \tag{8.6.18}$$

$$\hat{\alpha} \approx \frac{1}{2}\eta_{m0}\frac{\omega\tau_d}{1+\omega^2\tau_d^2} \tag{8.6.19}$$

These results, typical of small attenuation and dispersion in a relaxing medium having a single relaxation time, show the effects of both frequency and relaxation time. Thus, when $\omega\tau_d \ll 1$ (i.e., for waves whose period, $2\pi/\omega$ is much larger than the relaxation time τ_d), the attenuation vanishes and the sound speed equals the equilibrium speed $c_T(0)$. Clearly, the particles have plenty of time to adjust to the changes induced by the waves. At the other end of the frequency range, when $\omega\tau_d \gg 1$, the attenuation also vanishes and the sound speed becomes equal to the sound speed in the gas alone. Between those two limits, both speed and attenuation vary with the frequency, and Fig. 8.6.1 shows the variations of $[c_{Tf}^2/c_T^2(\omega) - 1]/\eta_{m0}$ and $\hat{\alpha}/\eta_{m0}$ as a function of the nondimensional frequency $\omega\tau_d$. We see that the effects of the particles on the propagation of sound waves in a very dilute aerosol are to slightly decrease the phase velocity from c_{Tf} and to produce a slight attenuation. This attenuation is largest at a frequency equal to $\omega = \tau_d^{-1}$, where the phase velocity curve has a point of inflection.

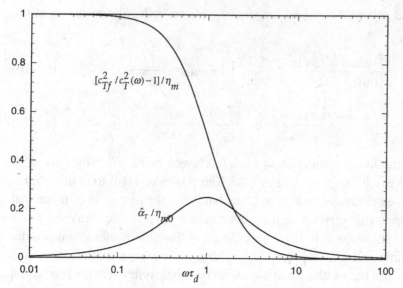

Figure 8.6.1. Scaled isothermal attenuation and dispersion in an aerosol composed of water droplets in air. $D = 100 \ \mu$m.

These features are directly related to the behavior of the relative velocity between particle and fluid. This may be seen by considering the particulate velocity. This is obtained from (8.6.10.b) and is

$$u_p = \frac{1}{1 - i\omega\tau_d} u_f \tag{8.6.20}$$

Although this refers to the velocity that the particulate phase has at some point x, it should be noted that it is the same as (4.4.5), which gives the velocity of a sphere in an oscillatory fluid when Stokes' law is used.

   Having obtained speed and attenuation, we may obtain any of the variables appearing in our system of equations. For example, the amplitude of the fluid-phase velocity is given by

$$u_f = U_0 \exp(-\alpha x)$$

where U_0 is the velocity amplitude at $x = 0$. We thus see that the fluid velocity decays exponentially. Given the small value of α, the decay is rather small. For example, at a frequency equal to $\omega = \tau_d^{-1}$, where $\hat{\alpha}$ has its maximum value, the amplitude of the fluid velocity in an aerosol having a mass loading equal to 0.004 decays about 1% in one wavelength. The same decay would apply to the other variables in the wave.

   This concludes our discussion of propagation in isothermal aerosols. The matter is pursued in Section 8.8, where we consider thermal effects in both aerosols and emulsions, and in considerably more detail in Chapter 9.

PROBLEMS

1. Determine $c_T(\omega)$ and $\hat{\alpha}$ when the mass loading is small, but finite.
2. Determine the variations in volume concentration, pressure, and number concentration of particles in a sound wave in an aerosol.

8.7 Isothermal Sound Propagation in a Bubbly Liquid

In the previous section, we considered isothermal propagation in aerosols, entirely disregarding whatever contributions might exist as a result of pressure changes inside the aerosol particles. The main reason that permitted us to do this was that the compressibility of the gas surrounding the particles is several orders of magnitude larger than that of the particles. In bubbly liquids, the opposite is true. Furthermore, as studied in Chapter 6, the remarkable response of a bubble to an external pressure fluctuation, makes it clear that the bubble pressure here must play a significant role. To study this role, we consider isothermal propagation in a bubbly liquid, neglecting, for simplicity, relative motion between the phases. This can be easily included, but merely obscures the derivation without affecting the results significantly. A far more important effect, to be considered later, is that produced by the temperature changes in a bubble. It should be added that the present derivation will also apply to propagation in isothermal emulsions whose droplets move with the fluid. This is so because droplets in liquids also sustain pulsational motions to some degree.

For future reference, we start by writing the system of equations retaining pulsational and translational effects. Thus, the two-phase equations in linearized form give for one-dimensional, isothermal motions

$$(1 - \phi_{v0})\left(\frac{1}{\rho_{f0}}\frac{\partial \rho'_f}{\partial t} + \frac{\partial u_f}{\partial x}\right) = \frac{\partial \phi'_v}{\partial t}$$

$$\phi_{v0}\left(\frac{1}{\rho_{p0}}\frac{\partial \rho'_p}{\partial t} + \frac{\partial u_p}{\partial x}\right) = -\frac{\partial \phi'_v}{\partial t} \qquad (8.7.1a,b)$$

$$(1 - \phi_{v0})\left(\rho_{f0}\frac{\partial u_f}{\partial t} + \frac{\partial p'_f}{\partial x}\right) = f'_p$$

$$\phi_{v0}\left(\rho_{p0}\frac{\partial u_p}{\partial t} + \frac{\partial p'_p}{\partial x}\right) = -f'_p \qquad (8.7.2a,b)$$

$$\rho'_f = \rho'_f(p'_f) = p'_f/c_{Tf}^2 \qquad \rho'_p = \rho'_p(p'_p) = p'_p/c_{Tp}^2 \qquad (8.7.3a,b)$$

Consider the particulate pressure equation. Without linearizing, it is given by (8.2.27) or

$$p_p = p_f + \rho_f R\frac{\partial^2 R}{\partial t^2} + \rho_f\left(\frac{4\nu_f}{R} + \frac{3}{2}\frac{\partial R}{\partial t}\right)\frac{\partial R}{\partial t} \qquad (8.7.4)$$

Putting $p_p = p_{p0} + p'_p$ and $R = a + \varepsilon$ in this equation, and dropping second-order terms, we get

$$p'_p = p'_f + \rho_{f0}a\frac{\partial^2\varepsilon}{\partial t^2} + 2\rho_{f0}\beta_\mu\frac{\partial\varepsilon}{\partial t} \tag{8.7.5}$$

where $\beta_\mu = 2\nu_f/a^2$ is the viscous damping coefficient introduced in Section 6.5. This coefficient was obtained for an incompressible pressure field outside the pulsating particle and does not, therefore, include acoustic radiation by the particle. So far as the pulsation is concerned, this radiation signifies a loss, like that produced by viscosity, and can be accounted for by adding to the viscous damping coefficient the acoustic damping coefficient $\beta_{ac} = \frac{1}{2}b\omega$. We denote the sum of viscous and acoustic damping coefficients by β_T, noting that, at a later stage, it can be modified to include thermal effects. Thus,

$$p'_p = p'_f + \rho_{f0}a\frac{\partial^2\varepsilon}{\partial t^2} + 2\rho_{f0}\beta_T\frac{\partial\varepsilon}{\partial t} \tag{8.7.6}$$

The radial displacement, ε, may be expressed in terms of the pressure in the bubble. Since the motions are isothermal, we have, using (6.4.7)

$$\varepsilon = -\frac{a}{3}\frac{p'_p}{c_{Tp}^2} \tag{8.7.7}$$

Using this in (8.7.6), we obtain

$$p'_p = p'_f - \frac{\rho_{f0}a^2}{3\rho_{p0}c_{Tp}^2}\left(\frac{\partial^2 p'_p}{\partial t^2} + 2\beta_T\frac{\partial p'_p}{\partial t}\right) \tag{8.7.8}$$

We are now ready to start our attempt to solve the system of equations, which is reduced to the two continuity and momentum equations, plus the particulate pressure equation. Rather than specifying the force, as was done with the aerosol case, we eliminate that need, momentarily, by adding the two momentum equations. We also eliminate ϕ'_v by adding the two linearized continuity equations, obtaining

$$(1 - \phi_{v0})\left(\frac{1}{\rho_{f0}c_{Tf}^2}\frac{\partial p'_f}{\partial t} + \frac{\partial u_f}{\partial x}\right) + \phi_{v0}\left(\frac{1}{\rho_{p0}c_{Tp}^2}\frac{\partial p'_p}{\partial t} + \frac{\partial u_p}{\partial x}\right) = 0 \tag{8.7.9}$$

$$(1 - \phi_{v0})\left(\rho_{f0}\frac{\partial u_f}{\partial t} + \frac{\partial p'_f}{\partial x}\right) + \phi_{v0}\left(\rho_{p0}\frac{\partial u_p}{\partial t} + \frac{\partial p'_p}{\partial x}\right) = 0 \tag{8.7.10}$$

Since this system has four unknowns (u_f, u_p, p'_f and p'_p), a fourth equation is needed. For this purpose, we could bring back the force equation because it involves the two velocities. The resulting system could then be solved using the same procedure as followed in the previous section – that is, we would assume a wave-like solution and reduce the system to a set of algebraic equations for four unknowns. This would result in a 4×4 matrix whose vanishing determinant would specify the dispersion relation for the assumed wavenumber. Instead of proceeding in that manner, we make the simplifying assumption that the two velocities are equal. This would leave only the most significant feature of the problem: the particulate pulsations. Thus, putting $u_f = u_p = u$ we obtain

$$\sigma_0 \frac{\partial u}{\partial x} + \frac{(1 - \phi_{v0})}{\rho_{f0} c_{Tf}^2} \frac{\partial p_f'}{\partial t} + \frac{\phi_{v0}}{\rho_{p0} c_{Tp}^2} \frac{\partial p_p'}{\partial t} = 0 \qquad (8.7.11)$$

$$\sigma_0 \frac{\partial u}{\partial t} + (1 - \phi_{v0}) \frac{\partial p_f'}{\partial x} + \phi_{v0} \frac{\partial p_p'}{\partial x} = 0 \qquad (8.7.12)$$

Taking cross-derivatives and subtracting gives

$$\sigma_0 \left(\frac{1 - \phi_{v0}}{\rho_{f0} c_{Tf}^2} \frac{\partial^2 p_f'}{\partial t^2} + \frac{\phi_{v0}}{\rho_{p0} c_{Tp}^2} \frac{\partial^2 p_p'}{\partial t^2} \right) = (1 - \phi_{v0}) \frac{\partial^2 p_f'}{\partial x^2} + \phi_{v0} \frac{\partial^2 p_p'}{\partial x^2} \qquad (8.7.13)$$

Before we substitute (8.7.8) into this to obtain a single equation for either pressure, we reduce it for a monochromatic time dependence. This gives, on using complex notation for the two pressures and writing the time dependence as $\exp(-i\omega t)$,

$$\left[1 - \frac{\rho_{f0} a^2}{3 \rho_{p0} c_{Tp}^2} (\omega^2 + 2i\beta_T \omega) \right] p_p' = p_f' \qquad (8.7.14)$$

Further simplification can be made by using the definition of the isothermal natural frequency, (6.5.4), but neglecting surface tension in that result. This shows that $\rho_{f0} a^2 / 3\rho_{p0} c_{Tp}^2 \equiv 1/\omega_{T0}^2$. Then, solving for p_p' we obtain

$$p_p' = \chi_T p_f' \qquad (8.7.15)$$

where χ_T is the isothermal response function given by (6.5.14),

$$\chi_T = \frac{1}{1 - (\omega/\omega_{T0})^2 - i\hat{d}_T} \qquad (8.7.16)$$

Here, $\hat{d}_T$ is the nondimensional, isothermal damping coefficient given by (6.5.15). We now return to (8.7.14). Using (8.7.16), we obtain

$$[1 + \phi_{v0}(\chi_T - 1)] \frac{\partial^2 p_f'}{\partial x^2} = \omega^2 \sigma_0 \left(\frac{1 - \phi_{v0}}{\rho_{f0} c_{Tf}^2} + \chi_T \frac{\phi_{v0}}{\rho_{p0} c_{Tp}^2} \right) p_f' \qquad (8.7.17)$$

Since we are looking for wavelike solutions, we take p_f' to be proportional to $\exp(ikx)$ and find that one exists if k satisfies

$$k^2 = \frac{\sigma_0 \omega^2}{\rho_{f0} c_{Tf}^2} \frac{1 + \phi_{v0}(\chi_T N_T - 1)}{1 + \phi_{v0}(\chi_T - 1)} \qquad (8.7.18)$$

where $N_T = \rho_{f0} c_{Tf}^2 / \rho_{p0} c_{Tp}^2$ is the ratio of particle to fluid materials compressibilities. The result may be used as is, but two more simplifications are useful. First, we note that the first fraction on the right-hand side of (8.7.18) can, with the aid of (7.3.5) and (7.3.6), be written as

$$\frac{\sigma_0 \omega^2}{\rho_{f0} c_{Tf}^2} = \frac{k_0^2}{1 + \phi_{v0}(N_T - 1)} \tag{8.7.19}$$

where $k_0 = \omega/c_T(0)$. Thus, retaining only terms of first order in ϕ_{v0} and rearranging, we obtain the final result:

$$\left(\frac{k}{k_0}\right)^2 = 1 + \phi_{v0}(\chi_T - 1)(N_T - 1) \tag{8.7.20}$$

This result clearly show that dispersive effects occur when the compressibility ratio and the response function differ from 1. Also, since χ_T is a complex function of the frequency, (8.7.20) shows that k is complex, implying that the phase velocity and the attenuation are frequency-dependent. Before considering this dependence, we first note that, if $\omega \to 0$, the response function is equal to 1, showing that k is real and equal to k_{T0}. The same result applies if the particulate and fluid phase have the same compressibilities, as may be the case with some droplets in liquids.

To obtain the speed and attenuation, we put $k = \omega/c_T(\omega) + i\alpha(\omega)$, and separate the resulting real and imaginary parts of (8.7.20), obtaining

$$\frac{c_T^2(0)}{c_T^2(\omega)} - \frac{\alpha^2 c_T^2(0)}{\omega^2} = 1 + \phi_{v0}\left[\Re\{\chi_T\} - 1\right](N_T - 1) \tag{8.7.21}$$

$$2\frac{\alpha c_T(0)}{\omega}\frac{c_T(0)}{c_T(\omega)} = \phi_{v0}(N_T - 1)\Im\{\chi_T\} \tag{8.7.22}$$

The real and imaginary parts of χ_T are, respectively,

$$\Re\{\chi_T\} = \frac{1 - (\omega/\omega_{T0})^2}{[1 - \omega/\omega_{T0}^2)]^2 + \hat{d}_T^2} \tag{8.7.23}$$

$$\Im\{\chi_T\} = \frac{\hat{d}_T}{[1 - \omega/\omega_{T0}^2)]^2 + \hat{d}_T^2} \tag{8.7.24}$$

In writing the first of these, we have put $1 + b_T^2 \approx 1$ because of the assumed smallness of b_T (see Section 6.5).

It is noted that, since χ_T vanishes at both ends of the frequency spectrum, the attenuation per wavelength, $\alpha c_T(0)/\omega$, also vanishes there. Now, (8.7.21) and (8.7.22) may be solved exactly for the phase velocity and the attenuation, but it is useful to note first that if ϕ_{v0} is of the order of 10^{-6} or smaller, the quantity $\phi_{v0}N_T$ is small, even for bubbly liquids, where N_T is large. The resulting attenuation is then also small, compared with unity and we may drop its square in (8.7.21). The speed changes and the attenuation are then both proportional to the volume concentration and are given by

$$\frac{c_T^2(0)}{c_T^2(\omega)} - 1 = \phi_{v0}(N_T - 1)\left(\frac{\omega}{\omega_{T0}}\right)^2 \frac{1 - (\omega/\omega_{T0})^2}{[1 - \omega/\omega_{T0}^2)]^2 + b_T^2(\omega/\omega_{T0})^4} \tag{8.7.25}$$

$$\frac{\alpha c_T(0)}{\omega} = \tfrac{1}{2}\phi_{v0}(N_T - 1)\frac{b_T(\omega/\omega_{T0})^2}{[1 - \omega/\omega_{T0}^2)]^2 + b_T^2(\omega/\omega_{T0})^4} \tag{8.7.26}$$

where $b_T = \omega a/c_{Tf}$. These results were first obtained by Kennard in 1941, using a different procedure. More recent derivations have been given by Carstensen and Foldy (1947) and by Wijngaarden (1972), also using different approaches.

As derived here, the equations are applicable to both isothermal bubbly liquids and emulsions. For bubbly liquids, the compressibility ratio, N_T, is rather large, whereas for emulsions, it is, at most, of order 10. Since the speed changes and the attenuation are, for $\phi_{v0} \to 0$, proportional to ϕ_{v0}, it is convenient to plot $[c_T^2(0)/c_T^2(\omega) - 1]/\phi_{v0}$ and $[\alpha c_T(0)/\omega]/\phi_{v0}$ versus the nondimensional frequency ω/ω_{T0}. Figure 8.7.1 shows those quantities for a bubbly liquid composed of water and a bubbly liquid composed of water and 100-μm diameter air bubbles and having a volume concentration equal to 10^{-6}. Consider first the attenuation. As pointed out previously, it vanishes at both ends of the frequency spectrum, indicating that the energy dissipated by acoustic radiation is then a negligible part of the energy of the incident wave. In between those two limits, $\alpha c_T(0)/\omega$ has a maximum at $\omega/\omega_{T0} \approx 1$. This is, of course, produced by resonance effects, as discussed in Chapter 6. It is seen that if, as assumed, $\phi_{v0} \to 0$, the maximum value of $\alpha c_T(0)/\omega$ is small. Hence, having dropped its square is warranted. The figure also shows that the phase velocity approaches the equilibrium sound speed at very low frequencies and a slightly different value at high frequencies. This high-frequency limit is essentially equal to the sound speed in the liquid and is due to the bubbles being unable, at those large frequencies, to pulsate. Finally, we also see that the phase velocity changes rapidly in the vicinity of the natural frequency of the particle, an effect due to resonance.

As the volume concentration is increased, the value of $\alpha c_T(0)/\omega$ remains small for emulsions. But, in bubbly liquids, significant attenuation is produced because $\phi_{v0}N_T$ can be large, even when ϕ_{v0} is small. For example, when ϕ_{v0} has a value of 10^{-4}, the attenuation per wavelength is of the order of 1. This means that its square cannot be neglected in (8.7.21). An indication of the error that such neglect would produce then is provided by (8.7.25) for the speed ratio. The equation predicts that just below resonance, the value of $c_T^2(0)/c_T^2(\omega)$ is of the order of $1 + \phi_{v0}N_T$, whereas just above resonance it becomes $1 - \phi_{v0}N_T$. It then follows that, for values of ϕ_{v0} of the order of N_T^{-1}, the square of the phase velocity becomes negative, which it cannot be. Thus, when the attenuation per wavelength is not very small, it is necessary to obtain the phase velocity and the attenuation from the complete equations, (8.7.21) and (8.7.22). We postpone discussion of this case until the next chapter.

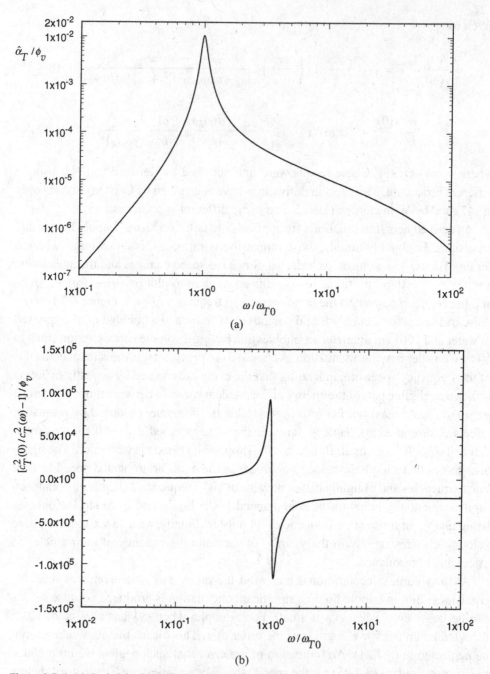

Figure 8.7.1. (a) Scaled isothermal attenuation in a bubbly liquid composed of air bubbles in water. $D = 100\,\mu m$. (b) Scaled isothermal dispersion in a bubbly liquid composed of air bubbles in water. $D = 100\,\mu m$.

8.8 Thermal Effects in Emulsions and Aerosols

So far we have studied the effects of particle translations and pulsations in isothermal conditions. We now use the two-phase model to study the effects of temperature changes. The simplest situation that would allow us to do so occurs in aerosols because, in them, the particle pressure can be disregarded. However, it is useful to consider those effects on sound propagation in less restrictive cases, and for that purpose we select emulsions. Here, the pressure fluctuations in the particle must be retained. This is done below using an approximate model based on the results obtained in Chapter 6. The final results also apply to aerosols composed of solid, elastic particles, or of droplets in gases.

The derivation of the dispersion equation is now necessarily more involved than in the previous two examples, because it involves the energy equations for both phases. We will, however, make use of some assumptions that will simplify the analysis. The first, also used in the previous section, is that the particles move with the fluid. The second relates to the pressure fluctuation in the droplets. Examination of Fig. 6.6.9 shows that, for a very wide range of frequencies below resonance, the pressure amplitude in a droplet does not differ significantly from its counterpart in the surrounding fluid and that as the droplet resonance frequency is reached, the pressure increase is relatively small. The more general theory presented in the next chapter allows for differences between the pressure fluctuations in the particles and the fluid, and shows that their disregard in the case of emulsions is warranted in a wide range of frequencies.

We thus consider the linearized, two-phase system of equations for a suspension where the two phases have the same velocity and pressure, but generally have different temperatures. To simplify the derivation, it is useful to start with those equations without imposing the pressure equality condition. Then, the continuity and momentum equation are given by (8.7.1) and (8.7.2). But, since the temperature now enters in the problem, we now require the two energy equations. In linearized form, these are

$$(1 - \phi_{v0}) \left(\rho_{f0} c_{pf} \frac{\partial T'_f}{\partial t} - \beta_f T_0 \frac{\partial p'_f}{\partial t} \right) = \dot{q}'_p$$

$$\phi_{v0} \left(\rho_{p0} c_{pp} \frac{\partial T'_p}{\partial t} - \beta_p T_0 \frac{\partial p'_p}{\partial t} \right) = -\dot{q}'_p \qquad (8.8.1a,b)$$

Similarly, the linearized equations of state are now

$$\rho'_f = p'_f / c^2_{Tf} - \rho_{f0} \beta_f T'_f \qquad \rho'_p = p'_p / c^2_{Tp} - \rho_{p0} \beta_p T'_p \qquad (8.8.2a,b)$$

The system thus contains nine equations. To reduce this number, we add the continuity and momentum pairs,

$$\frac{\partial u}{\partial x} + \frac{1 - \phi_{v0}}{\rho_{f0}} \frac{\partial \rho'_f}{\partial t} + \frac{\phi_{v0}}{\rho_{p0}} \frac{\partial \rho'_p}{\partial t} = 0 \qquad (8.8.3)$$

$$\frac{\partial u}{\partial t} + \frac{1 - \phi_{v0}}{\sigma_0} \frac{\partial p'_f}{\partial t} + \frac{\phi_{v0}}{\sigma_0} \frac{\partial p'_p}{\partial x} = 0 \qquad (8.8.4)$$

The velocity can be eliminated by taking cross-derivatives and subtracting. Doing this and setting $p'_f = p'_p = p'$, we obtain

$$\frac{1}{\sigma_0}\frac{\partial^2 p'}{\partial x^2} = \frac{1-\phi_{v0}}{\rho_{f0}}\frac{\partial^2 \rho'_f}{\partial t^2} + \frac{\phi_{v0}}{\rho_{p0}}\frac{\partial^2 \rho'_p}{\partial t^2} \tag{8.8.5}$$

Similarly, the energy equations become

$$(1-\phi_{v0})\left(\rho_{f0}c_{pf}\frac{\partial T'_f}{\partial t} - \beta_f T_0 \frac{\partial p'}{\partial t}\right) = \dot{q}'_p$$

$$\phi_{v0}\left(\rho_{p0}c_{pp}\frac{\partial T'_p}{\partial t} - \beta_p T_0 \frac{\partial p'}{\partial t}\right) = -\dot{q}'_p \tag{8.8.6a,b}$$

We also put $p'_f = p'_p = p'$ in the equations of state and use the result to eliminate the densities from (8.8.5), obtaining,

$$\frac{1}{\sigma_0}\frac{\partial^2 p'}{\partial x^2} = K_T(0)\frac{\partial^2 p'}{\partial t^2} - \left[\beta_f(1-\phi_{v0})\frac{\partial^2 T'_f}{\partial t^2} + \beta_p\phi_{v0}\frac{\partial^2 T'_p}{\partial t^2}\right] \tag{8.8.7}$$

where $K_T(0) = (1-\phi_{v0})K_{Tf} + \phi_{v0}K_{Tp}$ is the suspension's isothermal compressibility, as given by (7.3.5).

To complete the system, we require the linearized heat transfer rate, $\dot{q}'_p$. This is related to the local heat transfer rate, $\dot{Q}_p(x,t)$, by means of $\dot{q}'_p = -n_0\dot{Q}_p(x,t)$. Now, $\dot{Q}_p(x,t)$ was obtained in Section 6.6 for a monochromatic time dependence. That result is given by (6.6.17). When the pressures are equal, the complex pressure ratio Π is equal to 1, so that, for a *monochromatic* time dependence, we have

$$\dot{q}'_p = -\phi_{v0}\frac{\rho_{p0}c_{pp}}{\tau_t}\frac{1-\xi}{F(q,q_i)}T'_f \tag{8.8.8}$$

were τ_t is the particle temperature relaxation time defined by (3.2.20).

Having reduced our system to three differential equations, (8.8.6a,b) and (8.8.7), we look for wavelike solutions of those equations, assuming that our variables depend on position and time as $\exp(ikx - i\omega t)$. The resulting equations will be three algebraic equations that may be written in matrix form as $\underline{A}\underline{X} = 0$, where $\underline{X}$ is a one-column matrix whose components are p', T'_f, and T'_p, and where the coefficient matrix $\underline{A}$ is

$$\begin{pmatrix} k^2/\sigma_0 - K_T(0)\omega^2 & \omega^2\beta_f(1-\phi_{v0}) & \omega^2\beta_p\phi_{v0} \\ i\omega\beta_f T_0(1-\phi_{v0}) & \phi_{v0}\frac{1-\xi}{F(q,q_i)}\frac{\rho_{p0}c_{pp}}{\tau_t} - i\omega(1-\phi_{v0})\rho_{f0}c_{pf} & 0 \\ i\omega\beta_p T_0 & -\frac{1-\xi}{F(q,q_i)}\frac{\rho_{p0}c_{pp}}{\tau_t} & -i\omega\rho_{p0}c_{pp} \end{pmatrix}$$

The system has a solution provided the determinant of this matrix vanishes. Expanding the determinant results in an equation for k^2, which after some algebra may be expressed as

$$\left(\frac{k}{\omega}\right)^2 = \sigma_0 K_T(0) - \frac{\sigma_0}{\rho_{f0}} \frac{\beta_f^2 T_0}{c_{pf}} \frac{1 + \phi_{v0}\frac{\beta_p}{\beta_f}\left[\xi + i\frac{\beta}{\beta_f}\frac{1-\xi}{\omega\tau_t F(q,q_i)}\right]}{1 + i\phi_{v0}\frac{\rho_{p0}c_{pp}}{\rho_{f0}c_{pf}}\frac{1-\xi}{\omega\tau_t F(q,q_i)}}$$ (8.8.9)

where $\beta = (1 - \phi_{v0})\beta_f + \phi_{v0}\beta_p$. This equation defines the complex wavenumber k, whose real and imaginary parts define the phase velocity and attenuation coefficient. The separation into real and imaginary parts presents no difficulty, but produces involved equations whose meaning is difficult to asses. However, since the volume fraction has to be small for the analysis to apply, we may simplify (8.8.9) by taking $\phi_{v0} \ll 1$. Thus, with $\beta/\beta_f \approx 1 + \phi_{v0}\beta_p/\beta_f$, we can write

$$\left(\frac{k}{\omega}\right)^2 = \sigma_0 K_T(0) - \frac{\sigma_0}{\rho_{f0}} \frac{\beta_f^2 T_0}{c_{pf}}\left[1 - \phi_{v0}\frac{\rho_{p0}c_{pp}}{\rho_{f0}c_{pf}} - i\phi_{v0}\frac{\rho_{p0}c_{pp}}{\rho_{f0}c_{pf}}\frac{(1-\xi)^2}{\omega\tau_t F(q,q_i)}\right]$$ (8.8.10)

We now make use of several identities to simplify this result. First, remembering that $\sigma_0 \equiv \rho_0$, we see from (7.3.6) that $\sigma_0 K_T(0) = 1/c_T^2(0)$, where $c_T(0)$ is the suspension's isothermal sound speed. Second, the thermodynamic relation (7.3.32) gives $\beta_f^2 T_0/c_{pf} = (\gamma_f - 1)/c_{sf}^2$. Finally, when $\phi_{v0} \ll 1$, $\sigma_0/\rho_{f0} \approx 1 + \eta_{m0}$, where η_{m0} is the mass loading. This quantity, it may be remembered, may be finite, even if the volume fraction is small, because the density ratio can be large. Thus,

$$\left(\frac{k}{\omega}\right)^2 = \frac{1}{c_T^2(0)} - (1 + \eta_{m0})\frac{\gamma_f - 1}{c_{sf}^2}\left[1 - \eta_{m0}\frac{c_{pp}}{c_{pf}} - i\eta_{m0}\frac{c_{pp}}{c_{pf}}\frac{(1-\xi)^2}{\omega\tau_t F(q,q_i)}\right]$$ (8.8.11)

But when the volume fraction is so small that the mass loading can also be taken as small, further simplification is possible. Thus, neglecting quantities of order η_{m0}^2, we can write

$$\left(\frac{kc_{sf}}{\omega}\right)^2 = \gamma_f\frac{c_{Tf}^2}{c_T^2(0)} - (\gamma_f - 1)\left[1 + \eta_{m0} - i\eta_{m0}\frac{c_{pp}}{c_{pf}}\frac{(1-\xi)^2}{\omega\tau_t F(q,q_i)}\right]$$ (8.8.12)

Now, consider the sound speed ratio appearing on the right-hand side of this equation. Its exact value is given by (7.3.7), which because of the assumed smallness of the volume fraction can be written as

$$\frac{c_{Tf}^2}{c_T^2(0)} \approx (1 + \eta_{m0})(1 + \phi_{v0}N_T)$$ (8.8.13)

where N_T is the fluid to particle isothermal compressibility ratio. For droplets in gases and in liquids, the product $\phi_{v0}N_T$ is normally small. Hence, to a good approximation, we may put that speed ratio equal to $1 + \eta_{m0}$. This gives,

$$\left(\frac{kc_{sf}}{\omega}\right)^2 = \gamma_f(1 + \eta_{m0}) - (\gamma_f - 1)\left[1 + \eta_{m0} - i\eta_{m0}\frac{c_{pp}}{c_{pf}}\frac{(1-\xi)^2}{\omega\tau_t F(q,q_i)}\right]$$ (8.8.14)

Separating this into real and imaginary parts and putting $k = \omega/c_s(\omega) + i\alpha$, we obtain

$$\frac{c_{sf}^2}{c_s^2(\omega)} - \hat{\alpha}^2 = 1 + \eta_{m0} - \eta_{m0}(\gamma_f - 1)\frac{c_{pp}}{c_{pf}}\frac{(1 - \xi)^2}{\omega\tau_t}\Im\left\{\frac{1}{F(q, q_i)}\right\} \qquad (8.8.15)$$

$$2\hat{\alpha}\frac{c_{sf}}{c_s(\omega)} = (\gamma_f - 1)\eta_{m0}\frac{c_{pp}}{c_{pf}}\frac{(1 - \xi)^2}{\omega\tau_t}\Re\left\{\frac{1}{F(q, q_i)}\right\} \qquad (8.8.16)$$

where we have put $\hat{\alpha} = \alpha c_{sf}/\omega$.

As with the isothermal cases treated in the previous sections, these two equations may be solved, exactly, for the attenuation and the phase velocity, but anticipating that the attenuation might be small for sufficiently small volume concentrations, we neglect its square on the left-hand side of (8.8.15) to obtain

$$\frac{c_{sf}^2}{c_s^2(\omega)} = 1 + \eta_{m0} - \eta_{m0}(\gamma_f - 1)\frac{c_{pp}}{c_{pf}}\frac{(1 - \xi)^2}{\omega\tau_t}\Im\left\{\frac{1}{F(q, q_i)}\right\} \qquad (8.8.17)$$

Similarly, on the left-hand side of the second equation, we put the speed ratio equal to 1, noting that retaining the $O(\eta_{m0})$ term would produce a correction to the attenuation of still higher order. Thus, to this approximation, the attenuation is given by

$$\hat{\alpha}_{th} = \frac{1}{2}\eta_{m0}(\gamma_f - 1)\frac{c_{pp}}{c_{pf}}\frac{(1 - \xi)^2}{\omega\tau_t}\Re\left\{\frac{1}{F(q, q_i)}\right\} \qquad (8.8.18)$$

This result for the attenuation, due to Isakovitch, (1948) and to Epstein and Carhart (1952), is referred to as the thermal attenuation coefficient and applies to very dilute emulsions and to aerosols.

Aerosols.

Here, the phase velocity and the attenuation are obtained from the above result simply by putting $\xi = 0$. This produces a result that is valid in a very wide range of frequencies, including very high ones. Also, for aerosols, it is sufficient to approximate the function $F(q, q_i)$ for frequencies that are not too large. In that limit, it is sufficient to retain only leading terms in it. Thus, with reference to (3.5.15) and keeping in mind that for aerosols k_f/k_p is negligible, we have on account of the expansions of the function $G(q_i)$ noted in Section 3.5,

$$F(q, q_i) \approx 1 + \frac{i}{hz_f^2} = -\frac{1 - i\omega\tau_t}{i\omega\tau_t}$$

where use was made of (3.3.4). Thus,

$$\frac{c_{sf}^2}{c_s^2(\omega)} = 1 + \eta_{m0} + \eta_{m0}(\gamma_f - 1)\frac{c_{pp}}{c_{pf}}\frac{1}{1 + \omega^2\tau_t^2} \qquad (8.8.19)$$

$$\hat{\alpha}_{th} = \frac{1}{2}\eta_{m0}(\gamma_f - 1)\frac{c_{pp}}{c_{pf}}\frac{\omega\tau_t}{1 + \omega^2\tau_t^2} \qquad (8.8.20)$$

These results give the thermal contributions to $(c_{sf}/c_s)^2 - 1$ and $\hat{\alpha}/\eta_{m0}$ for aerosols and are shown in Fig. 8.8.1. The figure also shows the corresponding translational results derived in Section 8.6. Because these results apply to very low volume concentrations, these contributions can be added to those resulting from the translational motion that were derived in Section 8.6. We thus obtain (Temkin and Dobbins, 1966a)

$$\frac{c_{sf}^2}{c_s^2(\omega)} = 1 + \eta_{m0}\frac{1}{1+\omega^2\tau_d^2} + \eta_{m0}\frac{c_{pp}}{c_{pf}}\frac{\gamma_f - 1}{1+\omega^2\tau_t^2} \qquad (8.8.21)$$

$$\hat{\alpha} = \frac{1}{2}\eta_{m0}\frac{\omega\tau_d}{1+\omega^2\tau_d^2} + \frac{1}{2}\eta_{m0}(\gamma_f - 1)\frac{c_{pp}}{c_{pf}}\frac{\omega\tau_t}{1+\omega^2\tau_t^2} \qquad (8.8.22)$$

For the case chosen, the thermal effects are more pronounced, as is seen in Fig. 8.8.1. As shown in the next chapter, these results have been confirmed experimentally, thus giving support to the validity of the two-phase model.

Emulsions. Here, $\xi \neq 0$ and the function $F(q, q_i)$ must be used without approximation because the thermal properties of the two phases have similar values. Owing to this, the results cannot be expressed in a more compact manner than those given by (8.8.17) and (8.8.18), and are shown graphically in Fig. 8.8.2. These results have also received experimental confirmation, as will be shown later. It should nevertheless be observed here that, in the above formulation particle, compressibility is entirely due to thermal effects, as described by the coefficient of thermal expansion of the particles, β_p. This enters in the formulation only through the factor ξ. Hence, the results shown in the figure differ from those corresponding to a rigid particle only by a constant factor, namely, $(1 - \xi)^2$. Pressure differences between the two phases, which were neglected in the derivation, produce further changes, but these cannot be simply added to the thermal and dynamical effects because the pressure and the temperature in the particle are coupled. We postpone until the next chapter the discussion of such effects through a more complete theory that includes the three effects treated separately in this chapter.

8.9 Flow of a Dusty Gas Across a Shock Wave

The previous applications of the two-phase model have dealt with certain aspects of the propagation of monochromatic sound waves that illustrate, separately, some of the mechanisms that affect the propagation. A common feature in those applications was that the amplitudes of the waves treated in them were small. This enabled us to drop all the nonlinear terms in the equations of motion for both phases, a step that greatly simplified the solution.

While linear sound waves are the main focus of this book, it is useful to consider a flow in which the nonlinear terms must be retained. The simplest example of such a

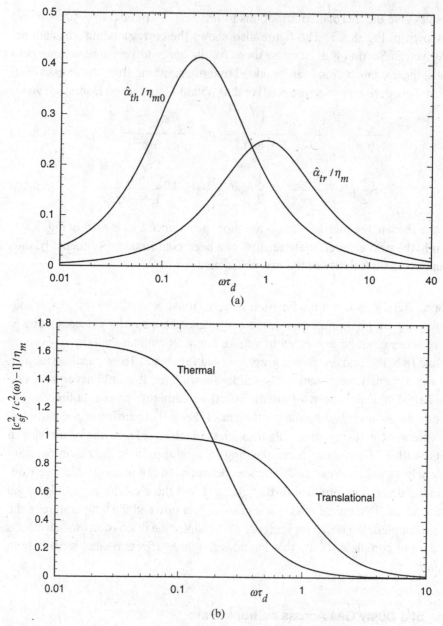

Figure 8.8.1. (a) Scaled thermal and translational attenuations in an aerosol. $D = 100 \ \mu$m. (b) Scaled thermal and translational dispersions in an aerosol. $D = 100 \ \mu$m.

flow is that of a dusty gas through a normal shock wave. The simplicity is due to the availability of an equation of state and the near rigidity of the particles.

Of course, a shock wave is a finite-amplitude sound wave that, owing to nonlinear distortion, has developed a very sharp front. In a viscous, thermally conducting gas, the variations across the front are rapid, but continuous. However, except for very weak shock waves, the variations are so rapid that for many purposes they may be regarded

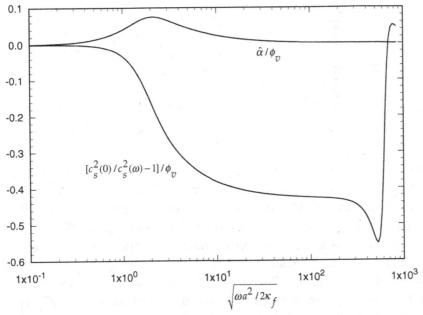

Figure 8.8.2. Scaled thermal attenuation and dispersion in an emulsion of toluene droplets in water. $D = 100\ \mu\text{m}$.

as having zero thickness, in which case the velocity, pressure, and temperature of the gas vary discontinuously across the front.

Now, the wave profile behind and ahead of the discontinuous front, called a shock from now on, generally varies with distance from it in a manner that depends on the initial profile of the sound wave. However, it is useful to consider, as a limiting form, a wave that has a flat profile behind the front (i.e., a wave where the pressure, velocity, and temperature have constant values, as sketched in Fig. 8.9.1). As depicted in the figure, the wave is compressive (i.e., the wave increases the pressure of the fluid it overtakes). Waves that decrease the pressure (i.e., waves of expansion), cannot develop sharp fronts because regions having lower pressure travel with smaller speeds and do not overtake those in front of them, as it happens with compressive waves.

Relative to a fixed observer, the motion indicated in the figure is unsteady. To make it steady, we superimpose a uniform velocity of magnitude $u_1 = c_{shock}$. The resulting flow is then a steady, uniform flow traveling from left to right, and crossing a shock front that is *fixed* in space. In gas dynamics, the motion is referred to as flow through a normal shock.

The Normal Shock in a Perfect Gas

We would like to study the one-dimensional, steady flow of a dusty gas across a normal shock. But, before we discuss the effects of particles, it is useful to consider the motion

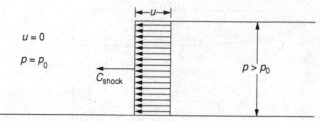

Figure 8.9.1. Wave with a rectangular profile.

when no particles are present. The motion is depicted in Fig. 8.9.2, with the double line representing the fixed shock. The gas approaching the shock with supersonic velocity u_1 is assumed to be a perfect gas with constant specific heats. Furthermore, in the spirit of the two-phase model, the viscosity and heat conductivity of the gas are assumed to be zero. On crossing the shock, the pressure and temperature of the gas increase, while its velocity is decreased to a subsonic value. We denote these quantities by $p_{f,2}$, $T_{f,2}$, and $u_{f,2}$, respectively. Their values follow from the conservation equations for the gas alone. When these equations are reduced for steady flow conditions, and the perfect gas assumption is made, they can be integrated to give expressions that relate $p_{f,2}$, $T_{f,2}$, and $u_{f,2}$ to the corresponding quantities ahead of the shock. These expressions also follow from (8.5.8) to (8.5.10) when we disregard the particles and note that, for a perfect gas, $\beta_f T = 1$. Thus,

$$\dot{m}_f = \rho_{f,1} u_1 = \rho_{f,2} u_{f,2} \tag{8.9.1}$$

$$\dot{m}_f u_1 + p_1 = \dot{m}_f u_{f,2} + p_{f,2} \tag{8.9.2}$$

$$c_{pf} T_1 + \tfrac{1}{2} u_1^2 = c_{pf} T_{f,2} + \tfrac{1}{2} u_{f,2}^2 \tag{8.9.3}$$

The perfect gas equation for the gas, $p_f = \rho_f \Re_f T_f$, is also available.

The solution to the system of algebraic equations may be expressed in various forms (Landau and Lifshitz, 1959; Liepmann and Roshko, 1959). A form suitable for our purposed expresses the solution in terms of the Mach number of the incoming

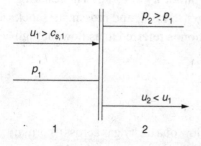

Figure 8.9.2. Standing shock wave.

flow, $M_1 = u_1/c_{sf,1}$, and on the specific heat ratio, γ_f of the gas:

$$\frac{u_{f,2}}{u_1} = \frac{\rho_{f,1}}{\rho_{f,2}} = \frac{2 + (\gamma_f - 1)M_1^2}{(\gamma_f + 1)M_1^2} \tag{8.9.4}$$

$$\frac{T_{f,2}}{T_1} = \frac{[2\gamma_f M_1^2 - (\gamma_f - 1)][2 + (\gamma_f - 1)M_1^2]}{(\gamma_f + 1)M_1^2} \tag{8.9.5}$$

The pressure ratio may also be expressed in the same manner, but in view of the perfect gas equation of state, and of (8.9.4) for the density ratio, we also have

$$\frac{p_{f,2}}{p_1} = \frac{T_{f,2}/T_1}{u_{f,2}/u_1} \tag{8.9.6}$$

Since $M_1 > 1$ and $\gamma_f \geq 1$, these show that the gas is decelerated as it crosses the shock, while its temperature and pressure increase. It is to be noted that no further changes occur downstream of shock. That is, the pressure, temperature, and velocity remain equal to $p_{f,2}$, $T_{f,2}$, and $u_{f,2}$, respectively.

The Normal Shock in a Dusty Gas

Let us now consider the flow of a dusty gas through a normal shock, assuming that the shock exists. For this purpose, we take an incoming equilibrium flow of gas and particles toward the shock – that is, the gas and the particles approach the shock with the same velocity, pressure, and temperature. Since the particles are solid, we could take them to be rigid. But to ensure complete upstream equilibrium, we will assume that their material is perfectly elastic, so that $p_p = p_f$ at all times. This assumption can be dispensed with, but its use enables us to use the equilibrium relations derived in Chapter 7. We will also limit the discussion to dusty gases having very small volume concentrations.

Now, consider the gas and the particles approaching the shock together, at conditions p_1, T_1, and u_1. The density of the dusty gas is $\sigma_1 = \sigma_{f,1} + \sigma_{p,1}$, but the densities of the two phases are, of course different, being $\sigma_{f,1} = \rho_{f,1}(1 - \phi_{v,1}) \approx \rho_{f,1}$ for the gas and $\sigma_{p,1} = \rho_{p,1}\phi_{v,1}$ for the particles.

With reference to Fig. 8.9.3, let the conditions *immediately* downstream of the shock be denoted with the subindex 2. On traversing the shock, the flow variables for the gas jump discontinuously; but, given that the dusty gas is very dilute, we expect that the gas will have, immediately after the shock, the same conditions as when no particles are present (i.e., $p_{f,2}$, $T_{f,2}$, and $u_{f,2}$). This expectation is confirmed later. The particle's variables, on the other hand, are not changed as they cross the shock – that is, since the shock wave thickness is of the order of the mean free path in the gas, the time taken by a particle to traverse it is much too short for the particle to experience

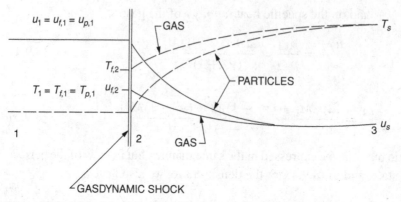

Figure 8.9.3. Schematic of the dusty flow through a normal shock (after Marble, 1963).

significant changes. Thus, on the downstream of the shock, the particle temperature and velocity are those in the approach flow, $T_{p,2} = T_1$ and $u_{p,2} = u_1$, respectively. These differ from those corresponding to the gas at that location. These velocity and temperature differences, or slips, destroy the equilibrium that existed in the suspension before the shock. It then follows that beyond the shock, there follows a region where the temperatures and velocities of the particles and the fluid undergo equilibration processes, with the particle force and heat transfer coming into play to decrease those differences.

The rates at which the temperature and velocity differences decrease are generally different, meaning that the distance (or time) required for temperature equalization is generally different from that required by the velocity. However, sufficiently far downstream of the shock, a new equilibrium condition exists (denoted by subindex 3 in Fig. 8.9.3). Both the equilibration process and the new equilibrium condition are of interest, and may be studied with the two-phase model, though the equilibration process is more difficult to calculate because it involves the force and the heat transfer rates. Below we first give equations relating the flow variables to one another anywhere in the flow. The equations are then solved to obtain the solution in the downstream equilibrium region. Finally, some remarks are then given that relate to the equilibration process $2 \rightarrow 3$.

Two-Phase Equations for Steady Motion

We begin by considering the two-phase equations of motion applicable to the one-dimensional flow through a standing, normal shock wave. To keep the flow one-dimensional, we shall neglect the body force. Also, by assumption, the flow is steady throughout. The incoming flow velocity is still u_1, but this time the material flowing across the shock is a two-phase medium made of a gaseous phase and a particulate phase. Thus, its sound speed and density are different from those of the gas alone. The gas will be assumed to be a perfect gas having constant specific heats, whereas particulate material will be regarded as an elastic solid, whose density, ρ_p, is much larger than that of the gas, and whose specific heats will be taken to be constant. For a rigid material, the two specific heats are equal, but our regarding the particles as elastic solids implies that they are generally different, although the differences are very small. On the other hand, we shall assume that the pressure, p_p, in the particulate phase is equal, at all instants, to that of the gas, p_f. Finally, the motion is inviscid, and there is no heat added from the exterior.

Let us start by writing the two-phase equations as they apply to steady, one-dimensional flow, which is in the direction of the positive x-axis. These are, with $p_f = p_p = p$,

$$\frac{d(\sigma_f u_f)}{dx} = 0 \qquad \frac{d(\sigma_p u_p)}{dx} = 0 \qquad \text{(8.9.7a,b)}$$

$$\sigma_f u_f \frac{du_f}{dx} = -\frac{\sigma_f}{\rho_f}\frac{dp}{dx} + f_p \qquad \sigma_p u_p \frac{du_p}{dx} = -\frac{\sigma_p}{\rho_p}\frac{dp}{dx} - f_p \qquad \text{(8.9.8a,b)}$$

$$\sigma_f u_f \left(c_{pf}\frac{dT_f}{dx} - \frac{\beta_f T_f}{\rho_f}\frac{dp}{dx} \right) = \dot{q}_p + f_p(u_p - u_f)$$

$$\sigma_p u_p \left(c_{pp}\frac{dT_p}{dx} - \frac{\beta_p T_p}{\rho_p}\frac{dp}{dx} \right) = -\dot{q}_p \qquad \text{(8.9.9a,b)}$$

$$p = \rho_f \mathfrak{R}_f T_f \qquad \text{(8.9.10)}$$

These are complemented by equations for the force and the heat transfer rate. Our first goal is to reduce this system of equations to another that can be integrated, so that we can write relations similar to those applying to the gas alone. The first pair can be integrated as it is. Thus, integrating the two equations between station 1, and any other point in the flow, we obtain $\sigma_{f,1}u_1 = \sigma_f u = \dot{m}_f$ and $\sigma_{p,1}u_1 = \sigma_p u = \dot{m}_p$, respectively. The last relations show that the mass flow rates of particles and gas are separately constant. The sum of these two flow rates is the total mass flow rate, $\dot{m}$, also constant. In view of the definition of the mass loading, $\eta_m = \dot{m}_p/\dot{m}_f$, we may write $\dot{m} = (1 + \eta_m)\dot{m}_f$, noting that η_m is also constant.

Let us now consider the momentum and energy equations. These contain the particle force and heat transfer; but, if we add the two momentum equations, the force term drops out, yielding

$$\dot{m}_f \frac{du_f}{dx} + \dot{m}_p \frac{du_p}{dx} + \frac{dp}{dx} = 0 \qquad \text{(8.9.11)}$$

Adding the two energy equations eliminates the heat transfer rate, giving

$$\dot{m}_f c_{pf}\frac{dT_f}{dx} + \dot{m}_p c_{pp}\frac{dT_p}{dx} - [(1 - \phi_v)\beta_f T_f u_f + \phi_v \beta_p T_p u_p]\frac{dp}{dx}$$
$$= f_p(u_p - u_f) \qquad \text{(8.9.12)}$$

To eliminate the force from this, we multiply the fluid and particulate momentum equations by the corresponding velocities, respectively, and obtain

$$f_p u_f = \frac{1}{2}\dot{m}_f \frac{du_f^2}{dx} + (1 - \phi_v)u_f \frac{dp}{dx} \quad \text{and} \quad f_p u_p = \frac{1}{2}\dot{m}_p \frac{du_p^2}{dx} + \phi_v u_p \frac{dp}{dx}$$

$$(8.9.13a,b)$$

Substituting these expressions in (8.9.12) gives

$$\dot{m}_f \left(c_{pf}\frac{dT_f}{dx} + \frac{1}{2}\frac{du_f^2}{dx} \right) + \dot{m}_p \left(c_{pp}\frac{dT_p}{dx} + \frac{1}{2}\frac{du_p^2}{dx} \right)$$

$$+ \left[(1 - \phi_v)(1 - \beta_f T_f)u_f + \phi_v(1 - \beta_p T_p)u_p \right]\frac{dp}{dx} = 0 \qquad (8.9.14)$$

This can be further reduced. First, since the gas is perfect, $\beta_f T_f = 1$, so that the first term in the square bracket vanishes. Next, since the volume concentration is very small, we may drop the remaining term in the square bracket, noting that the second term on the left-hand side must be retained because the product $\rho_p \phi_v$, contained in $\dot{m}_p$, is not negligible. With these assumptions, (8.9.14) reduces to

$$\dot{m}_f \frac{d}{dx}\left(c_{pf}T_f + \frac{1}{2}u_f^2 \right) + \dot{m}_p \frac{d}{dx}\left(c_{pp}T_p + \frac{1}{2}u_p^2 \right) = 0 \qquad (8.9.15)$$

Equations (8.9.11) and (8.9.15) can now be integrated. Thus, adding the mass conservation equation, we obtain

$$\dot{m} = \sigma_f u_f(1 + \eta_m) = \text{Constant}. \qquad (8.9.16)$$

$$\dot{m}_f u_f + \dot{m}_p u_p + p = \text{Constant}. \qquad (8.9.17)$$

$$c_{pf}T_f + \frac{1}{2}u_f^2 + \eta_m\left(c_{pp}T_p + \frac{1}{2}u_p^2 \right) = \text{Constant}. \qquad (8.9.18)$$

These are complemented by the equation of state, $p = \rho_f \Re_f T_f$, and apply anywhere in the flow.

Conditions Immediately After the Shock

Let us apply the previous equations to station 1, upstream of the shock, and 2, immediately after the shock, where the variables have been labeled with the subindex 2. This gives

$$\sigma_{f,2}u_{f,2} + \sigma_{p,2}u_{p,2} = \sigma_{f,1}u_1 + \sigma_{p,1}u_1 \qquad (8.9.19)$$

$$\dot{m}_f u_{f,2} + \dot{m}_p u_{p,2} + p_2 = \dot{m}_f u_1 + \dot{m}_p u_1 + p_1 \qquad (8.9.20)$$

$$c_{pf}T_{f,2} + \frac{1}{2}u_{f,2}^2 + \eta_m\left(c_{pp}T_{p,2} + \frac{1}{2}u_{p,2}^2 \right) = c_{pf}T_1 + \frac{1}{2}u_1^2 + \eta_m\left(c_{pp}T_1 + \frac{1}{2}u_1^2 \right)$$

$$(8.9.21)$$

But, as argued before, the particles experience no change of velocity and temperature on crossing the shock (though their pressure remains equal to that of the gas). Then, $T_{p,2} = T_1$ and $u_{p,2} = u_{p1}$. Substituting these values in the above system, we find that the particle terms on both sides of each equation cancel out. Furthermore, since $\phi_v \ll 1$, we have $\sigma_f \approx \rho_f$. Hence,

$$\dot{m}_f = \rho_{f,2} u_{f,2} = \rho_{f,1} u_1 \tag{8.9.22}$$

$$\dot{m}_f u_{f,2} + p_2 = \dot{m}_f u_1 + p_1 \tag{8.9.23}$$

$$c_{pf} T_{f,2} + \tfrac{1}{2} u_{f,2}^2 = c_{pf} T_1 + \tfrac{1}{2} u_1^2 \tag{8.9.24}$$

Comparison with (8.9.1)–(8.9.3) shows that the gas in the dusty gas suspension obeys the same jump conditions as a gas crossing, alone – a normal shock. This is due to the very small volume concentration of the particles and to the assumed negligible thickness of the shock. Thus, if the dusty gas approaches the shock with velocity u_1, pressure p_1, and temperature T_1, its pressure, velocity, and temperature in condition 2, p_2, $T_{f,2}$, and $u_{f,2}$, respectively, are given by equations (8.9.4)–(8.9.6). Thus, as anticipated previously, the gas state right ahead of the shock is the same as it would be without particles. As those equations show, the values of the flow variables depend only on the approach Mach number and on the gas specific heat ratio.

Values of $u_{f,2}/u_1$, $T_{f,2}/T_1$, and p_2/p_1 corresponding to a given incident Mach numbers may be obtained from (8.9.4) to (8.9.6). We note that, on crossing the shock, the gas experiences significant changes. For example, at an approach Mach number, $M_1 = 1.5$ in air, that velocity decreases to about one-half of its initial value, meaning that its kinetic energy per unit mass has decreased to about one-fourth of its initial value. Some of that energy goes into thermal heating. Thus, for the same Mach number, the temperature increases by a factor of about 1.3, also in air.

Since the particle conditions at station 2 are equal to those at 1, we may also write $u_{f,2}/u_1$ as $u_{f,2}/u_{p,2}$, and $T_{f,2}/T_1$ as $T_{f,2}/T_{p,2}$. We therefore see that, in crossing the shock, significant velocity and temperature differences between particles and gas are found. These differences bring in the equilibration mechanisms into play, namely particle force and heat transfer. Eventually, a new equilibrium state is reached. This is considered next.

Equilibrium State Downstream of the Shock

We now wish to determine the conditions of the gas and the particles in the equilibrium state that exists sufficiently far downstream of the shock. While we do not know how far this region is from the shock, the conditions may be obtained from the equilibrium relations obtained earlier. The derivation is as above, but now the particulate phase contribution to the conservation equations does not drop out.

Thus evaluating (8.5.8)–(8.5.10) at a point in region 3, we obtain

$$\dot{m} = \text{Constant} \tag{8.9.25}$$

$$\dot{m}u_3 + p_3 = \dot{m}u_1 + p_1 \tag{8.9.26}$$

$$c_p T_3 + \tfrac{1}{2}u_3^2 = c_p T_1 + \tfrac{1}{2}u_1^2 \tag{8.9.27}$$

where we have used the expression (7.3.26) for the dusty gas' specific heat at constant pressure.

We thus find that the flow variables in the two equilibrium conditions are connected by equations of the same type as those that apply to a gas flowing thorough a shock wave, with the downstream condition being labeled 3, instead of 2. Hence, the solution is again given by (8.9.4)–(8.9.6), provided we make note of the differences between the flows. Thus, (8.9.4)–(8.9.6) refer to the flow of a perfect gas, approaching the shock at speed u_1 and acquiring, on crossing the shock, the conditions specified by those equations. Presently, we have a dusty gas approaching the shock at the same speed and eventually reaching equilibrium conditions 3. Thus, in addition to the obvious change in subindices for the downstream condition, the following ought to be kept in mind. First, since u_1 is the approach velocity with or without particles, the approach Mach numbers differ because the sound speed in a dusty gas is different from that in a gas alone. Thus, the approach Mach number should be based on the equilibrium sound speed corresponding to the dusty gas in condition 1, [i.e., $c_{s,1}$, where $c_{s,1}$ is the equilibrium sound speed, $c_s(0)$, evaluated at 1]. Hence, the approach Mach number that should be used in the solution is $u_1/c_{s,1}$. To avoid confusion with the Mach number that applies when no particles are present, $M_1 = u_1/c_{sf,1}$, we denote $u_1/c_{s,1}$ by $\overline{M}_1$ noting that it is equal to $M_1(c_{sf,1}/c_{s,1})$. Second, the specific heat ratio appearing in those equations is that corresponding to the dusty gas, $\overline{\gamma}$.

The values of $\overline{M}_1$ and $\overline{\gamma}$ can be significantly different from M_1 and γ_f, even when the volume concentration is small. A numerical example is useful. Take a dusty gas composed of silica particles in air. Suppose that the volume concentration of the particles is $\phi_{v,1} = 10^{-4}$. Then, $\eta_m \approx 0.18$. The corresponding sound speed ratio, $c_{sf,1}/c_{s,1}$, is about 1.12 so that a value of M_1 equal to 1.5 corresponds to a value of $\overline{M}_1$ equal to 1.68. The specific heat ratio of the dusty gas is given by (7.3.35) and is approximately equal to $\overline{\gamma} = 1.33$. With these remarks in mind, we have

$$\frac{u_3}{u_1} = \frac{\sigma_1}{\sigma_3} = \frac{2 + (\overline{\gamma} - 1)\overline{M}_1^2}{(\overline{\gamma} + 1)\overline{M}_1^2} \tag{8.9.28}$$

$$\frac{T_3}{T_1} = \frac{\left[2\overline{\gamma} M_1^2 - (\overline{\gamma} - 1)\right]\left[2 + (\overline{\gamma} - 1)\overline{M}_1^2\right]}{(\overline{\gamma} + 1)\overline{M}_1^2} \tag{8.9.29}$$

$$\frac{p_3}{p_1} = \frac{2\overline{\gamma}\,\overline{M}_1^2}{(\overline{\gamma} + 1)} - \frac{\overline{\gamma} - 1}{\overline{\gamma} + 1} \tag{8.9.30}$$

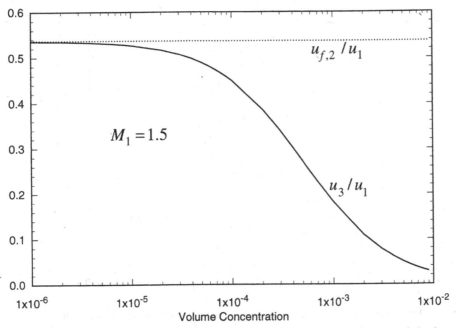

Figure 8.9.4. Downstream equilibrium velocities in a dusty gas flow through a normal shock.

It should be noted that the size of the particles does not appear in these equations. This is as it should be because conditions at 1, like those at 3, refer to an equilibrium state. However, the particles' presence affects the gas. To see this, we may evaluate the above ratios in those conditions as a function of the particulate concentration, for a given approach Mach number. The results then give those ratios in different flows, each one having a different concentration.

Figures 8.9.4 and 8.9.5 show the values of the velocity and temperature ratios, respectively, as a function of the volume concentration. The specific example chosen is that of a dusty gas composed of silica particles and air, for an approach Mach number, $M_1 = 1.5$. The ratios plotted give the velocity and temperature of the gas and the particles in equilibrium region 3, scaled with their corresponding values in condition 1. Because the state of the particles in condition 1 is the same as that in 2, the curves labeled u_3/u_1 and T_3/T_1 also show the changes in temperature and velocity that the particles must undergo between conditions 2 and 3. The corresponding changes for the gas can be inferred from the straight lines that are tangential to the above curves when the volume fraction is very small.

Consider first u_3/u_1. To appreciate the changes implied by Fig. 8.9.4, take a flow having a volume concentration equal to $\phi_v = 10^{-3}$. For that concentration, we see that $u_3/u_1 \approx 0.17$. Thus, the particles must lose about one-sixth of their initial momentum, and about one-thirty-sixth of their kinetic energy in going from state 2 to equilibrium condition 3. For the gas, we observe that its velocity must experience, in the equilibration region an additional decrease. This decrease reduces the gas kinetic energy to about one-tenth of its value in state 2.

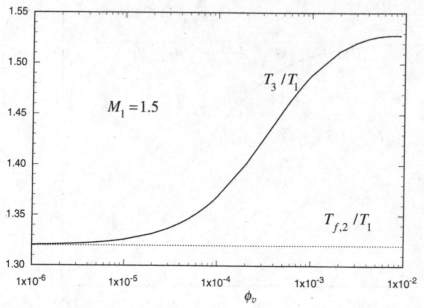

Figure 8.9.5. Downstream equilibrium temperatures in a dusty gas flow through a normal shock.

The particle loss of momentum was to be expected because immediately after the shock their velocity is larger than that of the gas. The decrease of the gas velocity, on the other hand, needs further clarification because it might have been expected that it would have increased owing to momentum transfer from the particles. However, as our derivation of the energy equation shows, the particle is unable to do this. All of the "extra" kinetic energy it loses is dissipated, thus producing additional heating. Similarly, the gas, in reaching equilibrium with the particle must spend additional energy, and this results in the additional velocity decrease that occurs between conditions 2 and 3.

Remarks About the Nonequilibrium Region

The overall changes taking place in the dusty gas in reaching equilibrium conditions 3 were obtained above making use of various simplifying that enabled us to reduce the equations of motion for a dusty gas to those of a medium satisfying the normal shock relations between equilibrium states. While the information we have gathered about the downstream equilibrium condition has elucidated some important features, the approach to that condition remains unknown. For example, we do not know where the downstream equilibrium condition is located in relation to the shock. Or stated differently, the length of the relaxation regions, where velocity and temperature readjustment takes place, is unknown. Such information requires that the pressure, velocities, and temperatures be determined at every point in the transition region.

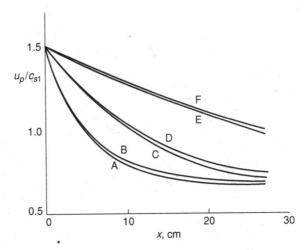

Figure 8.9.6. Particulate velocity in the equilibration region. [After Rudinger (1969). © 1969, Marcel Dekker.]

Now, to determine the conditions in the transition region, we first write the equations that need to be solved. First, we have the conservation conditions

$$\dot{m} = \dot{m}_f(1 + \eta_m) = \text{Constant} \tag{8.9.31}$$

$$\dot{m}_f u_f + \dot{m}_p u_p + p = \text{Constant} \tag{8.9.32}$$

$$c_{pf} T_f + \tfrac{1}{2}u_f^2 + \eta_m \left(c_{pp} T_p + \tfrac{1}{2}u_p^2\right) = \text{Constant} \tag{8.9.33}$$

In addition, we have the equation of state and the particle momentum and energy equations, (8.9.8) and (8.9.8), respectively, which involve the particle force and the heat transfer. When the volume fraction is very small, those quantities are $f_p = -nF_p$ and $\dot{q}_p = -n\dot{Q}_p$, so that

$$\dot{m}_p \frac{du_p}{dx} = nF_p \tag{8.9.34}$$

$$\dot{m}_p c_{pp} \frac{dT_p}{dx} = n\dot{Q}_p \tag{8.9.35}$$

Thus, to close the system, we require specific forms for the single-particle force and heat transfer rate. As we have seen, these depend on the relative velocity and temperature, and do not involve additional variables. Thus, the system consists of eight equations for the eight unknowns appearing in them: u_f, u_p, T_f, T_p, ρ_f, p, F_p, and $\dot{Q}_p$.

It is evident that a solution of the system for realistic conditions will require numerical calculations regardless of the model used for the particle force and heat transfer rates. Such calculations have been performed by several investigators since Carrier first studied the problem in 1958. For our purposes, which are meant to illustrate the approach to equilibrium in state 3, it is sufficient to show some of numerical results obtained by Rudinger.

Figures 8.9.6 and 8.9.7show the particulate phase velocity and temperature for the particulate phase in a dusty gas composed of 100-μm glass particles in air, at

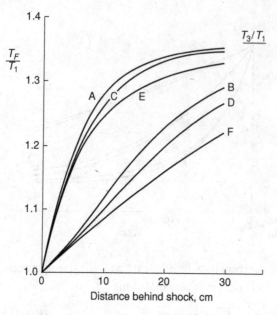

Figure 8.9.7. Particulate temperature in the equilibration region. [After Rudinger (1969). © 1969, Marcel Dekker.]

$M_1 = 1.5$, and Fig. 8.9.8 shows the corresponding variations in the gaseous phase. The mass loading was $\eta_m = 0.2$, corresponding to $\phi_v = 6 \times 10^{-4}$. The calculations were performed using for the particle force three models: Stokes' law, Ingebo's correlation, and the steady drag correlation (see Section 4.6), and using two models for the heat transfer rate: the pure conduction, quasi-steady heat transfer rate described by a Nusselt number equal to 2 (see Section 3.2), and by the Ranz-Marshall correlation given by (Rudinger, 1964)

$$Nu = 2 + 0.6 Re^{1/2} Pr^{1/3} \qquad (8.9.36)$$

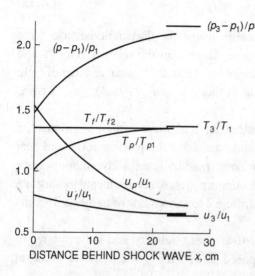

Figure 8.9.8. Gas pressure, velocity, and temperature in the equilibration region. [After Rudinger (1969). © 1969, Marcel Dekker.]

Table 8.9.1. *Drag and heat transfer correlations used for Figs. 8.9.6 and 8.9.7*

Drag coefficient	Heat transfer correlation	
	Equation (3.2.18)	Equation (8.9.3b)
Steady	A	B
Ingebo	C	D
Stokes' law	E	F

Each of the figures includes several curves that were obtained with three drag coefficients and the two heat transfer rates. Table 8.9.1 explains the labels appearing on the curves. Consider the particulate velocity first shown in Fig. 8.9.6 in terms of $u_p/c_{s,1}$. It is seen that the differences between the different heat transfer rate models are small. On the other hand, the drag influences the results significantly, with the standard drag producing the fastest approach to equilibrium. But, even then, velocity equilibration is relatively slow, requiring, as the figure implies, a considerable distance to be achieved.

Temperature equalization is even slower, as shown in Fig. 8.9.7. Here, as might be expected, different heat transfer rate correlations result in significantly different particulate temperature; but, as the figure shows, the drag force also plays a significant role. This is due to the different amounts of energy dissipated by viscous effects that each of the drag models implies.

Figure 8.9.8 shows Rudinger's results for the variations of the velocity, pressure, and temperature in both gas and particulate phase obtained for the same conditions as above, but obtained with Ingebo's correlation for the drag and (8.9.36) for the heat transfer correlation. It is seen that the velocity of the gas decreases monotonically in the relaxation zone. As mentioned previously, this is due to the inability of the particles to transfer momentum to the gas.

Results similar to those shown above have also been obtained by other investigators. However, it should be pointed out that the results obtained so far for the equilibration processes are based on steady correlations. The effects of acceleration must significantly affect those processes, but calculations involving such effects have not yet been performed.

There is, in addition, another important factor not taken into account by the model. Shock-induced relative motions are usually accompanied by collisions, as the experiments described in Section 4.12 show. These collisions may occur even if the particles are equal-sized and may also produce particle agglomeration. Such effects are important in a wide variety of situations ranging from industrial applications to astrophysical phenomena.

This concludes our discussion of a nonlinear, dusty gas flow. Many other examples have been treated in the literature on the subject, which continues to expand rapidly, particularly with regard to numerical simulations of two-phase flows.

8.10 Concluding Remarks

The chapter has presented the basic equations that describe the motions of suspensions as in terms of a two-phase model. The equations include thermal effects and apply to dilute aerosols, bubbly liquids, emulsions, and hydrosols. The model was used to discuss several basic motions, including specific cases of sound propagation and the flow through a normal shock. The bibliography section includes several references to additional examples.

9

Sound Propagation in Suspensions

9.1 Introduction

In the last chapter, we used the two-phase model of suspensions to consider several examples of sound propagation in suspensions. The examples were chosen to illustrate the main aspects of the propagation of sound waves, but they refer to some physically unrealistic conditions, such as the isothermal aerosol. Of course, the results obtained for those conditions may apply in limited portions of the frequency range; but, generally speaking, they leave out important effects. For example, as we saw in Chapter 6, the neglect of thermal effects on the pulsations of bubbles is not warranted at frequencies that are below resonance.

From a fundamental point of view, it is desirable to study sound propagation in suspensions without imposing such restricting conditions. To be sure, the two-phase model allows such a study, but the inclusion of all of the effects that play a role in the propagation would require the two complete sets of equations. While the system can be easily solved for linear motions, it is evident that its solution would generally require numerical computations and would produce results difficult to interpret. Furthermore, results valid for a wide range of frequencies would be obtained only if the particulate force and heat transfer rate are specified in their most general forms, thus increasing the complexity of the system.

In this chapter, we consider the propagation of sound waves in dilute suspensions using various techniques, models, and approximations. The goal is to obtain a theory, applicable to dilute suspensions, that simultaneously includes the various particle motions we have discussed, while retaining temperature changes. A theory that does this is developed in this chapter. That theory could be presented immediately, but it is better to approach the problem from various perspectives. In addition to confirming the results obtained with one another, this provides valuable additional understanding of the physical processes that take place in suspensions.

The initial discussion deals with sound propagation in real fluids that contain no particles in them. This is meant to clarify some of the concepts that follow later in the chapter, and to put in context the assumption that viscous and thermal effects in the main body of the fluid can be neglected in dilute suspensions.

9.2 Propagation in a Fluid Without Particles

The propagation of sound waves in fluids is determined by the conservation equations for mass, momentum, and energy, complemented by the fluid's equation of state. The simplest type of wave that can be considered is the plane, one-dimensional wave. It can be studied by means of the linearized, one-dimensional fluid dynamic equations for a viscous, heat-conducting, compressible fluid:

$$\frac{\partial \rho'_f}{\partial t} + \rho_{f0}\frac{\partial u_f}{\partial x} = 0 \tag{9.2.1}$$

$$\frac{\partial u_f}{\partial t} + \frac{1}{\rho_{f0}}\frac{\partial p'_f}{\partial x} = \tfrac{4}{3}\nu_f\frac{\partial^2 u_f}{\partial x^2} \tag{9.2.2}$$

$$\frac{\partial T'_f}{\partial t} - \frac{\beta_f T_0}{\rho_{f0}c_{pf}}\frac{\partial p'_f}{\partial x} = \kappa_f\frac{\partial^2 T'_f}{\partial x^2} \tag{9.2.3}$$

$$p'_f = c_{Tf}^2\rho'_f + \rho_{f0}\beta_f c_{Tf}^2 T'_f \tag{9.2.4}$$

When the viscosity and thermal conductivity vanish, these equations can be combined to yield the wave equation, showing that plane waves will travel unattenuated with speed c_{sf}, and that the temperature, density, and velocity fluctuations in the fluid are connected by means of the isentropic relations

$$P'_f = \rho_{f0}c_{sf}U_f, \quad \rho'_f = P'_f/c_{sf}^2 \quad \text{and} \quad \Theta'_f = (\beta_f T_0 c_{sf}/c_{pf})U_f \tag{9.2.5a,b,c}$$

It should be noted that P'_f, Θ'_f, and U_f refer to isentropic variations in a plane wave in a fluid without particles.

When the transport coefficients are nonzero, we can obtain modified wave equations for the pressure or the velocity. Alternately, we may seek for monochromatic, traveling-wave solutions, in which case all variables depend on time and distance through the exponential factor $\exp[i(kx - \omega t)]$, where k is a complex wavenumber, given by $k = k_r + i\alpha$. In order for the amplitude of the wave not to become infinitely large as x increases, it is necessary (for waves traveling toward larger values of x) that $\alpha(\omega)$ be nonnegative. Similarly, the real part of k – which is equal to $\omega/c(\omega)$, where $c(\omega)$ is the phase velocity of the waves – cannot be negative. The basic sound-propagation problem is to determine $c(\omega)$ and $\alpha(\omega)$.

Now, if a wavelike solution is assumed for all the variables appearing in (9.2.1) to (9.2.4), a system of algebraic equations will be obtained for the unknowns ρ'_f, T'_f, u_f, and p'_f. This system has a nontrivial solution if the determinant of the coefficient matrix vanishes, in which case it provides the complex wavenumber k. If this is expressed in a nondimensional manner by means of $L = kc_{sf}/\omega$, where

$$L = c_{sf}/c(\omega) + i\alpha(\omega)c_{sf}/\omega \tag{9.2.6}$$

then L satisfies (Temkin, 1981)

$$L^2(1 + i\omega\tau_\kappa L^2) - (1 + i\gamma_f\omega\tau_\kappa L^2)(1 + i\gamma_f\omega\tau_\nu L^2) = 0, \tag{9.2.7}$$

where

$$\tau_\nu = \frac{4}{3}\frac{\nu_f}{c_{sf}^2} \quad \text{and} \quad \tau_\kappa = \frac{\kappa_f}{c_{sf}^2}. \tag{9.2.8a,b}$$

These two quantities have dimensions of time, and are called the viscous and thermal relaxation times. They represent a measure of the rapidity with which molecular collisions restore mechanical and thermal equilibrium. Both time scales are very small, being of the order of 10^{-10} sec in gases and even smaller in liquids. Thus, if the frequency is less than 1 MHz, the products $\omega\tau_\nu$ and $\omega\tau_\kappa$ are rather small, so that (9.2.7) reduces to $L^2 \approx 1$. This corresponds to the ideal-fluid case, where $\alpha = 0$ and $c(\omega) = c_{sf}$, and represents a sound wave traveling with constant amplitude and with a constant speed equal to c_{sf} – the isentropic sound speed for the fluid.

Consider now larger frequencies. Here, $\omega\tau_\nu$ and $\omega\tau_\kappa$ are finite so that we must solve (9.2.7), which is a quadratic for L^2. The two values of L^2 satisfying that equation, represent monochromatic solutions of the full system, but only one of them represents a sound wave. The other represents highly attenuated viscous and thermal waves, of the type encountered in our discussion of particle velocity and temperature oscillations. As we saw there, these waves travel away from the boundaries that produce them with speeds proportional to $(\omega\tau_\nu)^{1/2}$, and with an amplitude that decreases exponentially according to $\exp(-x/\delta_\nu)$, in the case of velocity oscillations and similar expressions for the transverse temperature wave. Thus, at distances that are large, compared with the corresponding penetration depths, they may be ignored.

To obtain the values of L that correspond to the acoustic mode, we take advantage of the smallness of $\omega\tau_\nu$ and $\omega\tau_\kappa$, and look for solutions in the vicinity of $L = 1$, corresponding to the ideal case. To do this, we put $L^2 = 1$ inside each of the three parentheses in (9.2.7), so that

$$L^2 \approx \frac{(1 + i\gamma_f\omega\tau_\kappa)(1 + i\gamma_f\omega\tau_\nu)}{1 + i\omega\tau_\kappa} \tag{9.2.9}$$

The real and imaginary parts of this can be easily obtained, but since $\omega\tau_\nu$ and $\omega\tau_\kappa$ are less than 0.1 for frequencies as high as 10 MHz, we may neglect quantities proportional to their square or products and obtain

$$L^2 \approx 1 + \frac{1}{2}i[\omega\tau_\nu + (\gamma_f - 1)\omega\tau_\kappa] \tag{9.2.10}$$

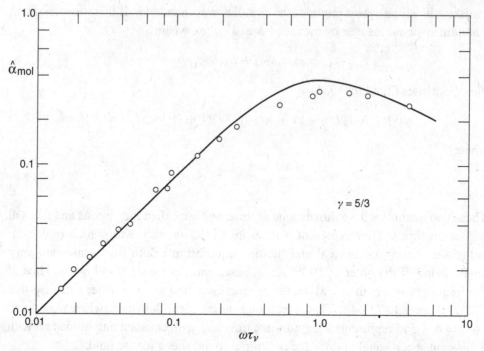

Figure 9.2.1. Molecular attenuation in helium. Solid line, from solution to (9.2.7); circle, experimental measurements by Greenspan (1956). [Reprinted with permission from Temkim (2001a). © Acoustical Society of America.]

Hence, in a very wide-frequency range, we have

$$c(\omega) = c_{sf} \tag{9.2.11}$$

$$\hat{\alpha} = \tfrac{1}{2}\omega\tau_v + \tfrac{1}{2}(\gamma_f - 1)\omega\tau_\kappa \tag{9.2.12}$$

where $\hat{\alpha} = \alpha c_{sf}/\omega$. Because the attenuation prescribed by (9.2.12) is due to molecular motions in the fluid, we shall denote this value by $\hat{\alpha}_{mol}$.

Equation (9.2.11) states that the phase velocity is independent of the frequency and equal to the ideal-fluid sound speed. Equation (9.2.12) gives the amplitude-attenuation coefficient for the sound waves in nondimensional form. Since the amplitude of the waves decay as $\exp(-\alpha x)$, and since both $\omega\tau_v$ and $\omega\tau_\kappa$ are very small, the equation states that, in one wavelength, the amplitude of the sound wave decays by a very small factor. At 10 MHz, this factor is equal to about 0.03. Thus, plane sound waves having frequencies smaller than a few mega-Hertz and traveling in unbounded fluids are largely unaffected by the effects of viscosity and heat conductivity. For higher frequencies, molecular relaxation can produce stronger effects, as the solution of (9.2.7) shows. That solution is displayed graphically in Fig. 9.2.1 for a monatomic gas, together with the experimental results of Greenspan (1956). It is seen that $\hat{\alpha}_{mol}$ can be large for higher frequencies than those for which $\omega\tau_v \approx 0.1$.

In a fluid having but a few particles, molecular absorption continues to dominate the attenuation; but, if the number of particles is sufficiently large, the particles take over as the main source of dissipation, at least if the frequency is not too high (as noted above). The concentrations required for this can only be determined after the various particulate sources of attenuation are evaluated, and this is done later. But, given that the molecular attenuation increases linearly with the frequency for frequencies as large as 10 MHz, and that, as we saw in the previous chapter, the particle attenuation per wavelength decreases at high frequencies, it is evident that the assumption that viscosity and heat conductivity in the fluid is negligible must break down at some frequency. In the next section, we consider particulate attenuation in very dilute suspensions, where we may anticipate that the attenuation is small. In addition to providing a comparative basis for the particulate attenuation and for the molecular attenuation, the calculations provide another useful means of probing suspensions.

9.3 Attenuation Coefficient

When the suspensions are very dilute, the attenuation produced by the particles, although normally much larger than that in the fluid alone, is generally small and can be determined from energy considerations alone. The method is similar to that used to compute the damping coefficient for the decay, in time, of oscillatory motions. But now we are interested in the decrease of the wave's amplitude with distance. Although for sound waves the temporal and spatial decay of the amplitude of a sound wave are related by a simple transformation, it is useful to derive here the basic equations that relate the attenuation to the energy dissipation rate.

Consider a plane sound wave traveling in the suspension. When it encounters a particle, it interacts with it in a variety of ways. For example, it makes the particle move. In general, the induced motion will be one in which the particle lags the fluid. As a result, viscous effects will come into play whose role is to eliminate this velocity difference, and, consequently, will result in some dissipation of energy. Furthermore, even in the frozen state, where the particles are at rest, they will scatter some of the energy in the incident wave. A fraction of this scattered energy is in the form of sound waves, but so far as the incoming wave is concerned, the scattered waves also represent dissipation.

Now, because energy is being removed from the incoming waves, the amplitude of the oscillations produced by them diminishes as the wave travels and, as we have seen, the decrease is exponential. Thus, if E_0 represents the acoustic energy in some portion of the traveling wave, for example the space between two wave crests, its value at subsequent time will be

$$E(t) = E_0 e^{-2\alpha x(t)} \tag{9.3.1}$$

Let the speed of the wave be c, then $x = ct$. Hence, the rate at which the energy of the wave changes is,

$$\dot{E} = -2\alpha c E(t) \tag{9.3.2}$$

This is an instantaneous energy loss, which we write as $-|\dot{E}_{lost}(t)|$ because $\alpha \geq 0$. Now, for monochromatic waves, it is useful to average (9.3.2) over one period. Thus, denoting the period $2\pi/\omega$ by τ, we have

$$\langle|\dot{E}_{lost}|\rangle = 2\alpha c E_0 \frac{1 - e^{-2\alpha c\tau}}{2\alpha c\tau} \tag{9.3.3}$$

If α is so small that $\alpha c\tau \ll 1$, terms of order $(\alpha c\tau)^2$ and higher can be neglected, so that the fraction on the right-hand side is, to that order, equal to 1, or

$$\alpha = \frac{\langle|\dot{E}_{lost}|\rangle}{2c E_0} \tag{9.3.4}$$

Since $c\tau = \lambda$, where λ is the wavelength, this equation requires that $\alpha\lambda$ be small.

Now, the determination of α via (9.3.4) requires that the quantities on its right-hand side be known. But, generally speaking, this is not the case because those quantities depend on the attenuation. Thus, as it stands, (9.3.4) is useless. However, in the limit in which it applies, namely, small $\alpha\lambda$, those quantities may be approximated by the nonattenuated pressure and velocity, which in this case means using the value of those quantities without particles.

Consider first the speed of propagation c. Since the waves have been taken to be monochromatic, that speed is the suspension's phase velocity, $c_s(\omega)$. But, since the concentration is very small, we may, for the purpose of computing α, put it equal to c_{sf}, the sound speed in the fluid alone. The same approximations apply to the calculation of E_0 (i.e., E_0 can be taken to be the energy in a sound wave in the same region, but without particles). For very dilute suspensions, this is given by the average acoustic energy in the fluid alone. Thus, if the volume of the region is denoted by δV, we have

$$E_0 = \frac{1}{2}\rho_{f0}\int_{\delta V}\langle U_f^2\rangle dV + \frac{1}{2\rho_{f0}c_{sf}^2}\int_{\delta V}\langle P_f'^2\rangle dV \tag{9.3.5}$$

For plane waves, the pressure and velocity appearing here are related by the isentropic relations (9.2.5), so that the two terms in (9.3.5) are equal. Thus, on taking the indicated time average, we obtain, for a monochromatic time dependence,

$$E_0 = \frac{1}{2}\rho_{f0}|U_f^2|\,\delta V \tag{9.3.6}$$

It remains to compute the rate at which energy is lost. This, too, can be calculated from the nonattenuated values of the fluid's velocity, temperature, and pressure. This does not mean that the energy loss rate vanishes. It means that the values for the variables that determine the energy loss rates should be those that would apply when no particles are present. The examples given below will help clarify these statements. Thus, in (9.3.4) we put c_{sf} instead of c and introduce the nondimensional attenuation

$\hat{\alpha} = \alpha c_{sf}/\omega$ to obtain

$$\hat{\alpha} = \frac{\langle|\dot{E}_{lost}|\rangle}{2\omega E_0} \tag{9.3.7}$$

An additional simplification is also possible on basis of the assumed smallness of $\hat{\alpha}$, and that is that the energy loss rates produced by the various motions can be calculated separately. Thus, $\hat{\alpha} = \hat{\alpha}_1 + \hat{\alpha}_2 + \cdots$.

Translational Attenuation

The simplest motion that results in attenuation is the translational motion in the particle. This produces attenuation because a certain amount of energy must be spent by the waves to move the particles relative to the fluid. Since the fluid velocity to be used in the calculation of $\hat{\alpha}$ is that which exists in the fluid without particles, we see that, in a unit time, a particle moves a relative distance $u_p - U_f$. The mechanism by which this takes place is, of course, the fluid force acting on each particle. Since the suspension is very dilute, that force is the same for each particle. If we denote that force by F_p, the rate at which work is done is $F_p(u_p - U_f)$. Thus, since the particle number concentration is $n = n_0 + n'$, we obtain, to second order in the fluctuations,

$$\langle|\dot{E}_{lost}|\rangle_{tr} = n_0\langle|F_p(u_p - U_f)|\rangle\delta V \tag{9.3.8}$$

Now, instead of using an explicit value for that force, we can, since the motion is monochromatic, use Newton's second law to write F_p as the real part of $-i\omega u_p$. Similarly, the velocity difference appearing above can also be written in complex form so that

$$\langle|\dot{E}_{lost}|\rangle_{tr} = -\frac{1}{2}n_0 m_p\omega|\Re\{iu_p(u_p - U_f)^*\}|\delta V \tag{9.3.9}$$

The term $iu_pu_p^*$ has no real part. For the second term, we put $u_p = VU_f$, where $V = u_p/U_f$ is the generally complex, but time independent, velocity ratio for one particle. This gives,

$$\langle\dot{E}_{lost}\rangle_{tr} = \frac{1}{2}\rho_{p0}\phi_{v0}\omega|\Im\{V\}||U_f|^2\delta V \tag{9.3.10}$$

Thus, substituting this and (9.3.6) into (9.3.7), and using the value of the mass loading applicable for dilute suspensions, $\eta_{m0} = \phi_{v0}/\delta$, we obtain the nondimensional attenuation resulting from the translational motion,

$$\hat{\alpha}_{tr} = \frac{1}{2}\eta_{m0}|\Im\{V\}| \tag{9.3.11}$$

This is the desired result. Despite its apparent simplicity, it describes the translational attenuation in dilute suspensions in a very wide-frequency range. The accuracy of the result depends, of course, on the accuracy with which the velocity ratio is calculated. This calculation was presented in Chapter 4, where the translational motion of a constant-radius sphere in a plane sound wave was studied using several models for the particle force. Before displaying the most general result, we reduce (9.3.11) for some important cases where specific results are known.

Velocity via Stokes' Law. First, we use the Stokes' velocity ratio given by (4.4.5) to obtain the translational attenuation in those suspensions where Stokes' law is applicable, namely aerosols. Thus, using (4.4.5), we get $\Im\{V^{(S)}\} = \omega\tau_d/(1 + \omega^2\tau_d^2)$. Hence,

$$\hat{\alpha}_{tr}^{(S)} = \frac{1}{2}\eta_{m0}\frac{\omega\tau_d}{1 + \omega^2\tau_d^2} \tag{9.3.12}$$

This is equal to the small-attenuation result obtained from the two-phase model in Section 8.6.

Unsteady Force in a Viscous, Incompressible Fluid. It will be remembered that Stokes' law leaves out the effects of unsteadiness and fluid compressibility. Unsteadiness produces significant effects for those suspensions whose particles have densities that are not much larger than that of the external fluid (e.g., emulsions and hydrosols). When unsteadiness is also taken into account, but still leaving out compressibility, the velocity ratio is given by (4.8.4), in which case the translational attenuation becomes

$$\hat{\alpha}_{tr}^{(V,I)} = 18\phi_{v0}\frac{|1 - \delta|(1 + y)y^2}{[2y^2(2 + \delta) + 9y\delta]^2 + 81\delta^2(1 + y)^2} \tag{9.3.13}$$

Here, $\delta = \rho_f/\rho_p$ and $y = \sqrt{\omega a^2/2v_f}$ are the ratios of particle radius, a, to viscous penetration depth $\delta_{vf} = \sqrt{2v_f/\omega}$.

Inviscid, Compressible Fluid. At very high frequencies, that is when $y \gg 1$, (9.3.13) shows that the effects of viscosity become insignificant, in the sense that the predicted attenuation then becomes negligible. However, when the frequency is so high that δ_v vanishes, other effects may come into play that produce attenuation (e.g., scattering). This occurs because the size of the particles is no longer small, compared with the wavelength.

At those frequencies, the motion of the particles, if any, is controlled by compressibility in the exterior fluid, in which case the velocity ratio is that given by (4.11.10). Therefore, the translational attenuation for an inviscid, compressible fluid is

$$\hat{\alpha}_{tr}^{(C)} = \frac{3}{2}\phi_{v0}\frac{1}{2 + \delta}\frac{|b\cos b - [1 - b^2/(2 + \delta)]\sin b|}{[1 - b^2/(2 + \delta)]^2 + b^2} \tag{9.3.14}$$

This result is not limited to small values of $b = \omega c_{sf}/a$.

Viscous, Compressible Fluid. The above expressions for $\hat{\alpha}_{tr}$ apply in limited frequency ranges and may be sufficient for specific suspensions. For example, (9.3.12)

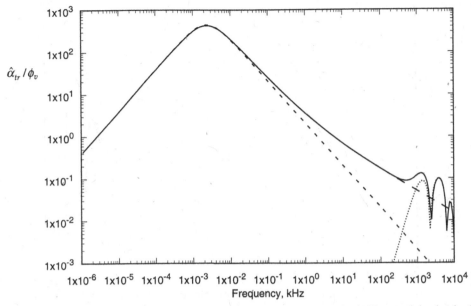

Figure 9.3.1. Scaled translational attenuation for an aerosol composed of silica particles in air. $D = 100\,\mu m$. Short, dashed line, Stokes; long, dashed lines, viscous, incompressible; dotted line, inviscid, compressible; solid line, viscous, compressible.

suffices for aerosols. It is, however, possible to obtain a theory for the translational attenuation by spherical particles in a viscous, compressible fluid taking the effects of unsteadiness into account. This can be also obtained from (9.3.11) using the more complete result for the velocity ratio given by (4.11.13). Thus,

$$\hat{\alpha}_{tr} = \frac{3}{2}\phi_{v0}\frac{|(GH - FI)\cos b - (FH + GI)\sin b|}{H^2 + I^2} \tag{9.3.15}$$

where the functions F, G, H, and I are given below (4.11.13). Although this result cannot be expressed concisely, it has the widest range of applicability as shown in the figures below.

Since the basic result, (9.3.11), shows that $\hat{\alpha}_{tr}$ is proportional to $|\Im\{V\}|$, its variations with frequency may be inferred from Figs. 4.11.1 and 4.11.2, where $|\Im\{V\}|$ was shown as a function of $\sqrt{\omega a^2/2\nu_f}$. However, so that we can later compare the translational attenuation with other attenuations, we show in Figs. 9.3.1 and 9.3.2 the variations of $\hat{\alpha}_{tr}/\phi_v$ with the frequency for an aerosol and a hydrosol, respectively, both containing silica spheres of the same size. The theory also applies to bubbly liquids and emulsions, but the corresponding results are shown later, after we discuss other attenuation coefficients.

Before discussing the figures in detail, we examine the validity of the results. As the derivation shows, they require that $\hat{\alpha}_{tr}$ be small. Figure 9.3.2 shows that this is the case in hydrosols having even modest values of the concentration. However, in the case of aerosols, Fig. 9.3.1, we see that a volume concentration equal to 10^{-3} results

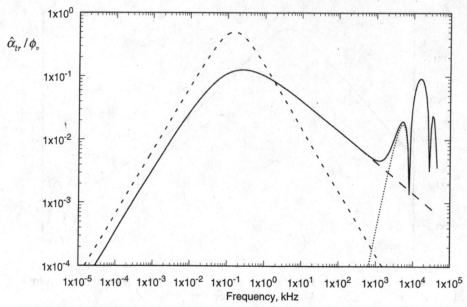

Figure 9.3.2. Scaled translational attenuation for a hydrosol composed of silica particles in water. $D = 100\,\mu\text{m}$. $D = 100\,\mu\text{m}$. Short, dashed line, Stokes; long, dashed line, viscous, incompressible; dotted line, inviscid, compressible; solid line, viscous, compressible.

in a translational attenuation that is close to 0.5. Thus, the dissipation-based result is not adequate for such high concentrations.

Now, consider the various results for $\hat{\alpha}_{tr}$ shown in the figures. First, it is seen that the result based on Stokes' law agrees closely with the viscous, compressible theory in the case of aerosols, at least for frequencies that are not large. Beyond these, differences appear that are more pronounced in hydrosols. This is because unsteady effects, left out in Stokes' law begins to affect the motion. On the other hand, (9.3.13) closely agrees with the viscous, compressible theory at most frequencies. However, it, too, fails when the frequency is very high, owing to compressibility effects. This is indicated by the agreement, in the high-frequency limit, between the inviscid theory and the viscous, compressible theory.

Pulsational Attenuation

We now consider the attenuation in suspensions caused by the particles ability to pulsate in response to the incident sound wave. This includes the important cases of bubbly liquids and emulsions, as well as aerosols whose particles are elastic, although here the losses caused by radiation are minimal. Now, to the extent that they exist, particle pulsations remove energy from the incident wave on two accounts: the radiation of acoustic energy resulting from the pulsations and the dissipation produced by irreversible heat transfer. The latter occurs even without pulsations; but, when these occur, it affects the acoustic radiation because the energy radiated by the particle depends on the amplitude of the pulsation, and this depends on the thermal conditions in the particle. Because of this, it is not generally possible to calculate these two effects separately. However, in the present case, where the attenuation is by assumption small, it is possible to do so. Below, we look at the acoustic radiation attenuation. The thermal attenuation is considered later.

Attenuation Due to Radiation. Let us first compute the average rate at which energy is lost from the wave owing to acoustic emission by the pulsating particles. This rate follows directly from the definition of the acoustic damping coefficient. When the damping is small, that rate is given by the second term in (6.5.11), multiplied by the number of particles per unit volume. Thus, in a volume δV of suspension, we have

$$\langle |\dot{E}_{lost}| \rangle_{ac} = n_0 2 \beta_{ac} e_0 \delta V$$

with the symbol e_0 in that equation now refers to the energy of pulsation, given by (6.5.10), or $^1/_2 M_0 |\dot{\varepsilon}|^2$. Thus, using the value of β_{ac}, namely $\beta = {}^1/_2 b\omega$, we obtain

$$\langle |\dot{E}_{lost}| \rangle_{ac} = \frac{3}{2} \phi_{v0} \rho_{f0} b\omega |\dot{\varepsilon}|^2 \delta V \tag{9.3.16}$$

In obtaining this, we have neglected b^2, compared with unity because, as it will be remembered from Chapter 6, the pulsational theory is limited to such values of b. We now need to express the amplitude, ε, of the radial pulsation in terms of the fluid's velocity. For this purpose, we first note that ε is given in terms of the response function by means of (6.6.25), with $P'_f = \rho_{f0} c_{sf} U_f$, as (9.2.5a) shows. Substitution in (9.3.16) yields,

$$\langle |\dot{E}_{lost}| \rangle_{ac} = \frac{3}{2} \phi_{v0} \rho_{f0} b\omega \frac{(\omega c_{sf}/a)^2}{\omega_{T0}^4} |\chi|^2 |U_f|^2 \delta V \tag{9.3.17}$$

where the response function, χ, is given by (6.6.28). Substitution of this into our defining equation for $\hat{\alpha}$, yields

$$\hat{\alpha}_{ac} = \frac{3}{2} \phi_{v0} \frac{(\omega/\omega_{T0})^4}{b} |\chi|^2 \tag{9.3.18}$$

This expresses the attenuation per wavelength in terms of known quantities and therefore provides the desired answer. But, in analogy with the translational attenuation, which was written in terms of the complex velocity ratio V, we would like to express (9.3.18) in terms of the complex pressure ratio Π, equal to p'_p/P'_f. To do this, we first use (6.6.32) for Π and (6.6.28) for χ. This gives

$$\Pi - 1 = -\frac{(\omega/\omega_{T0})^2 + i(\hat{d} - \hat{d}_{th})/\kappa_r}{1 - (\omega/\omega_0)^2 - i\hat{d}/\kappa_r}$$

Now, since the external fluid is, by assumption, inviscid, we have $\hat{a} - \hat{a}_{th} = \hat{a}_{ac}$. Thus, using $\hat{a}_{ac} = b(\omega/\omega_0)^2$, together with $(\omega_0/\omega_{T0})^2 = \kappa_r$ [equation (6.6.29)], we obtain

$$\Pi - 1 = -(\omega/\omega_{T0})^2 \chi (1 + ib) \tag{9.3.19}$$

Thus, again putting $1 + b^2$ equal to 1, we have the desired form:

$$\hat{\alpha}_{ac} = \frac{3}{2}\phi_{v0}\frac{|\Pi - 1|^2}{b} \tag{9.3.20}$$

Comparing this with (9.3.18) shows that, under the stated assumptions $|\Pi - 1| = |\chi|(\omega/\omega_{T0})^2$, which in turn shows that the pressure ratio increases very slowly as the frequency increases from low values. It should also be noticed that, if we assume that the pressure is uniform throughout, the acoustic radiation vanishes. But, as shown in the graphs below, acoustic radiation near resonance can produce significant attenuation, even in the case of emulsions, where the attenuation produced by other effects can be small

Thermal Attenuation. The derivation for the thermal attenuation is very similar to that given for the thermal damping coefficient in Section 6.7. In fact, all we need to do to determine the thermal dissipation rate is to multiply (6.7.2) by n_0, the mean number of particles in a unit suspension volume. This gives,

$$\langle |\dot{E}_{lost}| \rangle_{th} = \frac{n_0}{T_0} \langle \dot{Q}_p (\Theta'_f - \overline{T}'_p) \rangle \delta V \tag{9.3.21}$$

For the heat transfer rate, $\dot{Q}_p$, we also use (6.6.17). Thus, expressing the time average on the right-hand side of (9.3.21) as $^1/_2 \Re\{\dot{Q}_p(\Theta'_f - \overline{T}'_p)^*\}$, we obtain,

$$\langle |\dot{E}_{lost}| \rangle_{th} = \frac{1}{2}\phi_{v0}\frac{\rho_{p0}c_{pp}}{\tau_t T_0}\Re\left\{\frac{(1 - \xi\Pi)(1 - \mathrm{T})^*}{F(q, q_i)}\right\}|\Theta'_f|^2\delta V \tag{9.3.22}$$

where $\mathrm{T} = \overline{T}'_p/\Theta'_f$ and $\Pi = p'_p/P'_f$ are the dilute temperature and pressure ratios, and the quantities Θ'_f and P'_f are the temperature and pressure fluctuations in the fluid without particles. To put $|\Theta'_f|^2$ in terms of U_f, we use (9.2.5c) as well as the thermodynamic identity (7.3.32). This gives

$$|\Theta'_f|^2 = \frac{\gamma_f - 1}{c_{pf}}T_0|U_f|^2 \tag{9.3.23}$$

Putting this into (9.3.22) and substituting the result in the equation defining $\hat{\alpha}$, we obtain

$$\hat{\alpha}_{th} = \frac{1}{2}\phi_{v0}\frac{\rho_{p0}c_{pp}}{\rho_{f0}c_{pf}}\frac{\gamma_f - 1}{\omega\tau_t}\Re\left\{\frac{(1 - \xi\Pi)(1 - \mathrm{T})^*}{F(q, q_i)}\right\} \tag{9.3.24}$$

This result applies to dilute suspensions in general. But in Chapter 8, we obtained limited forms of this for aerosols, emulsions, and hydrosols from the two-phase model, making use of certain simplifying assumptions. Thus, the result we derived there for emulsions, (8.8.18), was obtained under the assumption that the pressure could be taken as uniform. That result is obtained from (9.3.24) by putting $\Pi = 1$. Similarly, the corresponding result for aerosols, (8.8.20), is obtained by putting $\xi = 0$ in the above equation. Thus, we find that the attenuation obtained with the energy method agrees with the corresponding results obtained by means of the two-phase model.

Total Attenuation

In the last three subsections, we calculated the various attenuations that particles produce when a plane sound wave travels in a suspension. These attenuations were calculated on the basis of average energy dissipation rates and on the assumption that they were small. Because of this, the resulting attenuation coefficients are additive. Thus, a total attenuation coefficient may be defined by

$$\hat{\alpha} = \hat{\alpha}_{tr} + \hat{\alpha}_{ac} + \hat{\alpha}_{th} \qquad (9.3.25)$$

When the attenuation is not small, the last two of these cannot, generally, be separated. It is therefore useful to refer to their sum as the pulsational attenuation, $\hat{\alpha}_{pul}$, keeping in mind that it includes thermal effects. Thus, in anticipation of that case, we put $\hat{\alpha}_{pul} = \hat{\alpha}_{ac} + \hat{\alpha}_{th}$, so that the total attenuation can be written as

$$\hat{\alpha} = \hat{\alpha}_{tr} + \hat{\alpha}_{pul} \qquad (9.3.26)$$

This form associates an attenuation coefficient with two of the basic particle motions that were introduced in Chapter 2. The third motion, rotational, also produces attenuation, but requires that the particle be nonspherical or that the external velocity be nonuniform around the particle. Neither condition is satisfied for dilute suspensions of spherical particles in a plane sound wave.

Now, the magnitude and frequency dependence of the attenuation coefficients depend on the type of suspension. Hence, it is useful to reduce the general, small-attenuation results for each type. Although it is possible to simplify the formulas taking advantage of the values of the relevant parameters applicable in each case, we simply show the results in the figures below, for an aerosol, a bubbly liquid, and an emulsion. To show the relative effects of each mechanism, the chosen suspensions have the same particle size and volume concentration, namely $a = 50\ \mu m$ and $\phi_v = 10^{-5}$, respectively. This small concentration was chosen so that we could compare the effects of each attenuation in across the various suspension types. But, as small as it seems to be, in bubbly liquids it results in an attenuation that is not very small.

With reference to Figs. 9.3.3–9.3.6, we first note that, except for the bubbly liquid case, each of the nondimensional attenuation coefficients are smaller than 1. Second, the acoustic and thermal attenuations in bubbly liquids are significantly larger than their counterparts in aerosols and emulsions, and, in fact, are not always small compared

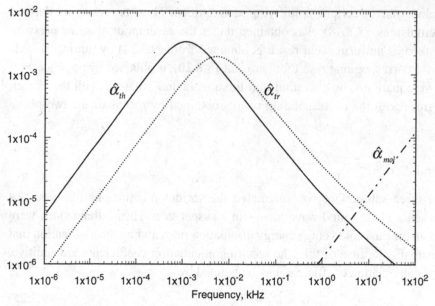

Figure 9.3.3. Attenuation coefficients in an aerosol composed of silica particles in air. $D = 100\,\mu m$, $\phi_v = 10^{-5}$.

with 1. This occurs because $\hat{\alpha}_{ac}$, and $\hat{\alpha}_{th}$ are proportional to $\phi_v N_s$, which is finite in bubbly liquids owing to their large effective compressibility. On the other hand, the translational attenuation in bubbly liquids is insignificant, compared with the other particle attenuations. This is due to the nearly negligible inertia of the bubbles.

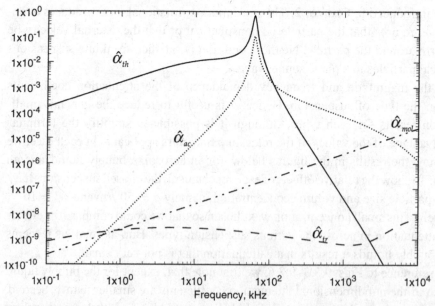

Figure 9.3.4. Attenuation coefficients in a bubbly liquid composed of air bubbles in water. $D = 100\,\mu m$, $\phi_v = 10^{-5}$.

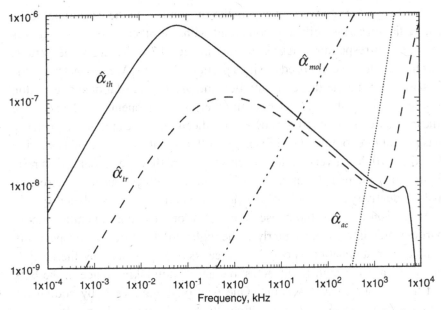

Figure 9.3.5. Attenuation coefficients in an emulsion composed of toluene droplets in water. $D = 100\,\mu m$, $\phi_v = 10^{-5}$.

In the case of emulsions, we note that the thermal component is the largest at most frequencies, except at very high frequencies where acoustic radiation overwhelms the other two. This larger value is produced by resonance effects, which as we have seen are also present for drops in liquids. This resonance effect also occurs in the

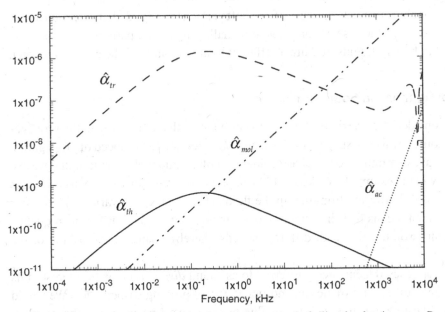

Figure 9.3.6. Attenuation coefficients in a hydrosol composed of silica droplets in water. $D = 100\,\mu m$, $\phi_v = 10^{-5}$.

thermal attenuation, although with much smaller magnitude. In the case of aerosols, the acoustic radiation is insignificant, compared with the other components, and is not shown in the corresponding figure. Finally, in Fig. 9.3.6, we show the various attenuations for a hydrosol composed of silica particles in water. We see that all of the attenuations are very small at this concentration and that the translational attenuation is several orders of magnitude larger than the thermal component. Incidentally, we note that the translational attenuation shows a significant increase at high frequencies. This increase is not due to pulsational resonance effects, but by acoustical scattering produced by the particles at very high frequencies, where they are essentially at rest.

The figures also include the attenuation per wavelength produced by viscosity and thermal conductivity in the bulk of the fluid. That attenuation is labeled $\hat{\alpha}_{mol}$ in the figures. It is seen that, for the chosen concentration, it is not generally smaller than the particle attenuations, particularly at the high end of the frequency spectrum. Here, the molecular attenuation is equal, or exceeds, that produced by the particle attenuation. Thus, at such high frequencies, the absorption taking place in the bulk of the fluid must be taken into account. One way of doing this is by adding the corresponding attenuation, $\hat{\alpha}_{mol}$, to total attenuation produced by the particles. It is also noted that, for the chosen concentration, some attenuations are smaller than $\hat{\alpha}_{mol}$ (e.g., the acoustic radiation attenuation produced in aerosols and hydrosols). In such cases, it would then be incorrect to neglect molecular attenuation while retaining acoustic radiation. However, as the concentration increases beyond the small value used in the above figures, most of the particulate attenuations become larger than the molecular contribution. This does not change with the particle concentration within the dilute suspension limit, and can thus be then ignored.

This concludes our discussion of the attenuation of sound that is produced in suspensions when the dissipation rates are small. Larger dissipations, such as those produced in bubbly liquids, require a different treatment and will be considered later.

9.4 Propagation via Sound Emission

Having considered the various attenuation coefficients that play a role in the propagation of acoustic waves in suspensions, we turn our attention to the speed of propagation of those waves, or rather, to the phase velocity of plane, monochromatic sound waves. As we saw in the examples worked out in Chapter 8, propagation in suspensions takes place at a speed different from that in the fluid alone. That speed varies with the frequency, which means that, in a nonmonochromatic sound wave, different frequency components would travel with different speeds, thereby changing the profile of the wave.

This frequency dependence – referred to as sound *dispersion* – is of interest in its own but, in addition, it provides an additional way of probing suspensions. We would therefore like to study this effect. For this purpose, we first note that the problem of obtaining the phase velocity of acoustic waves is generally equivalent to that of obtaining a complex wavenumber k in terms of known parameters and the frequency. As in Chapter 8, we may write that wavenumber as $k = \omega/c_s(\omega) + i\alpha(\omega)$, where ω is the circular frequency, $c_s(\omega)$ is the phase velocity, and $\alpha(\omega)$ is the attenuation coefficient. It is evident that some connection must exist between the attenuation and the phase velocity which, if known, might be used with our small-attenuation results. Such a connection does exist, and is based on certain requirements for the mathematical behavior of the functions $\alpha(\omega)$ and $c_s(\omega)$. This connection will be briefly discussed at the end of this chapter.

It is better, however, to first consider the problem from a more physical perspective. This can be done in several ways, each of which offers a different insight about the physical processes that take place during the propagation of acoustic waves in suspensions. Because of these differences, it is useful to regard the various methods as complementary.

The first method to be presented makes use of some aspects of the two-phase model and is based on the fact that, fundamentally, small particles in a suspension may be regarded as sources of sound. This was indeed the reason that attracted the attention of the earliest investigators of bubble pulsations, who associated them with the murmuring sounds of a brook.

But other types of particles also emit sound because particles in fluids are, essentially, point sources of volume, force, and heat. The strengths of these sources fluctuate in time, owing to the generally unsteady nature of suspension motions. However small the force and heat transfer sources might be, they affect the propagation of acoustic waves in suspensions. Therefore, if the strength and time dependence of these sources are known, the propagation problem can be reduced to one for the emission of sound into a fluid. This provides a method to investigate propagation in suspensions that is simpler than most.

Wave Equation with Sources of Mass, Force, and Heat

We begin by writing the linearized, one-dimensional equations of motion for the fluid phase. These are obtained from equation (8.3.1) to (8.3.4). For the present purposes, we write those equations as

$$\frac{\partial \rho'_f}{\partial t} + \rho_{f0}\frac{\partial u_f}{\partial x} = \rho_{f0}\dot{\phi}'_v(x, t) \tag{9.4.1}$$

$$\rho_{f0}\frac{\partial u_f}{\partial t} + \frac{\partial p'_f}{\partial x} = f'_p(x, t) \tag{9.4.2}$$

$$\rho_{f0}c_{pf}\frac{\partial T'_f}{\partial t} - {}^\bullet\beta_f T_0\frac{\partial p'_f}{\partial t} = \dot{q}'_p(x, t) \tag{9.4.3}$$

In these equations, f'_p and $\dot{q}'_p$ are the linearized force and heat transfer rate, respectively, and $\dot{\phi}'_v$ is the linearized time derivative of the particle volume concentration. The factor $(1 - \phi_{v0})$ appearing in the original equations has been set equal to 1 because, by assumption, $\phi_{v0} \ll 1$. As given, the equations are the linear versions of the equations of fluid mechanics with sources of mass, force, and heat. The system is complemented by the fluid's equation of state, which we now write as

$$\rho_{f0} c_{pf} T'_f = \beta_f T_0 \frac{\gamma_f}{\gamma_f - 1} \left(p'_f - c^2_{Tf} \rho'_f \right) \tag{9.4.4}$$

Since the fluid has been taken to be inviscid and nonheat conducting, we require that the volume concentration be sufficiently large so that the particle dissipation rates are much larger than those produced by the molecular motions in the fluid.

The system can be reduced to the wave equation for the pressure. Thus, eliminating the velocity from the first two equations by cross-differentiation gives

$$\frac{\partial^2 \rho'_f}{\partial t^2} - \frac{\partial^2 p'_f}{\partial x^2} = \rho_{f0} \frac{\partial^2 \phi'_v}{\partial t^2} - \frac{\partial f'_p}{\partial x} \tag{9.4.5}$$

For the last two equations, we use (9.4.4) in (9.4.3) and take the time derivative of the result, obtaining,

$$\frac{\partial^2 p'_f}{\partial t^2} - c^2_{sf} \frac{\partial^2 \rho'_f}{\partial x^2} = \frac{\gamma_f - 1}{\beta_f T_0} \ddot{q}'_p$$

Multiplying the first of these by c^2_{sf} and adding to the second gives the desired result

$$\frac{\partial^2 p'_f}{\partial t^2} - c^2_{sf} \frac{\partial^2 p'_f}{\partial x^2} = \rho_{f0} c^2_{sf} \ddot{\phi}'_v - c^2_{sf} \frac{\partial f'_p}{\partial x} + \frac{\gamma_f - 1}{\beta_f T_0} \ddot{q}'_p \tag{9.4.6}$$

This is an inhomogeneous wave equation, whose terms on the right-hand side represent the sources of sound. From the point of view of acoustic emission theory, the first and last terms are monopole sources. That is, the heat addition term produces sound in the same manner as the volume source term does. This happens because a fluctuating heat addition rate produces expansions and compressions in the fluid, which is what a fluctuating volume source does. The second term in (9.4.6), originating in the force on the particle, appears as a spatial derivative of that force. This is an acoustic dipole, a less efficient source of sound than the monopole.

Absent from (9.4.6) is another of fundamental source of sound: the quadrupole. This represents the limiting form of a force pair, or stress, and is absent here for two reasons. First, the shearing stresses, which are normally present in a viscous fluid, have been set equal to zero because of the inviscid-fluid assumption. Second, since in this model the particle force is represented by a dipole, we see that two close particles would, in fact, represent a quadrupole. Although this provides a means to study suspensions that are not dilute, that possibility is precluded here because of the large interparticle separation that exists in dilute suspensions.

Now, in sound emission studies, the basic problem is to determine the sound emitted by known source distributions. Here we are interested only in determining what happens to a plane sound wave traveling in a fluid with small particles, when the particles are being represented by point sources. To do this, we require explicit expressions of the source strengths.

Source Strengths

Let us first list the three source terms. These are:

Volume: $\quad \rho_{f0} c_{sf}^2 \ddot{\phi}_v'$

Heat: $\quad \dfrac{\gamma_f - 1}{\beta_f T_0} \ddot{q}_p'$

Force: $\quad -c_{sf}^2 \dfrac{\partial f_p'}{\partial x}$

These may be expressed in terms of the fluid and particle velocities, pressures, and temperatures. For example, the particulate force generally involves both u_p and u_f. The particulate velocity may be obtained in terms of its fluid counterpart, u_f. But this velocity is unknown because it depends on the wavenumber $k = \omega/c_s(\omega) + i\alpha$, and this is precisely what we want to determine (i.e., the source terms are generally unknown).

However, an approximate solution may be obtained by using, *in the source terms*, the fluid velocity, pressure, and temperature that apply when no sources are present. This assumes, *a priori*, that the effects produced by the particles (e.g., attenuation) are small. Thus, since the fluid is ideal, the pressure, density, velocity, and temperature appearing in the source terms are related by (9.2.5a)–(9.2.5c). Those equations imply, for example, that, for monochromatic waves, the derivative with respect to x appearing in the force term may be set equal to ik_0, where $k_0 = \omega/c_{sf}$.

In addition to the above approximation, we will also assume that the three types of sources act independently. This means, for example, that the force term is to be evaluated without taking into account the small variations of particle size that take place for pulsating particles.

Volume Source. We first compute $\ddot{\phi}_v'$. Putting $\phi_v = \phi_{v0} + \phi_v'$ in (8.3.8) and linearizing the result, we obtain,

$$\ddot{\phi}_v' = -\phi_{v0} \left(\frac{1}{\rho_{p0}} \frac{\partial^2 \rho_p'}{\partial t^2} + \frac{\partial^2 u_p}{\partial t \partial x} \right) \tag{9.4.7}$$

By the above assumptions, we have, for monochromatic time variations,

$$\ddot{\phi}_v' = \phi_{v0} \omega^2 \left(\frac{1}{\rho_{p0}} \rho_p' - \frac{1}{\rho_{f0} c_{sf}^2} p_f' \right)$$

We now eliminate the particle density by means of (8.3.4b), and multiply both sides by $\rho_{f0}c_{sf}^2$ to obtain

$$\rho_{f0}c_{sf}^2\ddot{\phi}_v' = \phi_{v0}\omega^2\left(\gamma_p\frac{\rho_{f0}c_{sf}^2}{\rho_{p0}c_{sp}^2}\frac{p_p'}{p_f'} - \beta_f\rho_{f0}c_{sf}^2\frac{T_p'}{p_f'} - 1\right)p_f'$$

The property ratio appearing in the first term of the right-hand side is, by definition, N_s. The second term inside the parentheses may be expressed as $(\beta_p/\beta_f)(\gamma_f - 1)(\overline{T}_p'/T_f')$. We now approximate p_p'/p_f' and T_p'/T_f' by their dilute values $\Pi = p_p'/P_f'$ and $T = T_p'/\Theta_f'$, respectively. Thus, we get

$$\rho_{f0}c_{sf}^2\ddot{\phi}_v' = \phi_{v0}\omega^2\left(\gamma_p N_s\Pi - \frac{\beta_p}{\beta_f}(\gamma_f - 1)T - 1\right)p_f' \qquad (9.4.8)$$

This specifies the volume source term in terms of the known quantities Π and T, and pressure fluctuation in the fluid.

Heat Source. This follows from our discussion in Section 8.8, where $\dot{q}_p'$ was calculated for the uniform pressure case (i.e., for $\Pi = 1$) and the function $F(q, q_i)$. In the present case, it is necessary to retain the pressure variations, in which case it is more convenient to express the local heat transfer rate in terms of the dilute pressure and temperature ratios. Thus, since $\dot{q}_p' = -n_0\dot{Q}_p(x, t)$, and since the fluctuations are monochromatic, we have

$$\dot{Q}_p(x, t) = -i\omega m_p c_{pp}\left[\overline{T}_p - (\beta_p T_0/\rho_{p0}c_{pp})\overline{p}_p'\right] \qquad (9.4.9)$$

Proceeding as with the volume source, we find

$$\frac{\gamma_f - 1}{\beta_f T_0}\ddot{q}_p' = \phi_{v0}\omega^2(\gamma_f - 1)\frac{\rho_{p0}c_{pp}}{\rho_{f0}c_{pf}}(T - \xi\Pi)p_f' \qquad (9.4.10)$$

Force Source. The last source term we require is that produced by the particle force, $-c_{sf}^2(\partial f_p'/\partial x)$. By our assumptions, this term is, for monochromatic time variations, equal to $-i\omega c_{sf}f_p'$, where f_p' is the local particulate force acting on the fluid. To obtain f_p', we note that the force on a particle is given by Newton's second law as $m_p(du_p/dt)$. The force on the fluid is not equal to this because the fluid also experiences a force due to its own acceleration. As discussed in Section 4.8, that force is $\rho_f v_p(du_f/dt)$. Hence, a small particle exerts on the fluid a net force equal to $v_p(\rho_p du_p/dt - \rho_f du_f/dt)$. The total particulate force is this amount multiplied by the particle number density n. Linearizing the result, we obtain, for monochromatic time variations,

$$f_p' = -i\omega\phi_{v0}\rho_{f0}\left(\frac{\rho_{p0}}{\rho_{f0}}\frac{u_p}{u_f} - 1\right)u_f \qquad (9.4.11)$$

We approximate u'_p/u'_f by its dilute value, $V = u'_p/U'_f$. Finally, we put $u_f = p'_f/\rho_{f0}c_{sf}$. Hence, the strength of the source term is

$$-c_{sf}^2 \frac{\partial f'_p}{\partial t} = \phi_{v0}\omega^2 \left(\frac{\rho_{p0}}{\rho_{f0}}V - 1 \right) p'_f \tag{9.4.12}$$

Complex Wavenumber

Having obtained the source terms, we add them and substitute the result in our inhomogeneous wave equation, (9.4.6), obtaining, since the time dependence is monochromatic

$$-c_{sf}^2 \frac{\partial^2 p'_f}{\partial x^2} - \omega^2 p'_f = \phi_{v0}\omega^2 \left\{ \frac{\rho_{p0}}{\rho_{f0}}V + \left[\gamma_p N_s - (\gamma_f - 1)\frac{\beta_p}{\beta_f} \right] \right.$$
$$\left. \times \Pi + (\gamma_f - 1)\left(\frac{\rho_{p0}c_{pp}}{\rho_{f0}c_{pf}} - \frac{\beta_p}{\beta_f} \right) T - 2 \right\} p'_f$$

This specifies the spatial variations of the pressure fluctuations produced by the sources. We are interested in wavelike solutions. Hence, assuming that p'_f depends on x as $\exp(ikx)$, where $k = \omega/c_s(\omega) + i\alpha(\omega)$, we find that acoustic waves can propagate in the suspension provided k satisfies

$$\left(\frac{k}{k_0} \right)^2 = 1 + \phi_{v0} \left\{ \frac{\rho_{p0}}{\rho_{f0}}V + \left[\gamma_p N_s - (\gamma_f - 1)\frac{\beta_p}{\beta_f} \right] \Pi \right.$$
$$\left. + (\gamma_f - 1)\left(\frac{\rho_{p0}c_{pp}}{\rho_{f0}c_{pf}} - \frac{\beta_p}{\beta_f} \right) T - 2 \right\} \tag{9.4.13}$$

This is the desired result. It includes both translational and pulsational motions, the latter including temperature changes, and applies to aerosols, bubbly liquids, emulsions, and hydrosols. It provides the answer to the question posed at the beginning of this section because the three complex ratios appearing on its right-hand side – namely V, Π, and T – are known. Thus, the value of the velocity ratio V is given by (4.11.13), whereas those for the pressure and temperature ratio are given by (6.6.32) and (6.6.15), respectively.

Before we examine the result in detail, we write it in a slightly different manner. For this purpose, we evaluate it in the low-frequency limit, $\omega \to 0$. Here, V, Π, and T are all equal to 1 and $\alpha = 0$, $k/k_0 = c_{sf}/c_s(0)$ so that (9.4.13) reduces to

$$\frac{c_{sf}^2}{c_s^2(0)} = 1 + \phi_{v0} \left[\frac{\rho_{p0}}{\rho_{f0}} + \gamma_p N_s + (\gamma_f - 1)\left(\frac{\rho_{p0}c_{pp}}{\rho_{f0}c_{pf}} - 2\frac{\beta_p}{\beta_f} \right) - 2 \right] \tag{9.4.14}$$

Comparison with the thermodynamic result, (7.4.6), shows that the two results are equal. This time, however, the result is obtained from the dynamic equations of motion. Next, we subtract (9.4.14) from (9.4.13), obtaining (Temkin, 2000)

$$\left(\frac{k}{k_0}\right)^2 = \frac{c_{sf}^2}{c_s^2(0)} + \phi_{v0}\left\{\frac{\rho_{p0}}{\rho_{f0}}(V-1) + \left[\gamma_p N_s - \frac{\beta_p}{\beta_f}(\gamma_f - 1)\right](\Pi - 1)\right.$$

$$\left. + (\gamma_f - 1)\left(\frac{\rho_{p0}c_{pp}}{\rho_{f0}c_{pf}} - \frac{\beta_p}{\beta_f}\right)(T-1)\right\} \tag{9.4.15}$$

This form, like that given by (9.4.13), contains all the information that is required to obtain the phase velocity and the attenuation dilute suspensions, and clearly displays the effects that determine the phase velocity and attenuation for a plane sound wave traveling in a suspension. For example, it shows that when V, Π, and T differ from 1 – that is when the particle velocity, pressure, or temperature differ from the corresponding quantity in the fluid – the propagation will be dispersive, in which case the wave's phase velocity will depend on the frequency, and their amplitude will decay with distance.

Let us now obtain the phase velocity and the attenuation. These follow from either of the above equations by means of

$$\left(\frac{k}{k_0}\right)^2 = \frac{c_{sf}^2}{c_s^2(\omega)} - \hat{\alpha}^2 + 2i\hat{\alpha}\frac{c_{sf}}{c_s(\omega)} \tag{9.4.16}$$

where $\hat{\alpha} = \alpha c_{sf}/\omega$. To save space, we write the real and imaginary parts of the right-hand sides of (9.4.15) as $X(\omega)$ and $Y(\omega)$, respectively. This gives.

$$\frac{c_{sf}^2}{c_s^2(\omega)} - \hat{\alpha}^2 = X(\omega) \tag{9.4.17}$$

$$2\hat{\alpha}\frac{c_{sf}}{c_s(\omega)} = Y(\omega) \tag{9.4.18}$$

Since $X(\omega)$ and $Y(\omega)$ are known, these two equations are sufficient to specify the phase velocity and the attenuation. Thus, the solution can be expressed as

$$\frac{c_{sf}^2}{c_s^2(\omega)} = \frac{1}{2}X\left[1 + \sqrt{1 + (X/Y)^2}\right], \qquad \hat{\alpha} = \sqrt{X/2}\left[\sqrt{1 + (Y/X)^2} - 1\right]^{1/2} \tag{9.4.19a,b}$$

A useful approximation of these, valid when $\hat{\alpha}$ is small, is

$$\frac{c_{sf}^2}{c_s^2(\omega)} \approx X \quad \text{and} \quad \hat{\alpha} \approx \frac{1}{2}|Y(\omega)| \tag{9.4.20a,b}$$

As shown below, this set is sufficient to treat those cases when the attenuation is small. Of course, given the rather complicated nature of the functions V, Π, and T, it is clear that, even then, the solution must be evaluated numerically. However, the simplified results may be used for comparison with previously obtained results.

For future reference, the symbols $X(\omega)$ and $Y(\omega)$, as well as the solution (9.4.19) of the two equations for $\hat{\alpha}$ and $c_s(\omega)$, will be also used in the forthcoming sections. However, it should be noted that those symbols will usually represent different functions.

Small Attenuation

Before we display the complete solution for specific cases, it is useful to consider it in the limiting case when the volume fraction is so small that the solution-pair (9.4.20a,b) is sufficient. In view of (9.4.18), it is clear that each of the terms in $Y(\omega)$ is proportional to an attenuation. The first of those terms is uncoupled to the remaining two and gives the small-volume fraction attenuation,

$$\hat{\alpha}_{tr} = \frac{1}{2}\phi_{v0}\frac{\rho_{p0}}{\rho_{f0}}|\Im\{V\}| \tag{9.4.21}$$

This is identical to the result obtained via the energy-dissipation method. The remaining two terms in $Y(\omega)$ are generally coupled because the pressure ratio depends on the temperature. But their sum represents the pulsational attenuation, $\hat{\alpha}_{pul}$, where

$$\hat{\alpha}_{pul} = \frac{1}{2}\phi_{v0}\left|\left[\gamma_p N_s - \frac{\beta_p}{\beta_f}(\gamma_f - 1)\right]\Im\{\Pi\} + (\gamma_f - 1)\left(\frac{\rho_{p0}c_{pp}}{\rho_{f0}c_{pf}} - \frac{\beta_p}{\beta_f}\right)\Im\{T\}\right| \tag{9.4.22}$$

Now, this quantity was also computed in the previous section, using the energy dissipation method. The result was expressed there as $\hat{\alpha}_{pul} = \hat{\alpha}_{ac} + \hat{\alpha}_{th}$, with $\hat{\alpha}_{ac}$ and $\hat{\alpha}_{th}$ given by (9.3.18) and (9.3.24), respectively. To see the correctness of the sound emission approach, we compare the two results for $\hat{\alpha}_{pul}$. We leave it to the reader to show that, when N_s and β_p/β_s are both very small, as is the case for aerosols and hydrosols, the above reduces to the thermal-dissipation attenuation obtained in the previous section. For bubbly liquids and an emulsions, it is simpler to compare the emission and dissipation results graphically. Thus, Figs. 9.4.1 and 9.4.2 show $\hat{\alpha}_{pul}$ as predicted by the two theories. Figure 9.4.1, pertaining to a bubbly liquid, was obtained with a volume concentration equal to 10^{-6} to ensure that the attenuation be small at all frequencies. For the emulsion case, the attenuation continues to be small. We therefore use the much larger value of 10^{-2}. It is seen that the emission and dissipation attenuations agree with one another at all frequencies.

Small Dispersion. Here, we look at the variations of the phase velocity $c_s(\omega)$ with the frequency. At very small frequencies, this is equal to $c_s(0)$. Hence, to see the effects of dispersion, it is useful to look at the difference between those two speeds. For simplicity, we introduce a dispersion coefficient, $\hat{\beta}$, by means of

$$\hat{\beta} = c_{sf}^2/c_s^2(0) - c_{sf}^2/c_s^2(\omega). \tag{9.4.23}$$

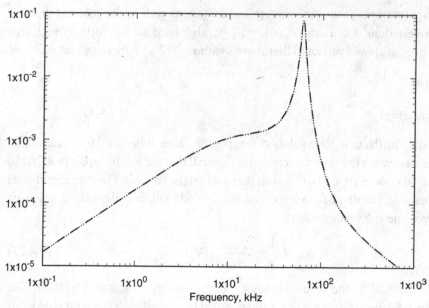

Figure 9.4.1. Pulsational attenuation in a bubbly liquid composed of air bubbles in water. $D = 100 \, \mu m$, $\phi_v = 10^{-6}$. Dotted line, $\hat{\alpha}_{ac} + \hat{\alpha}_{th}$; dashed line, $\hat{\alpha}_{pul}$.

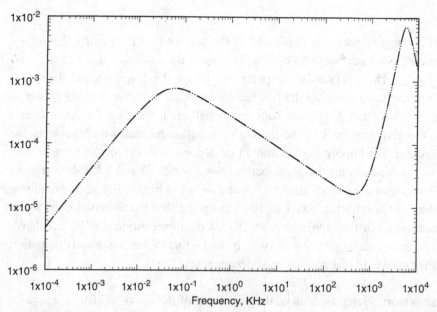

Figure 9.4.2. Pulsational attenuation in an emulsion composed of toluene droplets in water. $D = 100 \, \mu m$, $\phi_v = 10^{-2}$. Dotted line, $\hat{\alpha}_{ac} + \hat{\alpha}_{th}$; dashed line, $\hat{\alpha}_{pul}$.

This can be used whether the attenuation is small or not. But, when it is, the speed ratio $c_{sf}^2/c_s^2(\omega)$ is approximately equal to $X(\omega)$. Therefore, using (9.4.14), we obtain

$$\hat{\beta}/\phi_{v0} = \frac{\rho_{p0}}{\rho_{f0}}(1 - V) + \left[\gamma_p N_s - \frac{\beta_p}{\beta_f}(\gamma_f - 1)\right](1 - \Pi)$$

$$+ (\gamma_f - 1)\left(\frac{\rho_{p0}c_{pp}}{\rho_{f0}c_{pf}} - \frac{\beta_p}{\beta_f}\right)(1 - T) \qquad (9.4.24)$$

Thus, when $\hat{\alpha}$ is small, we see that $\hat{\beta}$ is given by the sum of translational, acoustic, and thermal effects. It thus may be expressed as $\hat{\beta} = \hat{\beta}_{tr} + \hat{\beta}_{ac} + \hat{\beta}_{th}$. Each of the separate contributions to $\hat{\beta}$ is generally small, except near resonance, particularly for bubbly liquids. Not all of these components contribute similar amounts. For bubbly liquids, the translational component, $\hat{\beta}_{tr}$, is negligible because $\rho_{p0}/\rho_{f0} \ll 1$. Similarly, $\hat{\beta}_{ac}$ is negligible for aerosols, and the same is true of $\hat{\beta}_{th}$ in hydrosols.

Figure 9.4.3 shows $\hat{\beta}_{ac}$, $\hat{\beta}_{th}$, and their sum for a bubbly liquid composed of 100-μm air bubbles in water, and having a volume concentration equal to 10^{-6}. As we see in the figure, the dispersion coefficients vary significantly in the region near resonance, with the acoustic component experiencing the widest swing. The thermal component also changes rapidly, although by smaller amounts, and from a positive value before resonance to a negative value beyond it. The figure also shows that $\hat{\beta}$ changes from a negative value to a positive one, which means that $c_s(\omega) < c_s(0)$ below resonance and $c_s(\omega) > c_s(0)$ beyond it. This is shown in Fig. 9.4.3b. Here, the speed ratios $c_s(\omega)/c_s(0)$ and $c_s(\omega)/c_{sf}$ are shown for the same bubbly liquid. It is seen that the speed changes are small everywhere, except at resonance. Once more, this is due to the small concentration used to reduce the theory.

The values of $\hat{\beta}_{tr}$, $\hat{\beta}_{ac}$, and $\hat{\beta}_{th}$ for emulsions are comparable in magnitude but generally very small. For hydrosols, the only component that is significant is the translational, but this is also small. Now, small values of the dispersion coefficient imply that the corresponding phase velocities are very close to the equilibrium sound speed in the suspension. This is seen in Fig. 9.4.4, where the speed ratios are shown for an emulsion and a hydrosol, both having a volume concentration equal to 0.01. It is observed that, even at this relatively large dilute concentration, the changes in sound speed in this emulsion are also very small. This is due to the close similarity between the properties of the particles and fluid. We also see that, except at the high end of the frequency range, the emulsion experiences smaller dispersion than the hydrosol. As with the attenuation, this is due to the emulsion particles having properties nearly equal to those of the exterior liquid. At high frequencies, resonance effects appear in the emulsion, producing a depression in the dispersion below resonance, followed by a significant increase beyond it. Here, self-scattering effects contribute in both cases.

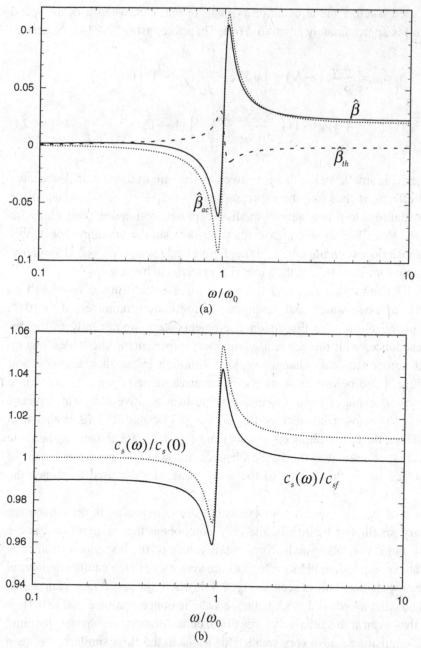

Figure 9.4.3. (a) Dispersion coefficients for a bubbly liquid composed of air bubbles in water. $D = 100\,\mu m$, $\phi_v = 10^{-6}$. (b) Sound speed ratios for a bubbly liquid composed of air bubbles in water. $D = 100\,\mu m$, $\phi_v = 10^{-6}$.

Finite Attenuation

As the volume concentration is increased from very small values, the attenuation coefficient increases proportionally, as indicated by the small attenuation results. For emulsions and hydrosols, the proportionality continues to hold for much larger values

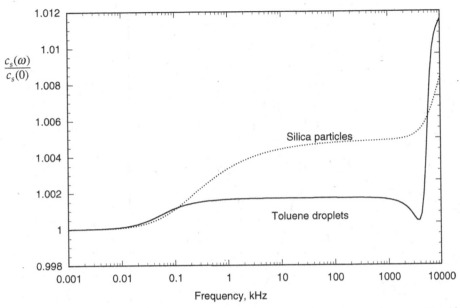

Figure 9.4.4. Sound speed ratios for an emulsion composed of toluene droplets in water. $D = 100\,\mu m$, $\phi_v = 10^{-2}$.

of ϕ_v than those used in the figure, because the attenuation remains small for them. But, for aerosols and bubbly liquids, the attenuation attains fairly significant values, meaning that the small-attenuation solution, (9.4.20a,b), cannot be used. Instead, we must use the solution-pair given by (9.4.19). Below, we consider the bubbly liquid case. Aerosols will be discussed at the end of the next section, where an additional effect – particle heating – is considered.

Examination of (9.4.13) shows that, in bubbly liquids, the magnitude of its imaginary part, denoted earlier by $Y(\omega)$, is proportional to the product $\phi_{v0}\gamma_p N_s$, where the isentropic compressibility ratio, $N_s = K_{sp}/K_{sf}$, is of the order of 10^4. Hence, the attenuation ceases to be small for quite small concentrations. This means that the ratio $\hat{\alpha}/\phi_{v0}$ will no longer be independent of the concentration, as it is when $\hat{\alpha} \ll 1$. This is shown in Fig. 9.4.5, where $\hat{\alpha}/\phi_{v0}$ is plotted as a function of the frequency for several small-volume concentrations. We see that some departures from the small attenuation result occur for values as small as $\phi_{v0} = 10^{-4}$. Most notable is the remarkable distortion of the bell-shaped curve that occurs for even very small concentrations. The distortion is due to a very marked increase in the dissipation rates that occurs at those concentrations and implies very large attenuations, as shown in Fig. 9.4.6, where $\hat{\alpha} = \alpha\lambda/2\pi$ is shown for several values of ϕ_{v0}. To get an idea of the significance of these large values, consider the amplitude of the waves, which, as we know, decreases with distance as $\exp(-2\pi\hat{\alpha}x/\lambda)$. Thus, in a distance equal to λ, the amplitude of the waves decreases by a factor equal to $1/\exp(2\pi\hat{\alpha})$. In the vicinity of resonance, this essentially implies that the wave has died.

The decay is even more drastic when we use the effective wavelength instead of that in the liquid alone, as used above. The effective wavelength when bubbles are present

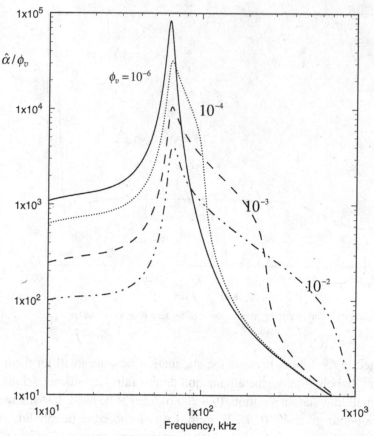

Figure 9.4.5. Scaled attenuation $\hat{\alpha}/\phi_v$ for a bubbly liquid composed of air bubbles in water. $D = 100\,\mu$m.

is given by λ multiplied by $c_s(\omega)/c_{sf}$. Hence, in a distance equal to one effective wavelength, the amplitude of the waves decreases by a factor equal to $\exp[-2\pi\hat{\alpha}c_s(\omega)/c_{sf})]$. The magnitude of the decrease can be assessed from Fig. 9.4.7, which shows the phase velocity as a function of the frequency for the same values of the volume concentration as used in Fig. 9.4.6. While the variations of $c_s(\omega)/c_{sf}$ are remarkable on their own, we see that when $\phi_v = 10^{-3}$, it is equal to about 10. It is evident that in such conditions, sound waves traveling in bubbly liquids die out in a very short distance.

Experimental Results

For completeness, we include in this section some experimental measurements of the attenuation coefficient and phase velocity obtained in three basic types of suspensions: aerosols, bubbly liquids, and emulsions. The experimental results shown here have been selected from a large number of studies available. Those selected were performed in suspensions having small-volume concentrations and well-defined particle sizes. Hence, the dilute, monodisperse theories given here are applicable.

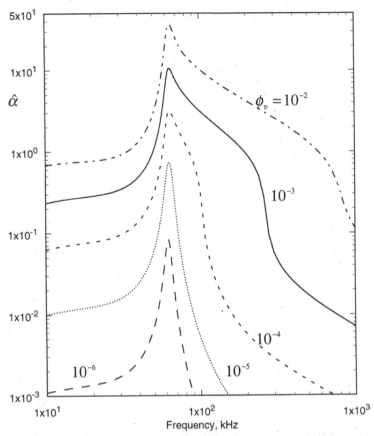

Figure 9.4.6. Attenuation $\hat{\alpha}$ for a bubbly liquid composed of air bubbles in water. $D = 100\ \mu m$.

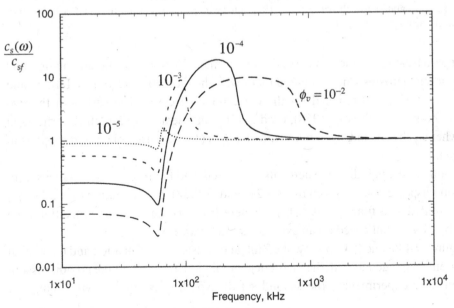

Figure 9.4.7. Speed ratio $c_s(\omega)/c_{sf}$ in a bubbly liquid.

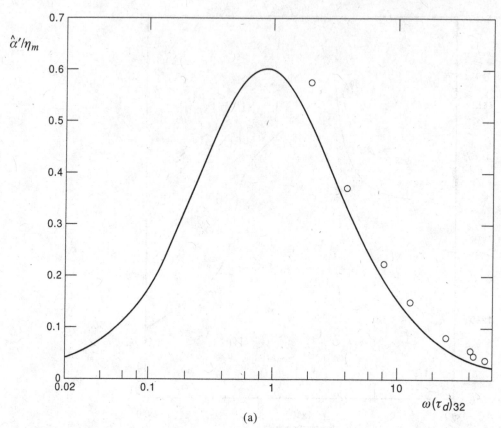

(a)

Figure 9.4.8. (a) Energy attenuation coefficient, $\hat{\alpha}' = 2\hat{\alpha}$, for an aerosol composed of alumina particles in air. Circle, Zink and Delsasso, 1958; solid line, (8.8.22), integrated over all sizes present. (b) Dispersion coefficient in an aerosol composed of alumina particles in air. Circle, Zink and Delsasso, 1958; solid line, (8.8.21) integrated over all sizes present. [Reprinted with permission from Temkin (2001a) © 2001, Acoustical Society of America.]

Aerosols. Below, we show two sets of experimental results for the attenuation and dispersion in aerosols, and compared them with the theoretical results of Temkin and Dobbins (1966a) and reduced for the corresponding cases. The earlier of the two sets, by Zink and Delsasso (1958), used a solid-particle aerosol (alumina particles in air), whereas the second, by Temkin and Dobbins (1966b), used oleic acid droplets in nitrogen.

In both cases, polydisperse aerosols were used, and the theory was obtained from the monodisperse results, given by (8.8.21) and (8.8.22), by integration over all sizes, and presented as a function of $\omega(\tau_d)_{32}$, where $(\tau_d)_{32}$ is the dynamic relaxation time given by (4.4.3), but based on an average droplet diameter.

Figures 9.4.8a and 9.4.8b show the Zink and Delsasso's attenuation and dispersion measurements, together with the theoretical prediction, reduced for the polydispersion used in their experiments. The second set of experiments cover a wider range of

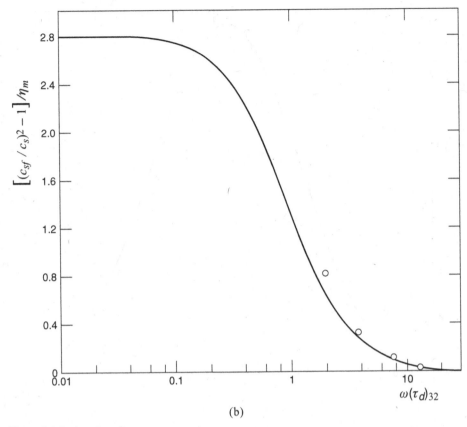

Figure 9.4.8. *(continued)*

frequencies, as shown in Figs. 9.4.9a and 9.4.9b. It is seen that both sets of experiments compare well with the theoretical results. In particular, this second set confirms the existence of an attenuation maximum that is predicted by the theory.

The experiments also imply a considerable support for two additional factors. One is for the two-phase model. The other is for the use of Stokes' law for aerosol particles when the frequency is not too large, as shown in Fig. 9.3.1.

Bubbly Liquids. For the attenuation in bubbly liquids, we select the measurements made by Silberman (1957) with air bubbles in water. These measurements were made with bubbles having a number of radii, and Fig. 1.1.3 in Chapter 1 shows his results for the attenuation, in dB/cm, for one of them, 0.264 cm, as a function of the frequency in kHz. It is seen that the finite-attenuation theory given in this section agrees well with the experiments, except in a band near resonance, where the theoretical value significantly exceeds the attenuation obtained experimentally. While the exact reason for the discrepancy is not known, it is noted that both experimental and theoretical attenuations are quite large in that region. Therefore, accurate attenuation measurements

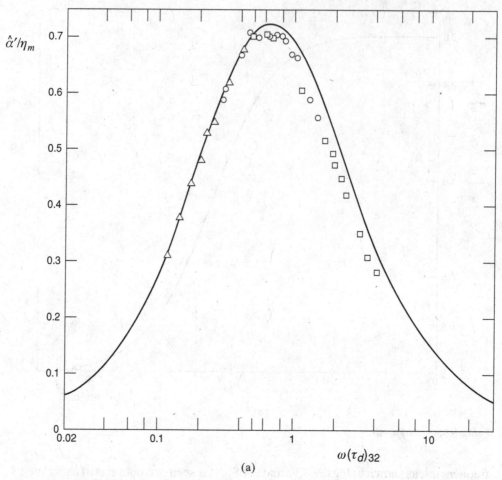

$\hat{\alpha}'/\eta_m$

$\omega(\tau_d)_{32}$

(a)

Figure 9.4.9. (a) Scaled energy-attenuation coefficient, $\hat{\alpha}' = 2\hat{\alpha}$, for an aerosol composed of oleic acid particles in nitrogen. Solid line, (8.8.22). The various symbols represent the experimental results of Temkin and Dobbins (1966b). (b) Dispersion coefficient in an aerosol composed of oleic acid particles in nitrogen. Solid line, (8.8.21). The various symbols represent the experimental results of Temkin and Dobbins (1966b). [Reprinted with permission from Temkin and Dobbins (1966b). © 1966, Acoustical Society of America.]

are difficult to obtain because of the rapid decay of the waves and the rapid variations of the phase velocity in that region.

Measurements of the phase velocity are also difficult to perform as a result of those factors indicated previously, but experiments exist that seem to have overcome the difficulties. These were performed by Cheyne et al. and were shown in Fig. 1.1.2, together with the theoretical predictions. As we see, the two results agree rather well.

Emulsions. As we saw previously, the dispersive effects produced in emulsions are quite small at small-volume concentrations because of the nearly equal properties

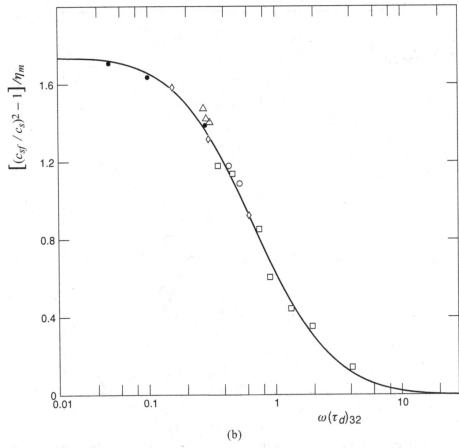

(b)

Figure 9.4.9. *(continued)*

of the particles and the fluid. Therefore, reliable measurements normally require concentrations that are not very small. Among the most accurate are the experimental measurements of Allegra and Hawley (1972). These were performed in a suspension of toluene particles in water made stable by the addition of a surfactant agent. The particle volume concentration used in the experiments had the rather large value of 0.2. Nevertheless, as Figure 9.4.10 shows, the theory agrees well with the experimental results.

This concludes our discussion of sound propagation in dilute suspensions that also have small values of the mass loading. In the next section, a different approach is used that is capable of dealing with such effects.

PROBLEM

Consider frequencies so high that $V = \Pi = \Theta = 0$. This state is known as the frozen state. Show that, for dilute suspensions, the isentropic sound speed in that state, $c_s(\infty)$,

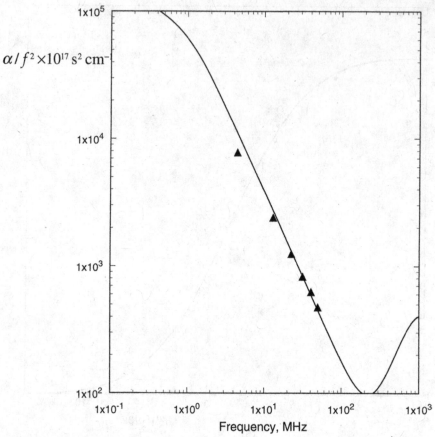

$$\alpha / f^2 \times 10^{17}\,\text{s}^2\,\text{cm}^{-1}$$

Figure 9.4.10. Attenuation in a toluene-water emulsion, $\phi_v = 0.2$. Solid line, (9.4.19); triangle, experimental results of Allegra and Hawley (1972). [Reprinted with permission from Temkin (2002). © 2002, Acoustical Society of America.]

is given by

$$c_{sf}^2 / c_s^2(\infty) \approx 1 - 2\phi_{v0}$$

9.5 Propagation via Compressibility

In the previous section, we considered sound propagation in suspensions from the point of view of a fluid having a large number of point sources of volume, force, and heat. In this section, we give back to the particles a more even role, and again consider a suspension as a two-phase fluid. As before, we limit the discussion to suspensions that are subject to small-amplitude pressures. Because of this, the suspension's volume element will normally change by small amounts. As a result, the change, or deformation, may be quantified it in terms of a linear, but time-dependent, compressibility. This approach also produces the phase velocity and the attenuation coefficient, showing that both quantities describe, in different ways, the energy dissipation that occurs in the suspension.

Dynamic Compressibility

We begin with the definition for the adiabatic compressibility of a medium. Let the medium – a suspension in our case – be enclosed by adiabatic walls and focus our attention on a volume element in it. At some time, the volume of the element is $\delta\tau$ and contains a certain number of particles that occupy a volume $\delta\tau_p$. The corresponding volume occupied by the fluid is $\delta\tau_f$, so that $\delta\tau = \delta\tau_f + \delta\tau_p$. When the suspension moves under the influence of an imposed pressure, the element is deformed, and the deformation can be quantified in terms of an adiabatic compressibility defined by

$$K_s = -\frac{1}{\delta\tau}\frac{d(\delta\tau)}{dp} \qquad (9.5.1)$$

where dp is the pressure change producing the deformation of the element. The subscript s in K_s is used to indicate that the deformation should be obtained in isentropic conditions. But as our discussion of the thermal attenuation in Section 9.2 shows, the total entropy of the element does not remain constant, but increases in accordance with the second law of thermodynamics. However, the increase is proportional to the squares of the variations of pressure and temperature. Therefore, in a linear theory, where only first-order quantities are retained, those increases may be disregarded, making the changes isentropic.

Now, (9.5.1) can be used in dynamic conditions, provided the physical quantities appearing in it can be defined for them. We note, however, that when the frequency is very small, K_s is given by the thermodynamic result obtained in Chapter 7, or

$$K_s(0) = \frac{1}{\sigma_0 c_s^2(0)} \qquad (9.5.2)$$

where $\sigma_0 = \rho_0$ is the suspension density and $c_s(0)$ is the equilibrium sound speed. The symbol 0 appearing in the sound speed has been used to remind us that this result applies in the low-frequency limit of the frequency-dependent phase velocity, $c_s(\omega)$. The quantity $K_s(0)$ thus defined is the adiabatic compressibility in, or very near, static conditions.

Let us now consider dynamic conditions. Here, the particle motions generally lag those in the fluid, which means that dissipation effects must enter in the picture. To allow for this lag, we first allow K_s and c_s to become complex functions of the frequency [i.e., $K_s \to \tilde{K}_s(\omega)$]; $c_s \to \tilde{c}_s(\omega)$], where the tilde is used to represent a complex quantity. We also postulate that $\tilde{K}_s$ and $\tilde{c}_s$ are related by the dynamic equivalent of (9.5.1). Thus

$$\tilde{K}_s(\omega) = \frac{1}{\sigma_0 \tilde{c}_s^2(\omega)} \qquad (9.5.3)$$

Because the main reason for a complex compressibility is attenuation, it is preferable to work with a complex wavenumber: $k = \omega/\tilde{c}_s(\omega)$. Thus, k is defined in the same manner as in the previous section and is given by $k = \omega/c_s(\omega) + i\alpha(\omega)$, where

$\alpha(\omega)$ is the amplitude-attenuation coefficient, and $c_s(\omega)$ is real and represents the phase velocity. In terms of this wavenumber, the compressibility can be expressed as $K_s(\omega)/K_s(0) = k^2 c_s^2(0)/\omega^2$, where, for simplicity, we have omitted the tilde on $\tilde{K}_s(\omega)$. Since the suspensions we are considering are dilute, it is better to scale this compressibility with that of the fluid alone, $K_{sf} = 1/\rho_{f0}c_{sf}^2$. Thus, we obtain a relationship between the complex wavenumber and the complex compressibility:

$$\left(\frac{k}{k_0}\right)^2 = \frac{\sigma_0}{\rho_{f0}} \frac{K_s(\omega)}{K_{sf}} \qquad (9.5.4)$$

In view of (9.4.16), this equation also connects the attenuation and the sound speed to the suspension compressibility. Now, as $\omega \to 0$, the attenuation vanishes and $c_s(\omega) \to c_s(0)$, so that we then have the identity $K_s(0)/K_{sf} = \rho_{f0}c_{sf}^2/\sigma_0 c_s^2(0)$. The value of this ratio was obtained in Chapter 7 and is given by (7.4.4).

The complex ratio $K_s(\omega)/K_{sf}$ differs from its equilibrium value, $K_s(0)/K_{sf}$ because of the various particle effects that exist in the suspension. For example, the particles can translate and pulsate. For small-amplitude deformations, these effects are independent of one another. They, therefore, contribute separate amounts to $K_s(\omega)/K_{sf} - 1$. That is,

$$K_s(\omega)/K_{sf} - 1 = [K_s(\omega)/K_{sf} - 1]_{tr} + [K_s(\omega)/K_{sf} - 1]_{pul} + \cdots \quad (9.5.5)$$

This is a valuable expression, but in order to obtain the attenuation and the phase velocity, we need compute each of the separate contributions. For this purpose, we first write (9.5.1) as

$$K_s(\omega) = -\frac{1}{\delta\tau}\frac{d(\delta\tau)/dt}{dp/dt} \qquad (9.5.6)$$

The derivatives appearing here need some explanation. First, d/dt is a Lagrangian derivative. It expresses the time derivative in a fixed frame of reference. Second, they must be evaluated while holding the entropy of the suspension element constant.

Translational Contribution

Let us first consider the translational contribution to the compressibility change. This is clearly due to the relative motion between particles and fluid. We wish to obtain this contribution for a plane, acoustic wave traveling in the suspension. To do this, it is advantageous to look at the suspension as a two-phase fluid, having a density $\sigma = \sigma_f + \sigma_p$, and an equilibrium compressibility $K_s(0)$, which is given by (7.4.6).

Now, $d(\delta\tau)/dt$ must be evaluated for an element having the same masses of particles and fluid. Although the particles and the fluid move relative to one another, we can, since the amplitude of the oscillatory motion is small, draw the volume element so that it always contains the same particles. Since the mass of the element is $\sigma\delta\tau$, we have

$$\frac{1}{\delta\tau}\frac{d(\delta\tau)}{dt} = -\frac{1}{\sigma}\frac{d\sigma}{dt} \tag{9.5.7}$$

Furthermore, since the motions are linear, we have $d/dt = \partial/\partial t$, so that on using (8.3.1), we have

$$\frac{1}{\sigma}\frac{d\sigma}{dt} \approx \frac{1}{\sigma_0}\frac{\partial\sigma'}{\partial t} = -\frac{1}{\sigma_0}\left(\sigma_{f0}\frac{\partial u_f}{\partial x} + \sigma_{p0}\frac{\partial u_p}{\partial x}\right) \tag{9.5.8}$$

Similarly, we write the time derivative of the pressure as $\partial p'/\partial t$, so that

$$K_s(\omega) = -\frac{\sigma_{f0}/\sigma_0}{\partial p'/\partial t}\left[\frac{\partial u_f}{\partial x} + \eta_{m0}\frac{\partial u_p}{\partial x}\right] \tag{9.5.9}$$

where $\eta_{m0} = \sigma_{p0}/\sigma_{f0}$. Also, since the wave is monochromatic, $\partial/\partial t = -i\omega$ and $\partial/\partial x = ik$. Hence,

$$K_s(\omega) = \frac{\sigma_{f0}}{\sigma_0}\frac{k}{\omega}\frac{u_f}{p'}\left(1 + \eta_{m0}\frac{u_p}{u_f}\right) \tag{9.5.10}$$

This expresses the compressibility in a suspension that is sustaining an acoustic wave in terms of the various quantities appearing on the right-hand side. If these are known, the equation provides the desired result. For example, suppose that $u_p = u_f = u$, then the pressure fluctuation is given by the plane wave relation $p' = \sigma_0 c_s(0)u$ and, furthermore, $k = \omega/c_s(0)$. Hence, using $\sigma_0/\sigma_{f0} = 1 + \eta_{m0}$, we obtain (9.5.2), which is the equilibrium value. Outside that equilibrium condition, not much can be done with the result, as it stands, because we do not know the value of the pressures, the velocities, and the wavenumber. These quantities are, in fact, part of the problem. However, keeping in mind that we are seeking changes in the compressibility that are small, we may approximate the quantities on the right-hand side as follows. First, since the wave is plane and the volume concentration is small, the acoustic pressure is approximately equal to $P'_f = \rho_{f0}c_{sf}U_f$, where P'_f and U_f are the pressure and velocity fluctuations without particles. Second, the wavenumber is, to the same order, equal to $k = \omega/c_{sf}$. Thus, we can write

$$\frac{K_s(\omega)}{K_{sf}} = \frac{\sigma_{f0}}{\sigma_0}(1 + \eta_{m0}V) \tag{9.5.11}$$

where $V = u_p/U_f$ is the single-particle velocity ratio. Hence, the translational contribution to the compressibility changes is given by

$$\left[\frac{K_s(\omega)}{K_{sf}(0)} - 1\right]_{tr} = \frac{\eta_{m0}}{1 + \eta_{m0}}(V - 1) \tag{9.5.12}$$

We shall return to this result after we compute the corresponding contribution due to the pulsational motion. It is noted, however, that the result was obtained by disregarding the pulsations of the particles, as well as whatever thermal effects exist in the fluid element.

Pulsational Contribution

We now consider the changes produced by the pulsational motion alone (i.e., without the translational effect considered above). The motion includes the effects of radial oscillations and temperature differences that may exist between particles and fluid, whether the particle pulsates or not. The calculation is necessarily more involved than that for the translational motion because the particles can affect the thermodynamic state of the fluid, even when the volume concentration is small. The reason for this is that, although the motions of the suspension element are isentropic, the entropies of the fluid and the particles change because of the heat transfer between particles and fluid. As we show below, this implies that the temperature of the fluid in the element is not equal to that in an adiabatic fluid. This means that the normal adiabatic relations that hold between pressure and temperature for a plane sound wave in a fluid without particles do not hold in a suspension. The differences between the adiabatic and nonadiabatic temperature are usually small for dilute suspensions, but are proportional to the volume concentration and can, therefore, be significant in some situations.

To obtain the relations that hold in isentropic conditions, we first note that since the masses of fluid and particles in the element are constant, the entropy changes of the fluid, ds_f, and of the particles, ds_p, in the suspension are well defined by (7.3.21). Thus, using (7.3.19b), we have

$$(1 + \eta_m)ds = c_{pf}\frac{dT_f}{T_f} - \frac{\beta_f}{\rho_f}dp_f + \frac{\rho_p\phi_v}{\rho_f(1 - \phi_v)}\left[c_{pp}\frac{dT_p}{T_p} - \frac{\beta_p}{\rho_p}dp_p\right] \quad (9.5.13)$$

where we have used $\eta_m = \rho_p\phi_v/\rho_{f0}(1 - \phi_v)$.

The variables appearing here are regarded as averages over the corresponding volumes. For example, T_f and T_p are average values of the fluid and particulate phase temperatures, respectively, in the element. When these variations are small, the total entropy remains constant, making $ds = 0$ in the linear approximation. We thus put $T = T_0 + T'$, $p = p_0 + p'$, with $p' \ll p_0$, $T' \ll T_0$, so that linearizing the result we get

$$T'_f = \frac{\beta_f T_0}{\rho_{f0}c_{pf}}p'_f - \frac{\phi_{v0}}{1 - \phi_{v0}}\frac{\beta_f T_0}{\rho_{f0}c_{pf}}\left(\frac{\rho_{p0}c_{pp}}{\beta_f T_0}T'_p - \frac{\beta_p}{\beta_f}p'_p\right) \quad (9.5.14)$$

If no particles were present, the second term would be absent and T'_f would equal Θ'_f, the adiabatic temperature fluctuation in a fluid without particles. Thus, the second term may be regarded as the temperature change induced in the fluid by the particles. To get an idea of this effect, we first note from (9.2.5) that the pressure change corresponding to Θ'_f is P'_f, which is equal to $(\rho_{f0}c_{pf}/\beta_f T_0)\Theta'_f$. Then, we put $p'_f = P'_f$, obtaining

$$\frac{T'_f}{\Theta'_f} = 1 - \frac{\phi_{v0}}{1 - \phi_{v0}}\left(\frac{\rho_{p0}c_{pp}}{\rho_{f0}c_{pf}}\mathrm{T} - \frac{\beta_p}{\beta_f}\Pi\right) \quad (9.5.15)$$

where $\mathrm{T} = T'_p/\Theta'_f$ and $\Pi = p'_p/P'_f$.

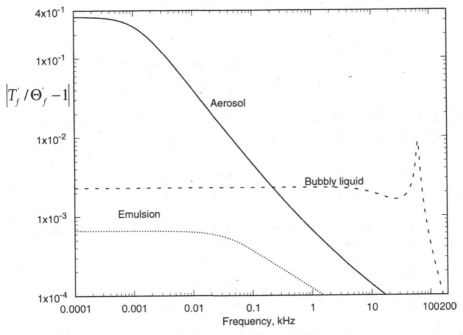

Figure 9.5.1. Fluid temperature change induced in the fluid by the particles. $D = 100\,\mu m$, $\phi_v = 10^{-4}$.

Figure 9.5.1 shows $|T'_f/\Theta'_f - 1|$ for three types suspensions, each having a volume concentration equal to 10^{-4}. We see that, generally, the heating term is small for such concentrations and larger at frequencies that are not too high. Two exceptions occur. One is in aerosols, where $\Pi \approx 0$, making that term proportional to the mass loading, η_m, which is equal to about 0.1 for the volume concentration used in the figure. The heating effect is particularly important at low frequencies, where the temperature ratio is not far from 1. The second exception occurs in bubbly liquids near resonance, where T is large. But, even then, the contribution at this very dilute concentration is small.

Although these changes affect the results only when the mass loading is finite, they do exist and may be included in the formulation without difficulty. To do this we will need expressions for T'_f/p'_f and T'_p/p'_f, both of which will appear in the calculations. These quantities may be approximated as follows. First, multiply both sides of (9.5.14) by $(1 - \phi_{v0})\beta_f$, use the identities $\beta_f^2 T_0 c_{sf}^2 = c_{pf}(\gamma_f - 1)$ and $K_{sf} = 1/\rho_{f0}c_{sf}^2$, and then divide by p'_f to obtain

$$(1 - \phi_{v0})\beta_f \frac{T'_f}{p'_f} \approx (\gamma_f - 1)(1 - \phi_{v0})K_{sf}$$

$$-\phi_{v0}(\gamma_f - 1)K_{sf}\left(\frac{\rho_{p0}c_{pp}}{\rho_{f0}c_{pf}}\frac{T'_p}{p'_f} - \frac{\beta_p}{\beta_f}\frac{p'_p}{p'_f}\right) \qquad (9.5.16)$$

Now add to both sides the quantity $\phi_{v0}\beta_f(T_p'/p_f')$ and factor out $(1 - \phi_{v0})\beta_f T_f'/p_f'$ to get

$$(1 - \phi_{v0})\beta_f \frac{T_f'}{p_f'} \approx (\gamma_f - 1)K_{sf} \frac{1 - \phi_{v0} + \phi_{v0}(\beta_p/\beta_f)(p_p'/p_f')}{1 + \eta_{m0}(c_{pp}/c_{pf})(T_p'/T_f')} \qquad (9.5.17)$$

Finally, approximate the pressure and temperature ratios on the right-hand side of this by their dilute limit, so that

$$(1 - \phi_{v0})\beta_f \frac{T_f'}{p_f'} \approx (\gamma_f - 1)K_{sf} \frac{1 - \phi_{v0} + \phi_{v0}(\beta_p/\beta_f)\Pi}{1 + \eta_{m0}(c_{pp}/c_{pf})\mathrm{T}} \qquad (9.5.18)$$

This reduces to (9.5.15) when the mass loading is very small. To the same degree of approximation, we obtain

$$\phi_{v0}\beta_p \frac{T_p'}{p_f'} \approx \frac{\phi_{v0}}{(1 - \phi_{v0})}(\gamma_f - 1)K_{sf} \frac{\beta_p}{\beta_f}\mathrm{T} \frac{1 - \phi_{v0} + \phi_{v0}(\beta_p/\beta_f)\Pi}{1 + \eta_{m0}(c_{pp}/c_{pf})\mathrm{T}} \qquad (9.5.19)$$

We shall make use of these relations later, keeping in mind the approximations that were used to obtain them.

Let us now return to the calculation of the suspension compressibility, starting with (9.5.6), written as

$$K_s = \frac{1}{\sigma_0} \frac{\partial\sigma'/\partial t}{\partial p'/\partial t} \qquad (9.5.20)$$

To take advantage of the fact that the mass concentration is constant, we evaluate $\partial\sigma'/\partial t$ by means of $1/\sigma = (1 - \phi_m)/\rho_f + \phi_m/\rho_p$. This gives, since the ϕ_m is constant, and since $\partial/\partial t = -i\omega$ for monochromatic time variations,

$$K_s = \frac{1 - \phi_{v0}}{\rho_{f0}} \frac{\rho_f'}{p'} + \frac{\phi_{v0}}{\rho_{p0}} \frac{\rho_p'}{p'} \qquad (9.5.21)$$

Now consider ρ_f' and ρ_p'. They are obtained from the corresponding linearized equations of state and are $\rho_f' = p_f'/c_{Tf}^2 - \rho_{f0}\beta_f T_f'$ and $\rho_p' = p_p'/c_{Tp}^2 - \rho_{p0}\beta_p T_p'$. Substituting these values above, we obtain

$$K_s(\omega) = \gamma_f(1 - \phi_{v0})K_{sf} + \gamma_p\phi_{v0}K_{sp} \frac{p_p'}{p_f'}$$

$$- \left[(1 - \phi_{v0})\beta_f \frac{T_f'}{p_f'} + \phi_{v0}\beta_p \frac{T_p'}{p_f'} \right] \qquad (9.5.22)$$

The quantities inside the square brackets are given by (9.5.18) and (9.5.19), so that their sum is given by

$$(1 - \phi_{v0})\beta_f \frac{T'_f}{P'_f} + \phi_{v0}\beta_p \frac{T'_p}{P'_f} = (\gamma_f - 1)K_{sf}(1 - \phi_{v0} + \phi_{v0}\beta_p/\beta_f)$$

$$\times \frac{1 - \phi_{v0} + \phi_{v0}(\beta_p/\beta_f)\Pi}{(1 - \phi_{v0})(1 + \eta_{m0}(c_{pp}/c_{pf})T}$$

Thus, putting the pressure ratio on the second term on the right-hand side of (9.5.22) equal to Π, and dividing by K_{sf} we obtain

$$\left[\frac{K_s(\omega)}{K_{sf}} - 1\right]_{pul} = -\phi_{v0} + \phi_{v0}\gamma_p N_s \Pi + (\gamma_f - 1)$$

$$\times \left[1 - \phi_{v0} - \frac{[1 - \phi_{v0} + \phi_{v0}(\beta_p/\beta_f)T][1 - \phi_{v0} + \phi_{v0}(\beta_p/\beta_f)\Pi]}{(1 - \phi_{v0})[1 + \eta_{m0}(c_{pp}/c_{pf})T]}\right] \quad (9.5.23)$$

This gives the pulsational contribution to the changes in the suspension compressibility. It is seen that the changes are affected by several property ratios, each representing a different mechanism. Thus, N_s measures the compressibility of the particles, relative to that of the fluid, β_p/β_f does the same for the thermal expansion and c_{pp}/c_{pf} for specific heat.

Total Changes

Having determined the various contributions to the compressibility changes, we may obtain the suspension dynamic compressibility by means of (9.5.12) and (9.5.23). Thus, adding the translational and pulsational contributions, we obtain

$$\frac{K_s(\omega)}{K_{sf}} = 1 - \phi_{v0} + \frac{\eta_{m0}(V - 1)}{1 + \eta_{m0}} + \phi_{v0}\gamma_p N_s \Pi + (\gamma_f - 1)$$

$$\times \left[1 - \phi_{v0} - \frac{[1 - \phi_{v0} + \phi_{v0}(\beta_p/\beta_f)T][1 - \phi_{v0} + \phi_{v0}(\beta_p/\beta_f)\Pi]}{(1 - \phi_{v0})[1 + \eta_{m0}(c_{pp}/c_{pf})T]}\right]$$

$$(9.5.24)$$

Before we reduce this result for small-volume concentrations, we note that, at very small frequencies, $T = \Pi = V = 1$, so that (9.5.24) reduces to

$$\frac{K_s(0)}{K_{sf}} = \gamma_f(1 - \phi_{v0} + \phi_{v0}N_s) - (\gamma_f - 1)\frac{[1 - \phi_{v0} + \phi_{v0}(\beta_p/\beta_f)]^2}{(1 - \phi_{v0})[1 + \eta_{m0}(c_{pp}/c_{pf})]}. \quad (9.5.25)$$

This is equal to the thermodynamic equilibrium result, given by (7.4.4). The corresponding equilibrium sound speed, $c_s(0)$ follows from this via $K_s(0)/K_{sf} = \rho_{f0}c_{sf}^2/\rho_0 c_s^2(0)$.

Let us return to the nonequilibrium result, (9.5.24). We wish to reduce it for small concentrations, where the particle-to-fluid ratios are known. The results, however, depend on the values of the property ratios appearing in the equation. As noted before, the mass loading, η_{m0}, can be finite for small-volume concentrations. The same is true of the product $\phi_{v0}N_s$ and may even be true for β_p/β_f for some particle-fluid combinations. Furthermore, the particle-to-fluid variable ratios can be large for some suspensions. Hence, in reducing the equation, we shall keep these facts in mind. But, to start, we note that, when ϕ_{v0} and η_{m0} are both very small, we obtain

$$\frac{K_s(\omega)}{K_{sf}} = 1 - 2\phi_{v0} + \frac{\phi_{v0}}{\delta}(V-1) + \phi_{v0}\left[\gamma_p N_s - (\gamma_f - 1)\frac{\beta_p}{\beta_f}\right]\Pi$$
$$+ (\gamma_f - 1)\phi_{v0}\left(\frac{\rho_{p0}c_{pp}}{\rho_{f0}c_{pf}} - \frac{\beta_p}{\beta_f}\right)T \qquad (9.5.26)$$

Multiplying this by the density ratio, σ_0/ρ_{f0}, yields (9.4.13). This is the value of $(k/k_0)^2$ obtained from the sound emission theory that was examined in Section 9.4 for both small and finite attenuations. Because of the agreement between the two theories, there is no need to examine the implications of (9.5.26).

Let us now consider dilute suspensions having finite values of η_{m0}, but with ϕ_{v0} remaining small, as required by the assumptions used in the derivation. Here, the attenuation and dispersion are affected by the heating effect mentioned previously and can be obtained from (9.5.24). Although this may be simplified for some specific suspensions, we shall use the complete expression. It should be remembered, however, that the theory is limited to values of ϕ_v such that the particle-fluid ratios, Π and T, are given by the single-particle results. Within this limitation, the attenuation and phase velocity are given by

$$\frac{c_{sf}^2}{c_s^2(\omega)} - \hat{\alpha}^2 + 2i\hat{\alpha}\frac{c_{sf}}{c_s(\omega)} = \frac{\sigma_0}{\rho_{f0}}\frac{K_s(\omega)}{K_{sf}} = X(\omega) + iY(\omega) \qquad (9.5.27)$$

where $X(\omega)$ and $Y(\omega)$ are now given by the real and imaginary parts of the right-hand side of $\sigma_0 K_s(\omega)/\rho_{f0}K_{sf}$, with $K_s(\omega)/K_{sf}$ given by (9.5.24). The attenuation and phase velocity are then obtained from these values of $X(\omega)$ and $Y(\omega)$ by means of (9.4.20a,b).

Now, our estimates of the temperature changes induced in the fluid by the particles, displayed in Fig. 9.5.1 for $\phi_v = 10^{-4}$, show that those effects are very small in emulsions, so that the attenuation and the sound speed are not affected by those changes. In bubbly liquids, the effect is slightly more noticeable, particularly near resonance. However, as Fig. 9.5.2 shows, the attenuation is not significantly affected.

On the other hand, the heating effect is important in aerosols. This is shown first in Fig. 9.5.3, where the attenuation is shown with and without the effect for a concentration equal to $\phi_v = 0.01$. This corresponds to a value of η_m equal to 8.3. As we see, particle heating decreases the attenuation at moderate frequencies by an order of magnitude. At smaller concentrations, the increase is not as large but it is, nevertheless, significant. To display the nonlinear dependence of the attenuation on η_m, we

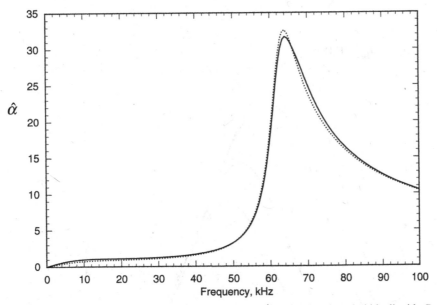

Figure 9.5.2. Fluid temperature change effects on the attenuation in a bubbly liquid. $D = 100\,\mu m$, $\phi_v = 10^{-2}$. Solid line, with temperature changes; dotted line, without temperature changes.

show in Fig. 9.5.4 the scaled attenuation, $\hat{\alpha}/\eta_{m0}$, as a function of the nondimensional frequency $\omega\tau_d$. As the figure shows, the most significant departures from linearity occur below the frequency at which the attenuation is the largest, and that the value of that frequency increases with η_m. Similar changes are observed for the phase velocity,

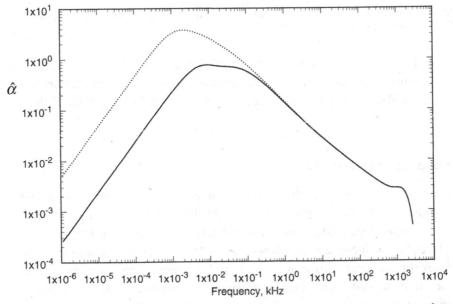

Figure 9.5.3. Temperature effects in the attenuation in an aerosol. $D = 100\,\mu m$, $\phi_v = 10^{-2}$. Solid line, with temperature changes; dotted line, without temperature changes.

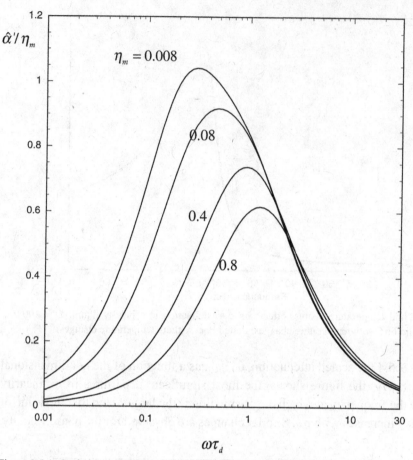

Figure 9.5.4. Scaled energy-attenuation coefficient, $\hat{\alpha}' = 2\hat{\alpha}$, in an aerosol for finite mass concentrations. Water droplets in air. $D = 100\,\mu m$.

as shown in Fig. 9.5.5. This shows the values of a dispersion coefficient, scaled with η_{m0}, also as a function of $\omega\tau_d$. These results, as well as those for the attenuation, agree with the results obtained on the basis of the two-phase model (Temkin and Dobbins, 1966a).

9.6 Propagation via Causality

It was mentioned in Section 9.4 that a connection exists between the phase velocity and the attenuation. That connection is presented below. First, we note that the results presented in the previous sections, as well as in Sections 8.6–8.8, show that the attenuation and the phase velocity are not independent of one another, but are related through the complex wavenumber $k(\omega) = \omega/c_s(\omega) + i\alpha$. This wavenumber was shown in the last section to be related to the complex adiabatic compressibility $K_s(\omega)$. This compressibility represents, in the complex plane, the response of the suspension element to a small-amplitude force having a frequency ω. It may, therefore, be regarded as

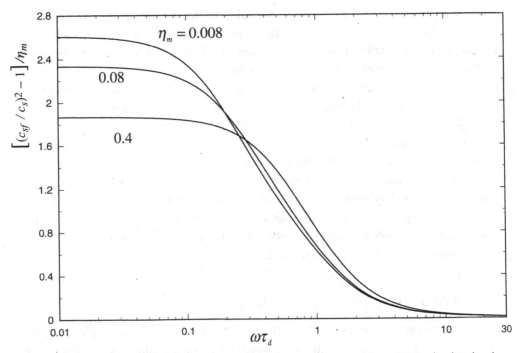

Figure 9.5.5. Scaled dispersion in an aerosol for finite mass concentrations. Water droplets in air. $D = 100\,\mu m$.

the Fourier transform of the suspension's response, or transfer function. Because this function represents the response of a physical system – a suspension element in this case, it must satisfy certain physical requirements. Particularly important is the requirement that the system response be causal (i.e., that it responds only after a force has been applied). This requirement imposes certain conditions on the mathematical behavior of the real and imaginary parts of the response function. The conditions are known as the Kramers-Kronig (K-K) relations. These provide yet another approach to suspension acoustics and are presented without derivation. Complete derivations may be found in the books cited in the bibliography.

The Kramers-Kronig Relations

Let $\zeta(\omega) = \zeta'(\omega) + i\zeta''(\omega)$ denote the response function of a linear system in the complex domain. Then, the real and imaginary parts of $\zeta(\omega)$ satisfy

$$\zeta'(\omega) - \zeta'(\infty) = \frac{1}{\pi}\fint_{-\infty}^{\infty} \frac{\zeta''(\Omega)d\Omega}{\Omega - \omega} \tag{9.6.1}$$

$$\zeta''(\omega) = -\frac{1}{\pi}\fint_{-\infty}^{\infty} \frac{\zeta'(\Omega) - \zeta'(\infty)d\Omega}{\Omega - \omega} \tag{9.6.2}$$

where the strokes through the integral symbol indicate principal parts and where the quantity $\zeta'(\infty)$ represents the value of the response function at a frequency where all the dissipative mechanisms included in ζ'' vanish. The symbol ∞ is used to represent such a frequency because it is normally very high.

The relations are supplemented by the condition that the response of a system be real at all frequencies, when the applied force is real. This requires that ζ' be an even function of the frequency and that ζ'' be an odd function – that is, $\zeta'(-\omega) = \zeta'(\omega)$ and $\zeta''(-\omega) = -\zeta''(\omega)$.

An important feature of the K-K relations is that they provide the connection between the real and imaginary parts of $\zeta(\omega)$ and show that if either is known, the other may be found by integration. This also implies that, for a given imaginary part, there can be only one function $\zeta'(\omega)$ that satisfies the relations and vice-versa. On the other hand, the response function is not unique. For example, multiplication of $\zeta(\omega)$ by a constant will also satisfy the relations.

The K-K Relations in Suspension Acoustics

For acoustic motions, the response function may be chosen in terms of various quantities; but, in what follows, we shall use the adiabatic compressibility, as this represents, in a very fundamental manner the response of the suspension to an applied pressure. However, since multiplication by constant is also a response function, we use, instead, $\zeta(\omega) = K_s(\omega)/K_s(0)$. By (9.4.26) and (9.5.4), this ratio can be written as

$$\zeta(\omega) = \frac{c_s^2(0)}{c_s^2(\omega)} - \overline{\alpha}^2(\omega) + 2i\overline{\alpha}\frac{c_s(0)}{c_s(\omega)} \tag{9.6.3}$$

where $\overline{\alpha} = \alpha c_s(0)/\omega$ is the attenuation based on $c_s(0)$. Thus, $\zeta'(\omega) = c_s^2(0)/c_s^2(\omega) - \overline{\alpha}^2$ and $\zeta'' = 2\overline{\alpha}c_s(0)/c_s(\omega)$. With these values, the K-K relations become (Temkin, 1990)

$$\frac{c_s^2(0)}{c_s^2(\omega)} - \overline{\alpha}^2(\omega) - \frac{c_s^2(0)}{c_s^2(\infty)} = \frac{2}{\pi}\fint_{-\infty}^{\infty} \frac{c_s(0)}{c_s(\Omega)}\frac{\overline{\alpha}(\Omega)}{\Omega - \omega}d\Omega \tag{9.6.4}$$

$$2\overline{\alpha}\frac{c_s(0)}{c_s(\omega)} = -\frac{1}{\pi}\fint_{-\infty}^{\infty} \frac{c_s^2(0)/c_s^2(\Omega) - \overline{\alpha}^2(\Omega) - c_s^2(0)/c_s^2(\infty)}{\Omega - \omega}d\Omega \tag{9.6.5}$$

In writing the last equation, we have again denoted by $c_s(\infty)$ the value of $c_s(\omega)$ in the frozen state. Here, the dissipation vanishes so that $\overline{\alpha}(\infty)$ is zero.

Although presented for suspensions sustaining acoustic motions, these relations are applicable to all acoustic motions in dispersive media. However, as given, the relations are of little use even if one of the two unknowns, $c_s(\omega)$ or $\overline{\alpha}(\omega)$, is known, because the remaining unknown appears both inside and outside the integrals. We may, nevertheless, take a hint from the previous sections and approximate the unknowns appearing on the right-hand side of the equations with known values. To do this, we proceed as follows. First, in the integral in (9.6.4), we approximate the speed ratio appearing there by 1, and use the small-dissipation attenuation obtained in Section 9.3. To avoid confusion, we will denote this approximate value of $\bar{\alpha}$ by $\bar{\alpha}_0$, noting that it is equal to the small dissipation value of $\hat{\alpha}$, multiplied by $c_s(0)/c_{sf}$. With these approximations, (9.6.4) becomes

$$\frac{c_s^2(0)}{c_s^2(\omega)} - \bar{\alpha}^2(\omega) = \frac{c_s^2(0)}{c_s^2(\infty)} + X(\omega) \tag{9.6.6}$$

where

$$X(\omega) = \frac{2}{\pi} \int_{-\infty}^{\infty} \frac{\bar{\alpha}_0(\Omega)}{\Omega - \omega} d\Omega \tag{9.6.7}$$

Since $\bar{\alpha}_0$ is known, the quantity $X(\omega)$ can be also regarded as known. Note that, if $\bar{\alpha}^2 \ll 1$ at all frequencies, then $c_s^2(0)/c_s^2(\omega) = X(\omega)$, thus completing the solution for that case.

Now consider the second of the K-K relations. Here, we neglect $\bar{\alpha}^2$ inside the integral and use, for the speed ratio appearing there, the small-attenuation result mentioned previously. This gives

$$2\bar{\alpha}\frac{c_s(0)}{c_s(\omega)} = Y(\omega) \tag{9.6.8}$$

where

$$Y(\omega) = -\frac{1}{\pi} \int_{-\infty}^{\infty} \frac{X(\Omega) - X(\infty)}{\Omega - \omega} d\Omega \tag{9.6.9}$$

Since $X(\omega)$ is known, $Y(\omega)$ can also be regarded as known. The solution is then given by solving the pair (9.6.6) and (9.6.8).

Although the approach seems straightforward, it assumes that the two improper integrals that define $X(\omega)$ and $Y(\omega)$ exist, and that they can be evaluated. The existence of $X(\omega)$, for example, depends on the behavior of $\bar{\alpha}_0$ at high frequencies. But, as our calculations in Section (9.3) show, $\bar{\alpha}_0$ decreases rapidly with ω, guaranteeing the existence of the integral. On the other hand, an exact evaluation of the integral may be difficult to obtain.

It should be added that the K-K approach can also be started with an expression for the sound speed, if this is known. In addition, the method can also be used with accurate experimental data for the attenuation, for example.

Sometimes the accuracy of the attenuation estimate/data can be assessed beforehand, but this requires information about the sound speeds at zero frequencies, and at such frequencies that the particles are frozen, called the frozen-equilibrium sound speed. If these are known, we obtain from (9.6.6) and 9.6.7

$$\frac{2}{\pi}\int_{-\infty}^{\infty}\frac{\bar{\alpha}_0(\Omega)}{\Omega}d\Omega = 1 - \frac{c_s^2(0)}{c_s^2(\infty)} \tag{9.6.10}$$

The result shows that the contributions of the attenuation at all frequencies is equal to difference appearing on the right-hand side. This is true regardless of the mechanisms that produce attenuation.

The difficulty, of course, is getting $c_s(\infty)$. One way of doing this is simply to set the three particle-fluid ratios equal to zero in the emission and in the compressibility theories, keeping in mind their approximate nature. If we limit ourselves to very dilute suspensions, then we obtain

$$\frac{c_{sf}^2}{c_s^2(\infty)} \approx 1 - 2\phi_{v0} \tag{9.6.11}$$

The zero frequency sound speed is given by (7.4.3). Thus, the ratio on the right-hand side of (9.6.10) may be regarded as known, so that the verification can be made. An example of this is presented below, where we use the K-K approach to reconsider the propagation in aerosols. Here, the integrals can be evaluated with relative ease, and the results for the attenuation and the phase velocity can be compared with those obtained before.

Propagation in an Aerosol

In aerosols, the attenuation is given by the sum of translational and thermal components. The small-dissipation values for these were obtained earlier, and their sum is given by (8.8.22). Thus, multiplying that sum by $c_s(0)/c_{sf}$ to get $\bar{\alpha}_0$, we obtain

$$\bar{\alpha}_0(\omega) = \frac{1}{2}\eta_{m0}\frac{c_s(0)}{c_{sf}}\left[\frac{\omega\tau_d}{1 + \omega^2\tau_d^2} + (\gamma_f - 1)\frac{c_{pp}}{c_{pf}}\frac{\omega\tau_t}{1 + \omega^2\tau_t^2}\right] \tag{9.6.12}$$

where $\eta_{m0} = \phi_{v0}/\delta$ is the dilute value of the mass loading. This result, it may be recalled, is based on the simplest forms for the particle force and heat transfer, and was obtained on the assumption that the suspension is dilute. We also note that $\bar{\alpha}_0(\omega)$ satisfies the assumptions used in the K-K system. Thus, $\bar{\alpha}_0(-\omega) = -\bar{\alpha}_0(\omega)$ and that $\bar{\alpha}_0(\omega \to \infty) = 0$.

Before proceeding, we assess the range of validity of $\bar{\alpha}_0$. Thus, substituting it on the left-hand side of (9.6.10), we obtain, after integration

$$\frac{2}{\pi}\int_{-\infty}^{\infty}\frac{\bar{\alpha}_0(\Omega)}{\Omega}d\Omega = \frac{\phi_{v0}}{\delta}\frac{c_s(0)}{c_{sf}}\left[1 + (\gamma_f - 1)\frac{c_{pp}}{c_{pf}}\right] \tag{9.6.13}$$

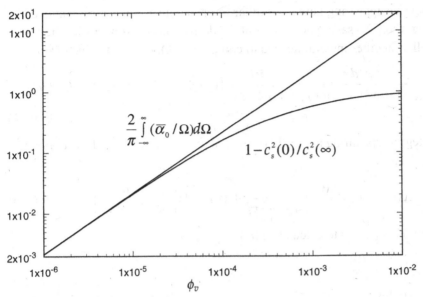

Figure 9.6.1. Thermodynamic and causal evaluations of (9.6.10) for water droplets in air.

On the other hand, the right-hand side of (9.6.10) can be obtained by using (9.6.11) for $c_{sf}^2/c_s^2(\infty)$ and from (7.4.7) for the value of $c_{sf}^2/c_s^2(0)$ applicable to aerosols. This gives

$$1 - \frac{c_s^2(0)}{c_s^2(\infty)} = \frac{(\gamma_f - 1)}{1 + \eta_{m0}} \frac{\eta_{m0} c_{pp}/c_{pf}}{1 + \gamma_f \eta_{m0} c_{pp}/c_{pf}}$$

When $\phi_v \to 0$, this becomes equal to (9.6.13), as may be easily verified. For larger, but still dilute concentrations, the value of $c_s(\infty)$ given by (9.6.11) is uncertain. But if use is of it is made to compute the above difference, we obtain the results shown in Fig. 9.6.1. This shows that differences occur for values of ϕ_v which are dilute, but for which the mass loading is not negligible. The main point of the figure is that, over a range of concentrations, the attenuation estimate is correct – which of course we knew before hand.

Having demonstrated that the initial approximation to $\bar{\alpha}$ is correct within a range of values of ϕ_v, we proceed to use the K-K system to obtain the phase velocity, as well as a closer approximation to the attenuation. We start the K-K procedure by computing $X(\omega)$. Substitution of (9.6.12) in (9.6.7) gives after making an obvious change of variables,

$$X(\omega) = \frac{1}{\pi} \eta_{m0} \frac{c_s(0)}{c_{sf}}$$

$$\times \left[\int_{-\infty}^{\infty} \frac{x\,dx}{(1+x^2)(x - \omega\tau_d)} + (\gamma_f - 1)\frac{c_{pp}}{c_{pf}} \int_{-\infty}^{\infty} \frac{x\,dx}{(1+x^2)(x - \omega\tau_t)} \right]$$

$$(9.6.14)$$

The two integrals appearing here are similar to one another and can be evaluated by contour integration. To save space, we omit the derivation and simply give the result. They, as well as another integral needed to compute $Y(\omega)$, can be obtained from

$$\pi I_n = \int_{-\infty}^{\infty} \frac{x^n dx}{(1+x^2)(x-\omega\tau)} = \frac{\pi i}{1+(\omega\tau)^2}\left[(\omega\tau)^n - i^{n-1}\omega\tau - i^n\right], \quad n = 0, 1$$

(9.6.15)

Thus, the integrals appearing in (9.6.14) are equal to $\pi/(1+\omega^2\tau_d^2)$ and $\pi/(1+\omega^2\tau_t^2)$, respectively. Hence,

$$X(\omega) = \eta_{m0}\frac{c_s(0)}{c_{sf}}\left[\frac{1}{1+\omega^2\tau_d^2} + (\gamma_f - 1)\frac{c_{pp}}{c_{pf}}\frac{1}{1+\omega^2\tau_t^2}\right] \qquad (9.6.16)$$

We now use this in (9.6.9) to calculate $Y(\omega)$,

$$Y(\omega) = -\frac{1}{\pi}\eta_{m0}\frac{c_s(0)}{c_{sf}}$$

$$\times \left[\int_{-\infty}^{\infty} \frac{dx}{(1+x^2)(x-\omega\tau_d)} + (\gamma_f - 1)\frac{c_{pp}}{c_{pf}}\int_{-\infty}^{\infty} \frac{dx}{(1+x^2)(x-\omega\tau_t)}\right]$$

(9.6.17)

Here, the integrals are equal to $-\pi\omega\tau_d/(1+\omega^2\tau_d^2)$ and $-\pi\omega\tau_t/(1+\omega^2\tau_t^2)$. Thus,

$$Y(\omega) = \eta_{m0}\frac{c_s(0)}{c_{sf}}\left[\frac{\omega\tau_d}{1+\omega^2\tau_d^2} + (\gamma_f - 1)\frac{c_{pp}}{c_{pf}}\frac{\omega\tau_t}{1+\omega^2\tau_t^2}\right] \qquad (9.6.18)$$

To obtain the values of $\bar{\alpha}$ and $c_s(\omega)$, we also need $c_{sf}^2/c_s^2(\infty)$. For $\phi_{v0} \ll 1$, those results give $c_{sf}^2/c_s^2(\infty) \approx 1$. Hence,

$$\frac{c_s^2(0)}{c_s^2(\omega)} - \bar{\alpha}^2 = \frac{c_s^2(0)}{c_s^2(\infty)} + \eta_{m0}\frac{c_s(0)}{c_{sf}}\left[\frac{1}{1+\omega^2\tau_d^2} + (\gamma_f - 1)\frac{c_{pp}}{c_{pf}}\frac{1}{1+\omega^2\tau_t^2}\right]$$

(9.6.19)

$$2\bar{\alpha}\frac{c_s(0)}{c_s(\omega)} = \eta_{m0}\frac{c_s(0)}{c_{sf}}\left[\frac{\omega\tau_d}{1+\omega^2\tau_d^2} + (\gamma_f - 1)\frac{c_{pp}}{c_{pf}}\frac{\omega\tau_t}{1+\omega^2\tau_t^2}\right] \qquad (9.6.20)$$

These can be combined to obtain an equation for either unknown whose solution provides the values of $c_s(\omega)$ and $\bar{\alpha}$. The results are not given here because they agree with those obtained from the compressibility model when the mass loading is finite, as may be easily verified. They also reproduce the phase-velocity results obtained in Section 9.4 for $\phi_v \ll 1$.

But, more importantly, the results show that the K-K system can be used to obtain the phase velocity if an approximate value of the attenuation is known. Furthermore, the method also produces a more accurate attenuation than initially estimated. Finally, the method is far more direct than those used before. On the other hand, the K-K procedure lacks the generality of the other methods presented in this chapter because each type of suspension requires a suspension-specific attenuation estimate, and, therefore, the evaluation of different integrals. Also, contrary to the aerosol case, the integrals appearing in other types of suspensions cannot be evaluated so easily.

Nevertheless, the K-K method does provide additional tools for the study of suspensions. One of those tools stems directly from the fact that the attenuation and the phase velocity are connected to one another. Hence, the accuracy of separate measurements of those two quantities can be assessed. This is possible because, to a value of the attenuation, there can correspond only one value of the phase velocity.

9.7 Concluding Remarks

This chapter has examined the propagation of sound in dilute suspensions. Several perspectives have been used that enabled us to examine the physical behavior of suspensions sustaining acoustic waves. Results have been presented for the attenuation and dispersion of sound in suspensions, which use certain particle-to-fluid ratios that were obtained in earlier chapters and that include the various mechanisms that come into play in dilute suspensions. These results apply to aerosols, bubbly liquids, emulsions, and hydrosols.

10

Applications and Extensions

10.1 Introduction

We conclude this work with a brief discussion of some topics that relate to the acoustic motions treated in Chapters 8 and 9. The topics selected are important examples of the use of acoustic waves – weak and strong – in suspension studies. The discussion is meant only to convey an idea about those applications, some of which are still being developed.

The examples that are addressed here relate to the use of acoustic theories to: (1) consider a simple acoustic problem; (2) measure particle sizes and concentrations in suspensions; and (3) to see the coalescence effects produced by sound waves. Some of these are important in those situations where a need arises to modify the sizes of the particles or to determine them. The number of techniques available to do the latter is rather limited, but optical scattering is perhaps the best known. However, optical beams are not capable to penetrate deeply into liquid media. The acoustic wave, on the other hand, can do that, and has therefore become another technique that can be used for that purpose.

The third example refers to the acoustic modification of particle size distributions in suspensions. In some applications – e.g., in trying to remove small particles from industrial fumes – it is desirable to increase that size, whereas in others the opposite effect is sought (e.g., in the production of finely dispersed droplets). Acoustic waves are capable of producing both effects, though strong waves are usually required. The first of these effects – coalescence – is considered in Section 10.5.

10.2 Reflection at a Fluid-Suspension Interface

Our study of sound propagation in suspensions has been limited to considering sound attenuation and dispersion in unbounded suspensions. Real suspensions have, of course, finite dimensions and therefore include interfaces separating them from other media, normally the fluid host without particles. From the acoustical perspective, these interfaces separate materials having different acoustical properties. Hence, a wave reaching the interface from either side will normally experience partial reflection and

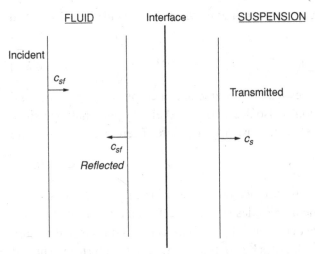

Figure 10.2.1. Schematic diagram for reflection at interface between two media.

transmission. These effects can be relevant in various settings (e.g., in the screening of sound by bubbles or by dust).

In what follows, we consider the simplest examples of reflection and transmission at such interfaces. In the first, we consider the reflection of a plane sound wave at an interface between a fluid and a suspension of the same fluid when equilibrium conditions hold. In the second example, we consider the same interface, this time at finite frequencies.

Equilibrium Conditions

In order to put the finite-frequency reflection in perspective, we first consider the reflection of sound waves from the surface separating two ideal media. This problem is usually stated in terms of two elastic materials, not necessarily fluids. Its solution also applies when the media are an ideal fluid and a suspension in equilibrium, and will be presented here for monochromatic waves, although it also applies to arbitrary plane waveforms.

In its simplest form, the incident wave is plane and travels in a direction perpendicular to the interface, which is a plane surface extending to infinity on the plane normal to the incoming wave. Both fluid and suspension extend to infinity, as sketched in Fig. 10.2.1. The interaction of the incident wave with the interface results in two additional waves, one reflected, traveling in a direction opposite to the incident, and one transmitted, traveling in the same direction. If the incident wave is monochromatic, the reflected and transmitted waves will also be monochromatic, and will have the same frequency as the incident wave. However, their amplitudes will generally be different. It is these quantities we wish to obtain. Since equilibrium conditions hold everywhere, and since the materials are inviscid and nonheat-conducting, the equations of motion for both reduce to wave equations, whose solution can be expressed in terms of a velocity potential. Thus, for region 1 in the figure, where both incident and reflected waves exist, we may write

$$\phi_1 = Ae^{-i(\omega t - k_1 x)} + Be^{-i(\omega t + k_1 x)} \tag{10.2.1}$$

Here, $k_1 = \omega/c_{s1}$, where c_{s1} is the isentropic sound speed corresponding to the fluid in region 1. In region 2, the motion is also described by a velocity potential; but, since region 2 extends to infinity, no reflected wave appears there. Therefore, we may write

$$\phi_2 = Ce^{-i(\omega t - k_2 x)} \tag{10.2.2}$$

where $k_2 = \omega/c_{s2}$ and c_{s2} is the isentropic sound speed in region 2. As stated, the problem reduces to obtaining the amplitudes of the reflected and transmitted waves, B and C, respectively. These are determined by the boundary conditions that require that the acoustic pressures and velocities be continuous across the interface. Thus, if the interface is placed at $x = 0$, we have

$$u_1(0, t) = u_2(0, t) \quad \text{and} \quad p_1'(0, t) = p_2'(0, t) \tag{10.2.3}$$

The velocity and pressure in either medium are given by their usual acoustic forms, $u = \partial\phi/\partial x$ and $p' = -\rho_0\partial\phi/\partial t$, where ρ_0 is the ambient density. This is equal to ρ_{10} for region 1 and ρ_{20} for region 2. Thus, using the above solutions, we obtain

$$u_1 = ik_1 e^{-i\omega t}\left[Ae^{ik_1 x} - Be^{-ik_1 x}\right]$$

$$\text{and} \quad p_1' = i\omega\rho_{10}e^{-i\omega t}\left[Ae^{ik_1 x} + Be^{-ik_1 x}\right] \tag{10.2.4a,b}$$

$$u_2 = ik_2 Ce^{-i(\omega t - k_2 x)} \quad \text{and} \quad p_2' = i\omega\rho_{20}Ce^{-i(\omega t - k_2 x)} \tag{10.2.5a,b}$$

Applying the boundary conditions and solving for B and C, we obtain

$$B/A = -\frac{1 - \rho_{20}c_{s2}/\rho_{10}c_{s1}}{1 + \rho_{20}c_{s2}/\rho_{10}c_{s1}} \tag{10.2.6}$$

$$C/A = \frac{2(c_{s2}/c_{s1})}{1 + \rho_{20}c_{s2}/\rho_{10}c_{s1}} \tag{10.2.7}$$

These are the amplitudes of the reflected and transmitted waves, relative to the amplitude of the incident wave. We note that both B and C are real, indicating that no lag exists between any of these waves.

Let us now consider the reflected wave. As (10.2.6) shows, its amplitude is different from that of the incident wave and could be used to quantify the reflective effects of the interface. However, a more appropriate measure of those effects is provided by the acoustic intensity in either the reflected wave or the transmitted wave, relative to the incident wave. By definition, the acoustic intensity, $\mathbf{I}$, in a wave traveling along a direction determined by the unit vector $\mathbf{n}$ is given by the product between the acoustic pressure and the acoustic velocity, or $\mathbf{I} = p'u\,\mathbf{n}$. Thus, the ratio of reflected to incident

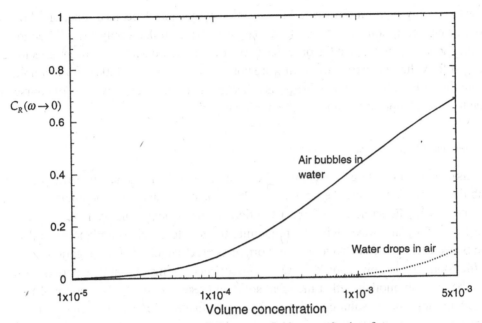

Figure 10.2.2. Equilibrium reflection coefficient at two fluid-suspension interfaces.

intensities defines the reflection coefficient, C_R. Using (10.2.4) and (10.2.6), we obtain

$$C_R = \left(\frac{1 - \rho_{20}c_{s2}/\rho_{10}c_{s1}}{1 + \rho_{20}c_{s2}/\rho_{10}c_{s1}}\right)^2 \tag{10.2.8}$$

The transmission coefficient can be similarly obtained, but since energy is conserved, it is simpler to compute it from C_R via $C_T = 1 - C_R$. Both depend on the ratio of the characteristic impedances, $\rho_0 c$, of the media. Special limiting forms of these results are worth mentioning here. Thus, at a water-air interface, where the impedance ratio is either very small or very large, the reflection coefficient is very nearly equal to 1, whereas the reflection coefficient for an interface between two fluids having nearly equal characteristic impedances, it is nearly equal to zero.

Let us now write (10.2.8) for the specific case of a fluid-suspension interface. That result applies here, as given, provided the variations in the incident wave occur so slowly that the suspension remains in equilibrium at all times. However, to avoid confusion, we write it using the notation applicable to suspensions in equilibrium. Thus, in region 1, we have a fluid whose ambient density and sound speed are density are ρ_{f0} and $c_{s1} = c_{sf}$, respectively, whereas in region 2 we have, $\rho_{20} = \rho_0$, and $c_{s2} = c_s(0)$. Therefore, (10.2.8) becomes

$$C_R = \left(\frac{1 - \rho_0 c_s(0)/\rho_{f0}c_{sf}}{1 + \rho_0 c_s(0)/\rho_{f0}c_{sf}}\right)^2 \tag{10.2.9}$$

If the host fluid in the suspension is the same as that in region 1, as it will be assumed, then we can use the equilibrium results obtained in Chapter 7 to evaluate the density and speed ratios appearing in (10.2.9). The results are shown in Fig. 10.2.2 for both a water-bubbly liquid interface and for a gas-dusty gas interface, as a function of the volume concentration. As we see, the dusty gas interface reflects only a small fraction of the incoming energy. On the other hand, the bubbly liquid acts as a strong reflector, at least for volume concentrations that are not too small. At $\phi_v = 0.001$, for example, more than 40% of the incident energy is reflected. This large reflectivity is, of course, due to the high compressibility of the bubbly liquid.

Nonequilibrium

Let us now consider finite frequencies. The geometry is still that sketched in Fig. 10.2.1, and the events that follow the interaction of the incident wave with the interface are, on the whole, the same as before: a fraction of the incident energy is reflected in the form of a plane wave, with the remaining transmitted, also as a plane wave. But the two situations differ in one very important aspect, namely that propagation in the suspension side is now dispersive. One consequence of this is that dissipation must also be present; another is that the ideal solution presented above does not apply. To obtain the appropriate solution, we make use of the fact that in a suspension sustaining a plane wave, the fluid pressure and velocity depend on time and space through the factor $\exp[-i(\omega t - \tilde{k}_2 x)]$, where the tilde over the wavenumber $\tilde{k}_2$ implies a complex value. This value was obtained in Chapter 9 and given by

$$\tilde{k}_2/k_1 = c_{sf}/c_s(\omega) + i\hat{\alpha} \tag{10.2.10}$$

where k_1 is the corresponding wave number for the fluid without particles. Now, although the spatial and time dependence of the pressure and velocity are identical, their amplitudes and phases, relative to one another, may differ. To take into account these differences and to obtain the velocity and pressure in the transmitted waves, we write those quantities as

$$u_{2f} = i\tilde{k}_2 C e^{-i(\omega t - \tilde{k}_2 x)} \quad \text{and} \quad p'_{2f} = i\rho_0 \omega C e^{-i(\omega t - \tilde{k}_2 x)} \tag{10.2.11a,b}$$

The coefficients of the exponential factors appearing here have been chosen so that they reduce to the equilibrium forms given earlier.

The field in region 1 is still given by (10.2.4) so that, as before, the problem reduces to obtaining B and C. To do this, we use the boundary conditions on the fluid at $x = 0$. These are that the *fluid* velocities and pressures on both sides of the interface be equal. That is,

$$u_1(0, t) = u_{2f}(0, t) \quad \text{and} \quad p'_1(0, t) = p'_{2f}(0, t) \tag{10.2.12a,b}$$

Applying these we obtain

$$1 - B/A = (\tilde{k}_2/k_1)C/A \tag{10.2.13}$$

$$1 + B/A = (\rho_0/\rho_{f0})C \tag{10.2.14}$$

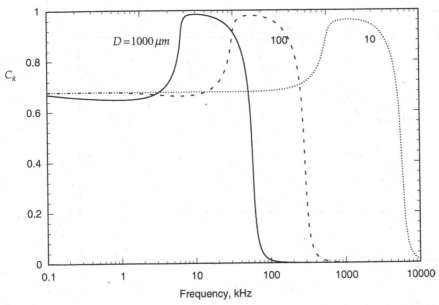

Figure 10.2.3. Finite-frequency reflection coefficient at three liquid-bubbly liquid (air bubbles in water) interfaces. $\phi_v = 0.005$.

Solving for B/A, we find

$$B/A = \frac{1 - \rho_{f1}\tilde{k}_2/\rho_0 k_1}{1 + \rho_{f1}\tilde{k}_2/\rho_0 k_1} \tag{10.2.15}$$

The appearance of a complex amplitude here implies a lag between incident and reflected waves. The reflection coefficient is equal to $|B/A|^2$. Thus, multiplication of (10.2.15) by its complex conjugate and using (10.2.10) for the complex ratio $\tilde{k}_2/k_1$ yields, after some rearrangement,

$$C_R(\omega) = \frac{\left[1 - \rho_{f1}c_{sf}/\rho_0 c_s(\omega)\right]^2 + (\rho_{f1}/\rho_0)^2 \hat{\alpha}^2}{\left[1 + \rho_{f1}c_{sf}/\rho_0 c_s(\omega)\right]^2 + (\rho_{f1}/\rho_0)^2 \hat{\alpha}^2} \tag{10.2.16}$$

This is the desired result; it applies to the reflection from a fluid-suspension interface at finite frequencies and obviously includes the effects of attenuation and dispersion. To evaluate it, we simply have to use the values of $c_{sf}/c_s(\omega)$ and $\hat{\alpha}^2$ obtained in Chapter 9. For bubbly liquids, however, a simpler result applies because, for them, the density ratio appearing in (10.2.16) can be set equal to 1. This gives

$$C_R(\omega) = \frac{\left[1 - c_{sf}/c_s(\omega)\right]^2 + \hat{\alpha}^2}{\left[1 + c_{sf}/c_s(\omega)\right]^2 + \hat{\alpha}^2} \tag{10.2.17}$$

This formula was obtained by Kennard in 1943 in an early study of sound propagation in bubbly liquids.

Figure 10.2.3 shows the reflection coefficient for three bubbly liquids, all having a volume concentration equal to 0.005. As a comparison with Fig. 10.2.2 shows, the

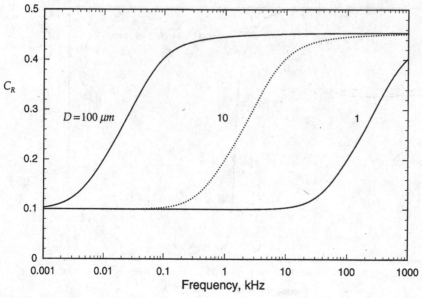

Figure 10.2.4. Finite-frequency reflection coefficient at three gas-dusty gas (water droplets in air) interfaces. $\phi_v = 0.005$.

low-frequency limits of this frequency-dependent result corresponds to the equilibrium results shown there. As the frequency increases, the value of C_R remains constant for a very wide range of frequencies below the frequency band corresponding to the resonance frequency for each chosen bubble diameter. Here, the bubbly reflection coefficient is close to 1, indicating that nearly all of the acoustical energy is then reflected back to the source. Beyond that frequency band, however, C_R diminishes rapidly, becoming very small soon after, showing the acoustical transparency of bubbly-liquid suspensions at such high frequencies.

For dusty gases (see Fig. 10.2.4), where no resonance band exists, we see that C_R increases gradually as the frequency increases and attains a fairly large, constant value at high frequencies. This limiting value is equal to $(1 + 2/\eta_m)^{-2}$, showing that C_R increases with the mass loading and can approach unity in some conditions. For example, in an aerosol composed of alumina particles in air and having a volume concentration equal to 0.01, η_m is about 32, so that the high frequency limit of C_R is 0.89.

The high acoustical reflectivity that applies to dusty gases at frequencies that are not small shows that layers of dust can be used to block undesirable sounds (although far more undesirable health effects can also result). Propagation through dusty layers of finite extent can be treated in a manner similar to that used above, but taking into account the reflected waves that will develop as a result of the layer being of finite extent. The interaction of the transmitted and reflected waves set up a standing field within the layer that can significantly the transmission. We shall not pose here to analyze this example, and refer the reader to work by Ishii (1983), who studied the reflection and transmission of sound through such a layer. Similar, but stronger effects occur with bubbly layers. Here, owing to the pulsations of the bubbles, the reflection and transmission of sound waves through such layers usually involve nonlinear effects. The problem has been studied by several investigators, including Miksis and Ting (1989), and Karpov, Prosperetti, and Ostrovsky (2003).

10.3 Extension to Polydisperse Suspensions

The propagation theories presented in the preceding chapters were obtained on the assumption that the suspensions are monodisperse (i.e., that the particles in them are all of a single diameter). Most suspensions are, of course, polydisperse and, furthermore, the distribution of sizes found in them can vary widely. It is therefore important to extend the monodisperse results obtained previously so that they can be applied to polydisperse suspensions. To do this, we consider suspensions so dilute that the attenuation and dispersion coefficients, $\hat{\alpha}$ and $\hat{\beta}$, respectively, are proportional to the volume concentration. Inclusion of larger, but still dilute, concentrations presents no additional difficulties. But, for the purposes of this section, it is simpler to use the very dilute results given in Section 9.4, in which case we may write

$$\hat{\alpha}(\omega, D) = \phi_v(D) f(\omega, D) \quad \text{and} \quad \hat{\beta}(\omega, D) = \phi_v(D) g(\omega, D) \quad (10.3.1a,b)$$

For example, in the case of aerosols, the functions $f(\omega, D)$ and $g(\omega, D)$ can be obtained from (8.8.21) and (8.8.22). When more than one particle size is present, the attenuation is given by the sum of the contributions of particles of each size present. If l denotes the number of different sizes, we get,

$$\hat{\alpha} = \sum_{n=1}^{l} \phi_v(D_n) f(\omega, D_n) \quad (10.3.2)$$

In practice, the sizes of the particles present do not divide themselves into a small number of different values, as assumed by the above formula. Rather, a very large number of particle sizes are encountered, and it is then advantageous to group together particles having sizes that fall within a small range of values. We refer to each range as a *section*. Consider, for example, the total number of particles in a unit suspension volume, N_0. This may be expressed as the sum of the number of particles in each section. Let that number be $N(D_n)$, then the total number of particles in a unit suspension volume is $N_0 = \sum_{n=1}^{l} N(D_n)$. The function $N(D_n)$ gives the distribution of particle sizes when the number of different sizes is discrete. In most suspensions, however, the number of sizes present is so large that the distributions are best fit by continuous functions. To get these, we first let $N(D_n) \Delta D_n$ be the number of particles having a diameter in the range D_n, $D_n + \Delta D_n$ and take the limit as $\Delta D_n \to dD$. Then, $N(D_n)$ approaches the continuous function $N_D(D)$ and the summations become integrals whose range of integration is chosen so that all particles are included. For this reason, the upper limit of integration is usually written as ∞. However, it is more convenient to write it as D_∞, with D_∞ representing the maximum diameter in the suspension. Then, $N_0 = \int_0^{D_\infty} N_D(D)dD$. Similarly, the volume fraction can be expressed as

$$\phi_v = \frac{\pi}{6} N_0 \int_0^{D_\infty} n_D(D)D^3 dD \tag{10.3.3}$$

where $n_D(D) = N_D(D)/N_0$ is called the particle size distribution function (PSDF). It satisfies the normalization condition, $\int_0^{D_\infty} n_D(D)dD = 1$, and plays a central role in determining the various quantities that depend on particle size. The attenuation is one such quantity, and can for spherical particles, be expressed as

$$\hat{\alpha} = \frac{\pi}{6} N_0 \int_0^{D_\infty} n_D(D)D^3 f(\omega, D)dD$$

where the symbol $\hat{\alpha}$ is used to represent the attenuation coefficient in polydisperse suspensions. Finally, using (10.3.3), we obtain

$$\hat{\alpha} = \phi_v \frac{\displaystyle\int_0^{D_\infty} n_D(D)D^3 f(\omega, D)dD}{\displaystyle\int_0^{D_\infty} n_D(D)D^3 dD} \tag{10.3.4}$$

Thus, if $n_D(D)$ is known, this equation provides a means of obtaining the attenuation coefficient in polydisperse suspensions.

A similar equation can be written for the dispersion coefficient. Thus,

$$\hat{\beta} = \phi_v \frac{\displaystyle\int_0^{D_\infty} n_D(D)D^3 g(\omega, D)dD}{\displaystyle\int_0^{D_\infty} n_D(D)D^3 dD} \tag{10.3.5}$$

The values of the integrals clearly depend on the PSDF. Sometimes, this may be available from optical scattering measurements, or from mathematical expressions that can be adjusted so as to fit typical suspension distributions; but, generally speaking, that information is lacking. Fortunately, as shown below, the exact value of $n_D(D)$ is not needed because $\hat{\alpha}$ and $\hat{\beta}$ are fairly insensitive to the shape of the distribution function.

Before evidence supporting these statements is presented by means of an example, we write our working equations in terms of the nondimensional diameter $\xi = D/D_\infty$. Thus,

$$\hat{\alpha}(\omega)/\phi_v = M_3^{-1} \int_0^1 n_D'(\xi)\xi^3 f(\omega, \xi)d\xi \tag{10.3.6}$$

$$\hat{\beta}(\omega)/\phi_v = M_3^{-1} \int_0^1 n_D'(\xi)\xi^3 g(\omega, \xi)d\xi \tag{10.3.7}$$

where $n'_D(\xi) = D_\infty n_D(\xi)$ satisfies the normalization condition $\int_0^1 n'_D(\xi)d\xi = 1$ and where $M_3 = \int_0^1 n'_D(\xi)\xi^3 d\xi$. This is an example of a *moment* of the PSDF, defined by

$$M_n = \int_0^1 n'_D(\xi)\xi^n d\xi \quad n = 0, 1, 2 \ldots \tag{10.3.8}$$

Thus, M_3 is called the third moment of the new PSDF. It should be noted that, because of the normalization condition on $n'_D(\xi)$, $M_0 \equiv 1$.

Several average diameters can be used to reduce the integrals. These are provided by averages of powers of D (i.e., by moments of the PSDF). More generally, the ratio of any two such moments also defines a mean diameter raised to some power. Thus, if n and m are integers, we define the $n - m$ power of the mean diameter D_{nm} by means of

$$D_{nm}^{n-m} = \frac{\int_0^{D_\infty} n_D(D)D^n dD}{\int_0^{D_\infty} n_D(D)D^m dD} \tag{10.3.9}$$

Although average values can be calculated through this definition when $n < m$, such average diameters would be larger than D_∞. Hence, we limit this definition to values of n that are larger than m.

In terms of the moments of $n'_D(\xi)$, these average diameters are given in nondimensional form by means of

$$(D_{nm}/D_\infty)^{n-m} = M_n/M_m \tag{10.3.10}$$

This relationship enables us to express D_∞ in terms of any chosen value of D_{nm}, or vice-versa. While any value of D_{nm} may be selected to reduce the integrals, it is evident that different choices may result in different values of the attenuation and the dispersion coefficients.

To clarify this statement, and to set the stage for subsequent discussion, it is useful to consider some simple examples of distributions and to obtain from them their first few moments. The chosen distributions have simple geometrical shapes that are not representative of any suspension. Mathematically, they can be expressed as:

Rectangular: $\quad n'_D(\xi) = 1, \quad 0 \le \xi \le 1$ \hfill (10.3.11a)

Parabolic: $\quad n'_D(\xi) = 6\xi(1 - \xi), \quad 0 \le \xi \le 1$ \hfill (10.3.11b)

Triangular: $\quad n'_D(\xi) = \begin{cases} 4\xi, & 0 \le \xi \le 0.5 \\ 4(1 - \xi), & 0.5 \le \xi \le 1 \end{cases}$ \hfill (10.3.11c)

It may be easily verified that the three distributions satisfy the normalization condition. Table 10.3.1 gives the values of the first five moments of $n'_D(\xi)$. These can be used to

Table 10.3.1. *First five moments of three distribution functions*

	M_1	M_2	M_3	M_4	M_5
Rectangular	0.5000	0.3333	0.2500	0.2000	0.1667
Parabolic	0.5000	0.3000	0.2000	0.1429	0.1071
Triangular	0.5000	0.2917	0.1875	0.1292	0.0938

evaluate a number of mean diameters via (10.3.10). Thus, for example, in a parabolic distribution, $D_{31}/D_\infty = \sqrt{M_3/M_1} = 0.632$.

Let us now return to the attenuation. Suppose first that the size distribution function obeys one of the distributions given above (e.g., the parabolic). This choice enables us to perform the integrations in (10.3.6). But, to carry out the integration, we must select a representative average diameter. To examine the choices we select D_{10}, D_{21}, and D_{32}. The first of these is the arithmetic mean, and the last is called the volume-to-surface mean diameter. Of course, we also need the function $f(\omega, D)$. This contains several effects, but to simplify the discussion, we consider only the translational contribution and furthermore limit the discussion to aerosols. Here, the functional dependence of $\hat{\alpha}$ on ω and D is simple and may be used to clarify the discussion. From (8.8.22) we see that the translational attenuation in dilute aerosols is

$$\hat{\alpha}_{tr}/\eta_{m0} = \frac{1}{2}\frac{\omega\tau_d}{1+(\omega\tau_d)^2} \tag{10.3.12}$$

where $\tau_d = D^2/18\nu_f\delta$ is the dynamical relaxation time. We shall use this as our function $f(\omega, D)$, noting that the attenuation is now scaled with the dilute value of the mass loading, $\eta_{m0} = \phi_{v0}/\delta$. Substitution of this in (10.3.6) yields

$$\hat{\alpha}_{tr}/\eta_{m0} = \frac{1}{2}(\omega\tau_d)_\infty M_3^{-1} \int_0^1 \frac{n_D'(\xi)\xi^5}{1+(\omega\tau_d)_\infty^2\xi^4}d\xi \tag{10.3.13}$$

where $(\omega\tau_d)_\infty$ can be expressed in terms of any of the above mean diameters by means of $(\omega\tau_d)_\infty = (\omega\tau_d)_{nm}(D_\infty/D_{nm})^2$. Figure 10.3.1 shows $\hat{\alpha}_{tr}/\eta_{m0}$ as a function of $(\omega\tau_d)_{nm}$, applicable to the three choices of D_{nm}. The figure may be best understood by considering each curve separately from the other two appearing in the it. Thus, for example, the abscissa corresponding to the curve labeled D_{10} is $(\omega\tau_d)_{10}$. We see that the three curves have similar shapes, which are reminiscent of the monodisperse attenuation. The peak value of all three results are equal, but the value, 0.227, is slightly smaller than the monodisperse maximum, that is equal to 0.25. Finally, the location of the maximum varies from one representation to another; it occurs at $(\omega\tau_d)_{10} = 0.48$ for D_{10}, $(\omega\tau_d)_{21} = 0.71$ for D_{21}, and $(\omega\tau_d)_{32} = 0.85$ for D_{32}. Which of these three representations is correct? The answer is simple: all three because the curves are simply shifted in proportion to the values of the corresponding ratios between the $(\omega\tau_d)_{nm}$. Thus, multiplying $(\omega\tau_d)_{10}$ by $(D_{21}/D_{10})^2$ shifts the D_{21} curve to the right so as to nearly match the next curve. But, because all choices yield the same results, we

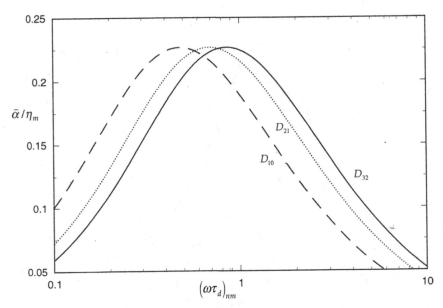

Figure 10.3.1. $\hat{\alpha}/\eta_m$ vs $(\omega\tau_d)_{nm}$ for three average diameters.

can choose any of them to represent the attenuation. However, it is desirable to find a mean diameter such that

$$\hat{\alpha}_{tr}/\eta_{m0} = \frac{1}{2}\frac{(\omega\tau_d)_{kl}}{1+(\omega\tau_d)^2_{kl}} \qquad (10.3.14)$$

Such a diameter would, in effect, reduce the size distribution to a monodisperse distribution. Representing the polydisperse result by means of this is advantageous for several reasons. First, the expression retains the physics of the problem, showing that dynamic relaxation controls the attenuation. Then, of course, is the simplicity of the expression, relative to (10.3.13).

Of course, we still have to determine which, if any, mean diameter can be used to express the attenuation in this manner. Two features in the monodisperse result are evident: the maximum value of the attenuation and its location. From the previous calculations, we already know that the polydisperse attenuation does not have the same peak value as the monodisperse one, though the differences are small. But, we still have at our disposal the location of the peak value, which defines $(\omega\tau_d)_{kl}$, and therefore D_{kl}. Thus, to determine this mean diameter we could reduce (10.3.13) for various mean diameters until we find one that produces the desired result. However, a more general approach is offered by examination of (10.3.13), which contains two integrals. One, denoted by M_3, appears in the denominator. The integral in the numerator is a bit more complicated because it depends on the frequency as well. But since $\xi \leq 1$, we see that for values of $(\omega\tau_d)_\infty$ that are not too large, the integral is controlled by the ξ^5 factor, which means that its value is proportional to M_5. Hence, for frequencies that include the

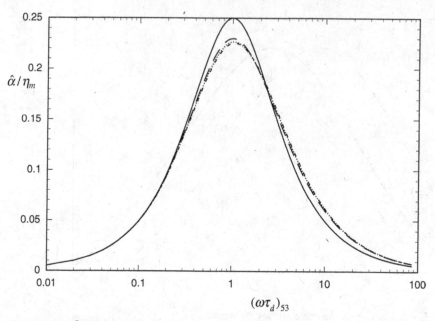

Figure 10.3.2. $\hat{\alpha}/\eta_m$ vs $(\omega\tau_d)_{53}$ for several particle distribution functions: solid line, monodisperse; long dashed line, rectangular; dotted line, parabolic; short, dashed line, triangular.

location of the maximum, but which are not much larger than this, $\hat{\alpha}_{tr}$ is proportional to M_5/M_3. This indicates that a suitable mean diameter is $D_{53}/D_\infty = \sqrt{M_5/M_3}$.

To see whether this is the case, we evaluate (10.3.13) with $(\omega\tau_d)_\infty = (\omega\tau_d)_{53}$ $(D_\infty/D_{53})^2$. Also, in order to determine how sensitive this choice is to the shape of the PSDF, we carry out the integration for the three simple size distributions given by (10.3.11). The results are shown in Fig. 10.3.2 as a function of $(\omega\tau_d)_{53}$. We see that this choice produces an attenuation curve that peaks at about $(\omega\tau_d)_{53} = 1$, as it was desired. Hence, the choice that mean diameter is correct. But because the maximum is not sharply defined, other mean diameters can probably be found that would also be suitable. Also shown in the figure is the monodisperse result, evaluated using (10.3.14) with $D_{kl} = D_{53}$. The figure makes it clear that the use of a well-selected mean diameter can be useful to represent the attenuation theory, provided we are not seeking very close agreement.

The same procedure may be used for the translational dispersion coefficient, $\hat{\beta}_{tr}/\eta_m$, whose monodisperse value can be obtained from (8.8.21). The polydisperse results are shown in Fig. 10.3.3 for the same three distributions as used for the attenuation. The figure includes the monodisperse theory, reduced for a diameter equal to D_{53}. Figures 10.3.2 and 10.3.3 also show $\hat{\alpha}_{tr}$ and $\hat{\beta}_{tr}$ for the rectangular, parabolic, and triangular particle size distributions introduced earlier. We see that, despite these distributions being very different, the results are nearly equal – that is, expressed in terms of D_{53}, the polydisperse attenuation and dispersion are fairly insensitive to the exact shape of the size distribution function. As we show below, this has important consequences.

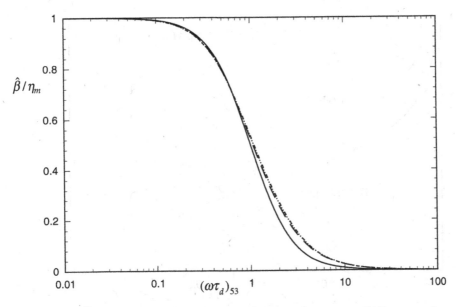

Figure 10.3.3. $\hat{\beta}/\eta_m$ vs $(\omega\tau_d)_{53}$ for several particle distribution functions: solid line, monodisperse; long, dashed line, rectangular; dotted line, parabolic; short, dashed line, triangular.

10.4 Suspension Characterization

Having extended the monodisperse results for the attenuation and dispersion, we now consider their use in determining several characteristics of a suspension of known particle and fluid materials (e.g., its particle size, or sizes, and its volume concentration). In static conditions, these could be obtained by weighting a known suspension volume and by sorting the sizes with a microscope. Can we do the same acoustically?

Volume Concentration

Consider first the volume concentration. Its value, for spheres, is formally given by

$$\phi_v = (\pi/6)N_0 \int_0^{D_\infty} n_D(D)D^3 dD \tag{10.4.1}$$

As shown below, the value of ϕ_v can be obtained by acoustic means that are independent of the size distribution of particles. To see this, we consider the sound speed in suspensions at *low* frequencies. The thermodynamic analysis given in Chapter 7, shows that this speed is independent of the size and shape of the particles, but depends on the volume concentration and on several particle-fluid property ratios. These ratios are regarded as known. Thus, for a given suspension, the equilibrium speed depends only on the volume concentration or vice-versa – that is, we may regard ϕ_v as a function of $c_s(0)$. Figure 10.4.1 shows that dependence for four types of suspensions, as described by (7.4.3). As stated in Chapter 7, that expression applies for all volume concentrations for which the suspension can be regarded as homogeneous.

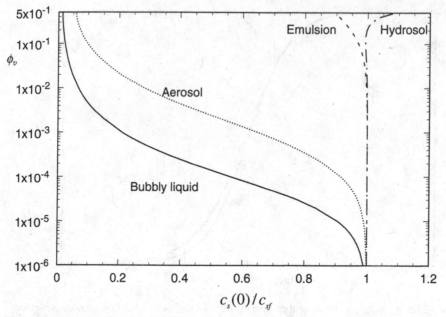

Figure 10.4.1. ϕ_v as a function of $\tilde{c}_s(0)/c_{sf}$ for typical suspensions.

Now, since the theory upon which the curves shown in the figure seems to describe well the low-frequency sound speed, we can use them to obtain the volume concentration from a single measurement of the sound speed at low frequencies. In the case of aerosols and bubbly liquids, the technique can yield accurate results because that speed differs considerably from its counterpart in the fluid without particles. On the other hand, as also shown in the figure, the curves applicable to the emulsions and hydrosol used to reduce the theory, do not differ significantly from unity for volume concentrations as large as 0.1. Hence, the procedure is far less accurate for these suspension types.

Particle Size

Let us now consider the acoustic measurement of particle size. One way of doing this is via the attenuation and dispersion coefficients, $\hat{\alpha}$ and $\hat{\beta}$, respectively, both of which depend on particle size. Because both of these vanish at low frequencies, it is clear that finite frequencies are necessary. As we show below, a variety of techniques can be used that relay on measurements at a single or multiple frequencies (Dobbins and Temkin, 1967). The procedures can be used with both monodisperse and polydisperse suspensions.

Monodisperse Suspensions. For simplicity, we shall use the attenuation and dispersion results in their small dissipation limits. Here, $\hat{\alpha}$ and $\hat{\beta}$ are proportional to ϕ_v and may be written as

$$\hat{\alpha}/\phi_v = f(\omega, D_0) \quad \text{and} \quad \hat{\beta}/\phi_v = g(\omega, D_0) \qquad (10.4.2a,b)$$

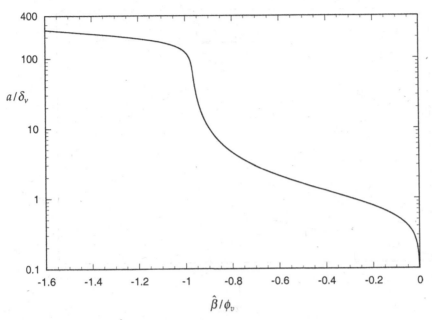

Figure 10.4.2. a/δ_v vs $\hat{\beta}/\phi_v$. Silica particles in water.

where D_0 is the diameter of all the particles and, for aerosols, the functions $f(\omega, D_0)$ and $g(\omega, D_0)$ are given by the right-hand sides of (8.8.21) and (8.8.22).

The problem at hand is the determination of D_0. If the volume concentration is known, then a measurement of the dispersion, together with the theories for that quantity is, in principle, sufficient. To illustrate the method we consider a suspension of silica particles in water, and show in Figs. 10.4.2 and 10.4.3 the variations of a/δ_v, where $\delta_v = \sqrt{2\nu_f/\omega}$, with the dispersion and attenuation coefficients, respectively. The graphs also show the variations of the frequency, via $y = a/\delta_v$ as a function of the attenuation and dispersion coefficients, and were obtained from previous graphs by transposing axes. Other frequency variables containing the particle size could, of course, have been used instead of y.

Suppose that we measure the phase velocity at some fixed-frequency ω. This speed, together with the concentration and the equilibrium sound speed, defines the value of $\hat{\beta}/\phi_v$, which combined with Fig. 10.4.2 yields y, and therefore the particle size. This value is unique. On the other hand, Fig. 10.4.3 shows that the particle size is not a single-valued function of $\hat{a}$. Hence, a measurement of the attenuation cannot, by itself, yield a unique particle size.

The above procedure uses single-frequency measurements. These are most easily conducted with a continuous wave. Finite-duration pulses can also be used. These contain an infinite number of frequencies, which offer additional means of obtaining particle size (e.g., by measuring the dispersion coefficient at several different frequencies). It is evident, however, that whatever measurements are performed, the ranges of sizes and frequencies where the methods will yield accurate results, depend on the

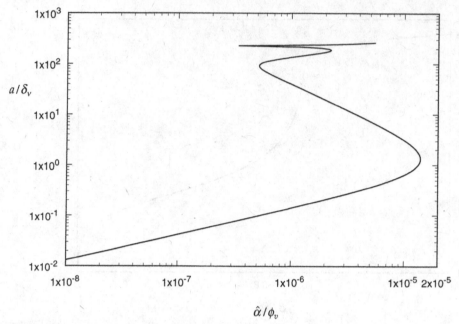

Figure 10.4.3. a/δ_v vs $\hat{\alpha}/\phi_v$. Silica particles in water.

type of suspension. To save space, we do not show every type. However, suspensions of highly resonant particles (e.g., gas bubbles in liquids) offer different possibilities for the determination of sizes and concentrations. This is due to the large attenuation that is produced at the resonance frequency of the bubbles. Given this large attenuation and the usually sharp nature of the resonance peak, the resonance frequency can be easily obtained. As (6.6.23) shows, that frequency depends on the bubble radius. Hence, its value plus that equation can, in principle, be used to determine the radius. However, although the frequency is a unique function of the radius, the converse is not true, as shown in Fig. 6.6.6. Thus, additional information is required to arrive at the correct value of the bubble radius.

Polydisperse Suspensions. The above procedures can also be used to obtain a number of mean diameters from acoustic measurements. To do this, we would integrate the attenuation and dispersion coefficients as was done in the previous section 10.3. It was shown there that these coefficients were fairly insensitive to the exact shape of the distribution function, and this enabled us to obtain them, even though the size distribution function is not known. Thus, with reference to Fig. 10.4.2, we see that a measurement of the dispersion coefficient at a single frequency is sufficient to obtain a mean diameter. Other mean diameters can also be obtained from the same measurement, by reducing the theory for that mean diameter, as done in Section 10.3, but as pointed out above, the attenuation coefficient is, by itself, not suitable to obtain a mean size because size is not a single valued function of the attenuation.

Size Distributions

The determination of the particle size distribution in a given suspension by acoustic means is, in principle, also possible, but presents considerable difficulties. These are caused by the inherent problems associated with the accurate measurement of attenuation or sound speed in suspensions, as well by the insensitivity of those quantities to the exact shape of the distribution function that was demonstrated earlier. In fact, the results of Section 10.3 show that widely different distribution functions produce very similar curves for those coefficients. Thus, obtaining the shape of an actual distribution function by inversion of the attenuation or the dispersion coefficient is not generally feasible. To illustrate the difficulties, we write below the set of equations that are available to obtain the distribution function. These are:

$$\phi_v = \frac{\pi}{6} N_0 D_\infty^3 M_3$$

$$\hat{\alpha}(\omega) = \frac{\phi_v}{M_3} \int_0^1 n'_D(\xi) \xi^3 f(\omega, D_\infty \xi) d\xi$$

$$\hat{\beta}(\omega) = \frac{\phi_v}{M_3} \int_0^1 n'_D(\xi) \xi^3 g(\omega, D_\infty \xi) d\xi$$

The idea would be to combine measurements of ϕ_v, $\hat{\alpha}(\omega)$ and $\hat{\beta}(\omega)$ with these equations to obtain $n'_D(\xi)$. This function appears in the integrals and in M_3, but the first of these equations shows that the ratio ϕ_v/M_3 is a constant for a given suspension. This constant can be evaluated through measurements of the attenuation at a well-defined frequency (e.g., the frequency where a maximum occurs). It may therefore be regarded as known. We still have at our disposal the equation for $\hat{\beta}(\omega)$. Absorbing ϕ_v/M_3 into $g(\omega, \xi)$, we can write the equation as

$$\hat{\beta}(\omega) = \int_0^1 K(\omega, \xi) y(\xi) d\xi$$

This is an integral equation whose unknown is $y(\xi) = n'_D(\xi) \xi^3$. Thus, to obtain n'_D, we need to invert the equation, assuming that the left-hand side is separately known, say from measurements of the phase velocity. But, as we saw earlier, very different distribution functions produce dispersion coefficients that are very similar. Experimental errors can easily produce uncertainties in the measurements of $\hat{\beta}$ that are larger than the differences between widely different distribution functions. This means that the solution of the equation, obtained by any means, might be inaccurate. Of course, given the nature of the functions involved, numerical solutions are required. These are based on a discrete, and therefore limited, number of data points. Even if the data are fairly accurate, the difficulties associated with the numerical inversion of the this type of equation are not insignificant. Nevertheless, inversion techniques are currently under development that appear to yield information on the shape of the distribution functions via acoustic measurements. The reader is referred to the bibliography where a number of works on this topic are listed.

10.5 Acoustical Coalescence

In the preceding chapters, we have studied the dynamic behavior of suspensions under the assumption that the particles are of equal size and do not interact with one another. As a consequence, the velocities acquired by each of the particles in a volume element are identical. This means that, although the particles move relative to the fluid, they remain fixed relative to one another.

The situation changes when the particles interact with one another, or when their size is not uniform. Particle interactions tend to modify their relative positions when the particles are sufficiently close to one another. Nonuniform particle size can produce strong relative motions at all interparticle distances because particles of different size respond differently to external forces.

Whatever the origin of this relative displacement, the particles may get so close to one another that collisions may occur. Furthermore, some of these collisions may be sticking, in the sense that the colliding particles do not rebound on touching but join to form a single particle having an aggregate mass and volume larger than those of either particle. For droplets, the process is referred to *coalescence*; otherwise, the term agglomeration is used. We shall use the first to refer to the joining of different particles, whether liquid or solid.

Coalescence effects obviously change the motions of both fluid and particles, reduce the number concentration of the particles in the suspension, modify the shape of the particle size distribution, and, therefore, all quantities that depend on size and concentration. These changes are important in many situations, and, in this section, we briefly consider particle coalescence in a simplified manner. The simplifications are needed because the phenomenon is a complex one, as we shall see. Among the more drastic simplifications that will be made, we will assume that *all* collisions lead to coalescence, and that regardless of their size, particles do not break up. This means that, whenever a collision occurs between particles of different volumes, a new particle having a volume equal to the sum of the volumes of the colliding particles will be formed, and that the new particle will not breakup, no matter how large it is. Both assumptions are incorrect for some collisions. Thus, collisions that occur at high relative velocities have kinetic energies that are sufficiently large to overcome bonding forces that would otherwise keep them together. Similarly, a very large particle is unstable and would tend to breakup into several smaller parts as a result of a collision with another particle, or by the action if some external force.

The Coagulation Equations

Given the large number of particles usually present in a unit volume of suspension, it is not feasible to study the trajectories of each of the particles in a suspension, and it is best to treat the problem in a statistical sense, by looking at the probability that collisions between certain particle pairs may occur. To do this, we first divide the distribution of sizes (or volumes) into a large number of sections. Each of the sections is associated with a given size or volume range. For example, if the volume width of the lth section is Δv_l, the particles in that section will have volume in the range $v_l - \frac{1}{2}\Delta v_l \leq v_l \leq v_l + \frac{1}{2}\Delta v_l$, and, for convenience, we will refer to them as the particles of volume v_l. Below, we use volume sections to describe the evolution of a particle distribution function. The reason for this choice is that particle volumes are additive, whereas sizes are not.

Suppose that we have a suspension whose particles have different sizes, as represented by some distribution function. We divide the distribution into L sections, each having a different size or volume. The width of each section, that is, the value Δv_l may vary from section to section, but their total number, L, is assumed to be sufficiently large so that a statistical description of the agglomeration process is meaningful. Let N_l denote the number of particles per unit volume in the lth section. Then, the total number of particles in a unit volume of the suspension is given by $N = \sum_{l=1}^{L} N_l$. When coalescence take place, these numbers depend on time in a manner dictated by the coalescence equations.

As the agglomeration process proceeds, the total number of particles decreases, but the number of particles in any given section may increase or decrease. To see this, consider the particles in one of the sections, say the kth. A collision between a particle in it with one belonging to *any* section, including the kth, will decrease the value of N_k because, by assumption, the collision is sticking. The resulting particle will have a larger volume, which means that section k, and possibly some other section, loses a particle, and that a larger volume section gains a particle.

But, on the other hand, the number of particles in the kth section may increase as a result of collisions between particles having smaller volumes than v_k, say v_l and v_m if their volumes are such that $v_l + v_m = v_k$. This sum represents a condition that must be taken into account when counting the number of collisions that increase N_k.

We also need a means of determining whether a collision between a particle pair occurs in the first place. In general, a deterministic solution to this question is not possible, although some numerical results exist for very simple relative motion between two particles that have yielded some information about the probability that a collision between two particles may occur. To compound matters, collision probabilities are only part of the problem, because an account must be made of whether a given collision is sticking. For the time being, however, we consider both issues together. Thus, we let $K(v_j, v_k)$ denote the probability that a sticking collision will occur between a particle of volume v_j and one of volume v_k in a unit time; $K(v_j, v_k)$ represents the collision frequency and for reasons that will be seen below, it is also called the coalescence kernel. In general, $K(v_j, v_k)$ depends on the specific details of the collision mechanism considered. Some particular mechanisms for which K is known will be considered later. For now, we note that the collision kernel is symmetric with respect to the volumes of the colliding particles, that is

$$K(v_j, v_k) = K(v_j, v_k) \tag{10.5.1}$$

We are now ready to determine the rate at which N_k changes owing to sticking collisions. To do this, we first consider those collisions that decrease the number of particles in section k. Since each of those collisions requires a particle in that section with other particles, denoted by the index j, whose number is N_j, the decrease in time dt of the number of particles in section k can be expressed as

$$N_j N_k K(v_j, v_k) dt$$

Thus, the number particles in section k that are lost due to collisions is obtained by adding $N_j N_k K(v_j, v_k) dt$ over j, or

$$(\Delta N_k)_{loss} = \sum_{j=1}^{L} N_j N_k K(v_j, v_k) dt = N_k \sum_{j=1}^{L} N_j K(v_j, v_k) dt \qquad (10.5.2)$$

Let us now consider those collisions that increase N_k. These only involve pairs of particles such that their combined volume is v_k. Consider particles having volume $v_i < v_k$. These may collide with particles having any volume, but only those collisions with particles of volume $v_k - v_i$ will result in a particle of volume v_k. As a result of this restricted class of collisions, there will be an increase in the number N_k equal to $N_i N_j K(v_i, v_j) dt$, with the above provision on v_i and v_j. To obtain the total increase in time dt owing to collisions of this type, we have to add all possible volume combinations that yield v_k upon addition. Thus,

$$(\Delta N_k)_{gain} = \frac{1}{2} \sum_{v_i + v_j = v_k} N_i N_j K(v_i, v_j) dt \qquad (10.5.3)$$

where the 1/2 factor is used to prevent collisions from being counted twice.

The difference between (10.5.3) and (10.5.2) gives the net change of N_k in time dt. Hence, the rate of change of the number of particles per unit volume in the kth section is given by

$$\frac{dN_k}{dt} = \frac{1}{2} \sum_{i+j=k} N_i N_j K(v_i, v_j) - N_k \sum_{j=1}^{L} N_j K(v_j, v_k), \quad k = 1, 2, \ldots L \qquad (10.5.4)$$

In the sums, we have used the indices i, j, and k to represent the sections with particles of volumes v_i and v_j and v_k, respectively. Thus, the constraint on the volumes is equivalent to the requirement that the sum of the indices i and j be equal to k. Noting that the *last* positive value of the integer i, which satisfies this condition, is $i = k - 1$, and that the value of j is restricted to $j = k - i$, we may write (10.5.4) as a sum over the index i. Thus,

$$\frac{dN_k}{dt} = \frac{1}{2} \sum_{i=1}^{k-1} N_i N_{k-i} K(v_i, v_{k-i}) - N_k \sum_{j=1}^{L} N_j K(v_j, v_k), \quad k = 1, 2, \ldots, L \qquad (10.5.5)$$

The total number of particles in a unit suspension volume at some instant can be obtained from this by summing over k. Thus, since by definition $N = \sum_{k=1}^{L} N_k$, we get

$$\frac{dN(t)}{dt} = \frac{1}{2} \sum_{k=1}^{L} \sum_{i=1}^{k-1} N_i N_{k-i} K(v_i, v_{k-i}) - N \sum_{j=1}^{L} N_j K(v_j, v_k) \tag{10.5.6}$$

Equations (10.5.5) and (10.5.6) are known as the coalescence equations. They represent $L + 1$ differential equations for the L sectional particle numbers N_k, and for the total number, $N(t)$, respectively. Given the initial values of these (i.e., given the initial particle distribution of volumes), the solution of these equations gives the evolution in time of the distribution function. However, analytical solutions are not generally possible, and the system is usually solved numerically.

It is instructive to write out some of the terms of some of the differential equations for the N_k. To do this, we first introduce the short-hand notation $K(v_i, v_j) = K_{i,j} = K_{j,i}$. Thus, since $N_\ell = 0$ if $\ell < 1$, we have

$$\frac{dN_1}{dt} = -N_1(K_{1,1}N_1 + K_{1,2}N_2 + \cdots + K_{1,L-1}N_{L-1} + K_{1,L}N_L)$$

$$\frac{dN_2}{dt} = \frac{1}{2}N_1^2 K_{1,1} - N_2(K_{2,1}N_1 + K_{2,2}N_2 + \cdots + K_{2,L-1}N_{L-1} + K_{2,L}N_L)$$

$$\frac{dN_3}{dt} = N_1 K_{1,2}N_2 - N_3(K_{3,1}N_1 + K_{3,2}N_2 + \cdots + K_{3,L-1}N_{L-1} + K_{3,L}N_L)$$

$$\frac{dN_4}{dt} = \frac{1}{2}N_2^2 K_{2,2} + N_3 K_{3,1}N_1 - N_4(K_{4,1}N_1 + K_{4,2}N_2 + \cdots + K_{4,L-1}N_{L-1}$$
$$+ K_{4,L}N_L)$$

Several interesting features about the coalescence process can be observed in these equations. One is that the number of particles in the first section cannot increase; this follows from the obvious fact that there are no smaller particles than those in v_1. Second, as the section number increases, more and more combinations of particle pairs can be found that increases its number of particles. Third, even if there were no particles in some of the sections at the beginning of the process, collisions may eventually populate them. This is particularly clear for distributions having, initially, no particles beyond the Lth section. Here, particles whose volumes are larger than v_L may eventually be produced. These larger particles can act as efficient collection centers and can have a significant fraction the total mass of particles in the suspension. They cannot, therefore, be ignored. Finally, it should be noticed that the equations include collisions among particles of the same size. These are represented by kernels having equal indices (i.e., $K_{1,1}$, $K_{2,2}$, etc.).

Continuous Distributions. Let us now consider the coalescence equations for the important case of size distributions that are continuous, rather than discrete as treated above. These may be obtained from those for the sectionalized distributions by letting the width of the volume sections become infinitesimal (i.e., by letting $\Delta v_j \to dv$ and by letting the number of sections become infinite). We can then write the number of particles in section j, say, as $N_j = N(v)dv$, where $N(v)$ is the number concentration, per unit volume range, of particles in the range v, $v + dv$ at time t. Its dimensions are those of $(\text{length})^{-6}$. In that limit, the sums in (10.5.5) and (10.5.6) become integrals over the respective ranges. For example, the sum giving the decrease in the number of particles becomes an integral extending to infinity. The gain summation, on the other hand, extends only to volume v, representing the particles that are produced by the combination of particles of volumes w and $v - w$. Thus, (10.5.5) becomes

$$\frac{dN(v)}{dt} = \frac{1}{2}\int_0^v N(w)N(v-w)K(w, v-w)dw$$

$$-N(v)\int_0^\infty N(w)K(v, w)dw \tag{10.5.7}$$

The time rate of change of the total number of particles per unit volume may be obtained from this by integration over v. Thus,

$$\frac{dN(t)}{dt} = \frac{1}{2}\int_0^\infty \int_0^v N(w)N(v-w)K(w, v-w)dwdv$$

$$-\int_0^\infty \int_0^\infty N(v)N(w)K(v, w)dwdv$$

We exchange the order of integration in the first term, noting that the range of w is between 0 and v, whereas that of v extends from 0 to ∞. Therefore, $\int_0^\infty dv \int_0^v N(w)N(v-w)K(w, v-w)dw = \int_0^\infty N(w)dw \int_0^\infty N(u)K(w, u)du$, where we have put $u = v - w$. Since u is a dummy variable of integration, we can denote it by any other symbol, say v, giving

$$\int_0^\infty dv \int_0^v N(w)N(v-w)K(w, v-w)dw = \int_0^\infty \int_0^\infty N(v)N(w)K(v, w)dvdw$$

where we have used the fact that the collision frequency is a symmetric function of its arguments. Thus, the total number of particles in a unit volume in the suspension decreases at a rate given by

$$\frac{dN(t)}{dt} = -\frac{1}{2}\int_0^\infty \int_0^\infty N(v)N(w)K(v, w)dwdv \tag{10.5.8}$$

Like the discrete equations, the continuous coagulation equations can, in principle, be used to describe the dynamic progress of a given distribution undergoing one or several simultaneous types of coagulation. The task of solving exactly either system of equations for realistic values of the coagulation kernel is, however, very difficult, and a variety of mathematical techniques have been developed. There is, however, one case where the equations have been solved, exactly, and this is considered next, as an illustration of the coalescence process.

Smoluchowski's Solution

Although the coagulation kernel is generally a strong function of the volumes of the colliding particles, we take it to be a constant. This was first by Smoluchowski in his study of coagulation resulting from the random displacements of small particles produced by fluid-molecules colliding with them, that is, by Brownian motion. Smoluchowski considered the coagulation equations in discretized form, for a sectionalized distribution that extends to infinity, of which only the first is initially populated – that is, the distribution is initially monodisperse. Thus, we consider the discrete coagulation equations with $K(v_i, v_j) = K_0$ for all possible volume combinations, and with the initial number of particles being equal to N_0, all of the same size. For this case, the discretized coagulation equations reduce to

$$\frac{dN_k}{dt} = \frac{1}{2}K_0 \sum_{i=1}^{k-1} N_i N_{k-i} - N_k K_0 \sum_{j=1}^{\infty} N_j, \quad k = 1, 2, \cdots \qquad (10.5.9)$$

The summation in the last term is, by definition, the total number of particles, N, in a unit volume. Hence,

$$\frac{dN_k}{dt} = \frac{1}{2}K_0 \sum_{i=1}^{k-1} N_i N_{k-i} - N_k K_0 N, \quad k = 1, 2, \cdots \qquad (10.5.10)$$

An equation for $N(t)$ may be obtained from this by summing this equation over all sections

$$\frac{dN}{dt} = \frac{1}{2}K_0 \sum_{k=1}^{\infty} \sum_{i=1}^{k-1} N_i N_{k-i} - K_0 N^2 \qquad (10.5.11)$$

Noticing that, in the inner sum in the first term $k \geq i + 1$, we can write

$$\sum_{k=1}^{\infty} \sum_{i=1}^{k-1} N_i N_{k-i} = \sum_{i=1}^{\infty} N_i \sum_{m=1}^{\infty} N_m = N^2$$

Hence,

$$\frac{dN}{dt} = -\frac{1}{2}K_0 N^2, \qquad (10.5.12)$$

a result that follows more directly from (10.5.8). We should note that this result applies whether the initial distribution is monodisperse or not, and that at $t = 0$, the rate of change of N is proportional to the square of the initial number.

Equation (10.5.12) can be easily integrated to yield the changes in the number concentration when $t > 0$. Thus,

$$\frac{N}{N_0} = \frac{1}{1 + \frac{1}{2} K_0 N_0 t} \qquad (10.5.13)$$

where N_0 is the initial value of N. This also applies regardless of the nature of the initial distribution of sizes, and shows that the total number concentration decreases, from its initial value, in inverse proportion to the elapsed time.

Having determined N, we return to (10.5.10) for the N_k and write that equation as

$$\frac{dN_k}{dt} + \frac{2\alpha_0}{1 + \alpha_0 t} N_k = \frac{\alpha_0}{N_0} \sum_{i=1}^{k-1} N_i N_{k-i}, \quad k = 1, 2, \cdots \qquad (10.5.14)$$

where we have put $\alpha_0 = K_0 N_0 / 2$. To solve these equations, we first note that the right-hand side vanishes for $k = 1$, and that for $k > 1$, it does not include any N_k larger than N_{k-1}. Thus, the equations may be integrated sequentially, starting with $k = 1$. The solution for this is then used in the equation for $k = 2$, and so on. Before giving some of these, we write (10.5.14) in terms of the nondimensional variables $n_k = N_k / N_0$ and $t' = \alpha_0 t$. Thus,

$$\frac{dn_k}{dt'} + \frac{2}{1 + t'} n_k = \sum_{i=1}^{k-1} n_i n_{k-i} \qquad (10.5.15)$$

This can be solved starting with $k = 1$, in which case the solution is

$$\frac{n_1}{n_1^{(0)}} = \frac{1}{(1 + t')^2} \qquad (10.5.16)$$

where is $n_1^{(0)} = N_1 / N_0$. But, for the case considered by Smoluchowski, where only one size of particles exists at time $t = 0$, it follows that $N_1 = N_0$ so that $n_1^{(0)} = 1$. Hence,

$$n_1 = \frac{N_1}{N_0} = \frac{1}{(1 + t')^2} \qquad (10.5.17)$$

Thus, the particles in the first section, which initially was the only one populated, decreases more rapidly with time than the total number of particles.

To obtain n_k for $k > 1$, we note again that the right-hand side of (10.5.15) is, for a given value of k, known in terms of the known values of $n_1, n_2, \ldots n_{k-1}$. We now write the equation as

$$\frac{d}{dt'}[(1 + t')^2 n_k] = \sum_{i=1}^{k-1} (1 + t')^2 n_i n_{k-i}$$

whose solution is

$$n_k = \frac{1}{(1 + t')^2} \sum_{i=1}^{k-1} \int_0^{t'} (1 + \zeta)^2 n_i(\zeta) n_{k-i}(\zeta) d\zeta \qquad (10.5.18)$$

This yields, upon integration, results for any value of k. When $k = 2$, this becomes

$$n_2 = (1+t')^{-2} \int_0^{t'} (1+\zeta)^2 n_1^2(\zeta) d\zeta$$

Using (4.17) for n_1, we get

$$n_2 = \frac{t'}{(1+t')^3}$$

Similarly, when $k = 3$, we obtain

$$n_3 = \frac{t'^2}{(1+t')^4}$$

We may continue this way, but the values of n_k thus far found show that the result for the kth number fraction is

$$n_k = \frac{t'^{(k-1)}}{(1+t')^{k+1}}, \qquad k = 1, 2, \cdots \tag{10.5.19}$$

These equations show several interesting features. First, while all sections, except the first, were initially empty, the result shows that they become populated at subsequent times. We also see that the number-fractions in those sections go through a maximum, first increasing as $t'^{(k-1)}$ and ultimately decreasing as $(1/t')^2$. For a given section k larger than $k = 1$, the maximum occurs at a nondimensional time given by $t'_m = (k-1)/2$. The time delay between the occurrence of two maxima in *consecutive* sections is $\Delta t_m = (K_0 N_0)^{-1}$ sec. This time delay provides a measure of the rapidity of the coagulation process. As might have been anticipated, this decreases as the collection efficiency, or the initial number concentration increase.

Some of these features are displayed in Fig. 10.5.1. This shows the variations of some of the first nondimensional section numbers, n_k, as a function of the nondimensional time $t' = \frac{1}{2} K_0 N_0 t$. The figure also shows that, by the time the particle population in the second section attains its maximum value, the number concentration in the first has decreased to about one-half of its initial value. This rapidity is due, in part, to the assumed uniform value of the collection efficiency. Let us now return to the solution, as given by (10.5.19) for the nondimensional numbers of particles in each section. We may use this solution in a variety of ways, the first of which is to study the evolution of the particle-volume distribution function, n_v. Since we are using discretized distributions, n_v is given by a series of vertical lines, of height n_k, and centered at v_k. Initially, of course, this consists of a single line at v_1, having a height equal to 1.0. At later times, more sections are populated and Fig. 10.5.2 shows the state of affairs for $0 < t' \leq 10$. As seen in the figure, the population numbers in sections having larger volumes are very small at this time. But, for larger times, these populations become nearly equal, so that the distribution flattens and shifts toward

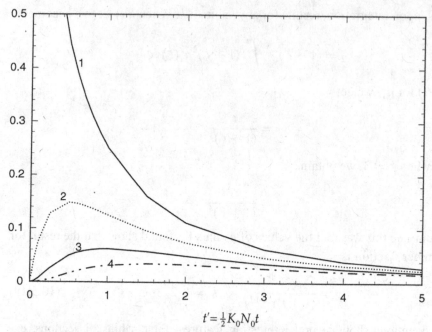

$$t' = \tfrac{1}{2}K_0 N_0 t$$

Figure 10.5.1. Nondimensional number concentrations, n_k, for the first four sections.

larger sizes, as shown in Fig. 10.5.3. One implication of this is that a proper account of the coalescence process generally requires a large number of sections. To clarify this statement, we consider again the variations for the number concentration. This was obtained before by direct integration of either (10.5.8) or (10.5.11), and is given

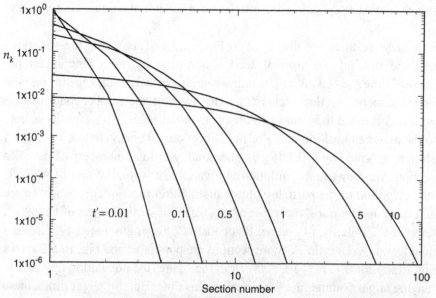

Figure 10.5.2. Distribution of sizes among the first 100 sections for $0.01 \le t' \le 10$.

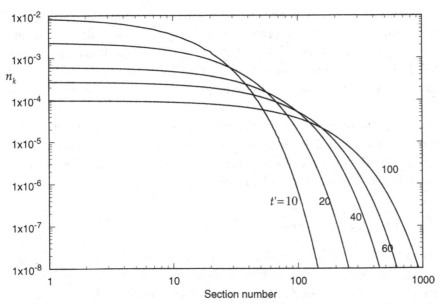

Figure 10.5.3. Distribution of sizes among the first 1,000 sections for $10 \leq t' \leq 100$.

by (10.5.13). In terms of our nondimensional time, it can be written as

$$\frac{N}{N_0} = \frac{1}{1 + t'}$$

We can also compute this quantity by adding up the numbers contained in each section at any desired time. Thus, using (10.5.19), we obtain

$$\frac{N}{N_0} = \sum_{k=1}^{\infty} \frac{t'^{(k-1)}}{(1 + t')^{k+1}} \tag{10.5.20}$$

These two results are equal, but the second requires that we add the particle numbers in an infinite number of sections. In numerical work having more realistic collection efficiencies, it is necessary to limit the number of sections, in which case some particles may be lost as the calculations proceed. This deficit can have a significant effect on the assessment of the coalescence process. In the present example, the deficit that results if only a finite number of sections is used is

$$\frac{\Delta N}{N} = 1 - \sum_{k=1}^{L} \frac{t'^{(k-1)}}{(1 + t')^k} \tag{10.5.21}$$

When $L \to \infty$, the sum in this equation is equal to 1, showing that no particles are lost. But, for finite values of L, the sum can differ from 1 when t' is large. For example, when $L = 100$, the sum is equal to 0.99993 at $t' = 10$, but only 0.630 when $t' = 100$. Thus, the unaccounted fractional number of particles in the latter case is about 37% of those remaining at that time. This is a rather large number of particles left unaccounted for, particularly since those particles are large. Larger particles are, of course, more massive and can also be more efficient in collecting other particles. To retain those particles, we need to increase the number of sections in the calculation. Thus, if $L = 1,000$, the above sum is 0.999952 at $t' = 100$ so that almost all particles are retained. However, the number of sections that has to be kept is rather large, and this increases the numerical effort for those cases when the collection efficiency is variable.

This concludes our review of Smoluchowski's solution for an initially monodisperse distribution. The coagulation equations have also been integrated for some cases, where the initial distribution is not monodisperse. These solutions are reviewed in the monograph by Drake, to which the reader is referred.

Collision Kernel for Rectilinear Motions

Smoluchowski's solution of the coagulation equations illustrates some of the characteristics of the coalescence process that occur when collisions between any two particles have the same probability of taking place. The assumption that the collision frequency is nearly constant appears to be satisfied for the coagulation induced by Brownian motion. Here, very small particles can display a random motion which is due to molecular collisions. This is relevant in liquids containing submicron particles. But, as the particle size increases, molecular collisions are less able to move the particles, so that they remain at rest if no external forces are applied to the suspension. Thus, coalescence effects produced by it diminishes significantly. Hence, noncolloidal suspensions that are at rest exhibit little, if any, coalescence.

For suspensions that are in motion, however, significant coalescence can take place among particles of all sizes, as a result of velocity differences that usually exist between particles and fluid. These differences are known to occur in a variety of motions that include shear flows and turbulence. The reader is referred to the book by Friedlander and to the monograph by Drake for excellent discussions of these and other types of coalescence.

Below, we consider the simplest type of externally induced coalescence. This occurs when the particles (and the fluid) move rectilinearly. An important example of this is gravitational settling of cloud particles. Here, the particles fall along the local vertical, and the fluid raises (to fill the volume vacated by the particles) also along the vertical. Another example is that produced when sound and shock waves propagate in a suspension. In both these cases, nonequilibrium regions are produced where collisions may take place.

Let us obtain the collision frequency for this type of motion. To do this, we need evaluate the probability that a collision between particles having different volumes and, therefore, different velocities may occur (i.e., we need the collision kernel). We need only consider one collision at a time because the moment our particle collides with another, their velocity changes owing to the collision. We thus consider two

Figure 10.5.4. Grazing trajectory ($E = 1$).

particles, whose volumes are represented by the indices i, and j, moving along the positive x-axis, say, with velocities u_i and u_j, respectively. For simplicity, we first take the centers of the particles to move along that axis. It then follows that if the ith particle is behind the jth and if $u_i > u_j$, a collision will occur. Thus, the probability that this collision takes place in time dt is proportional to $(u_i - u_j)dt$. Similarly, if the jth particle is behind the ith and if $u_j > u_i$, the corresponding probability is proportional to $(u_j - u_i)dt$. Thus, regardless of the approach side, we see that the probability that the collision occurs when the particles are aligned is $|u_j - u_i|dt$.

Now consider a particle pair that is not aligned, as sketched in Fig. 10.5.4. The figure represents the relative motion as seen by an observer fixed on the ith particle and shows that the collision depends on the trajectory of the j particle. Far from the ith particle, this j trajectory is usually straight, but becomes curved as the distance decreases. It is however useful to use as a reference trajectory, a horizontal one that grazes the ith particle, called a *grazing trajectory*. Since the center of that trajectory is at a distance $a_i + a_j$ from that of the ith particle, it follows that, on a plane perpendicular to the motion, the ith particle and the grazing jth particle combine to form a surface area – a collision cross-section – having a value equal to $\pi(a_i + a_j)^2$. If all trajectories were straight, then all particles having centers at a distance $R_c \le a_i + a_j$ from the x-axes would experience a collision, whereas those with $R_c > a_i + a_j$ would not. In actual conditions, the trajectories are curved and to assess the probability that a collision may occur between a given particle pair, we define a collision efficiency, $E(a_i, a_j)$, by means of

$$E(a_i, a_j) = \frac{R_c^2}{(a_i + a_j)^2} \tag{10.5.22}$$

Figure 10.5.5 shows, schematically, cases having $E(a_i, a_j) < 1$ and $E(a_i, a_j) > 1$. To understand these, we consider the gravitational sedimentation of a particle pair from the perspective of an observer fixed on the larger particle. For this observer, the large particle is fixed, smaller particles move towards it, and the fluid streams past it forming at small Reynolds numbers a well-defined streamline pattern. At the surface of the particle, of course, the fluid is at rest, which means that the streamlines in the vicinity of the fixed particle must be curved. Now consider Fig. 10.5.5a. The trajectory depicted there may occur when a small particle, approaches a larger particle. Not shown in the figure are the fluid streamlines. Far from the sphere, the small particle moves along a

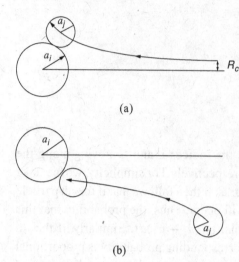

(a)

(b)

Figure 10.5.5. (a) Trajectory having $E < 1$. (b) Trajectory having $E > 1$.

streamline. Closer to it, the small particle initially follows the curved streamline, but because of its having a finite inertia, the small particle is generally not able to follow the fluid entirely, thereby colliding with the larger particle.

Now consider Fig. 10.5.5b. Here, we see another motion taking place at such Reynolds numbers that a wake is formed on the lee side of the sphere, thereby resulting in a region where other spheres experience a decreased drag (see Section 4.12). This may produce the trajectory sketched in the figure and results in a value of $E(a_i, a_j)$ larger than 1.

While the definition of the collision efficiency seems simple, the value of $E(a_i, a_j)$ depends intimately on the nature of the motion of the two particles and, more importantly, on the interaction among them. Particle interaction is, of course, a difficult matter, even in the simplest cases. Thus, to obtain $E(a_i, a_j)$ from first principles, one needs to solve the fluid mechanical problem of two particles in relative motion with respect to one another at all separation distances. This generally involves transient boundary conditions on the particles' surfaces, which even for the spherical case represent a difficult task. The problem is therefore tackled numerically, and some results are available in the literature (see, e.g., the book by Pruppacher and Klett, 1978) that provide an idea of the variations of $E(a_i, a_j)$ for given values of the particle size ratio, a_i/a_j, and the Reynolds number. For our purposes, we assume that $E(a_i, a_j)$ can be computed for any specific motion of interest. Thus, the probability that a collision between particles i and j will occur in time dt can be expressed as

$$\pi(a_i + a_j)^2 E(a_i, a_j)|u_i - u_j|dt.$$

In general, not all of the collisions produce the union agglomeration or coalescence between the colliding particles; and to account for those cases where a new particle is formed, we let $S(i, j)$ be the probability that the collision between particles i and j be sticking. Like $E(a_i, a_j)$, this *sticking* probability must be determined separately and will also assumed to be known.

A knowledge of $E(a_i, a_j)$, $S(i, j)$ and the relative velocity, $|u_i - u_j|$, determines the collection frequency, or kernel. Thus, since the particles are spherical, we have

$$K(v_i, v_j) = \pi (3/4\pi)\left(v_i^{1/3} + v_j^{1/3}\right)^2 E(v_i, v_j)S(v_i, v_j)|u_i - u_j| \tag{10.5.23}$$

This probability refers to a pair of particles of volumes v_i and v_j, and can be used in the coalescence equations when the distribution of sizes consists of a number of sections, each having a given value.

For continuous distributions, we are interested in the rate of collisions between particle of volumes v and w, respectively, and this is given by

$$K(v, w) = \pi (3/4\pi)^{2/3}\left(v^{1/3} + w^{1/3}\right)^2 E(v, w)S(v, w)|u_v - u_w| \tag{10.5.24}$$

where u_v and u_w are the velocities of particles of volumes v and w, respectively. It is noted that, under this model, no collisions are produced when the relative velocity vanishes.

Nondimensional Equations

Before addressing acoustic coalescence, we express the variables appearing in the above kernel and in the coalescence equations in nondimensional form. Consider first (10.5.24). To scale the particle velocities, we use a characteristic velocity in the imposed flow (e.g., the fluid velocity in a sound wave). Denoting that velocity by U_0, we put $u_v' = u_v/U_0$. Next comes the length scale. This is needed to scale both radii and volumes. A suitable scale that can be used in a variety of situations is the average interparticle distance for the suspension at rest. That length was introduced earlier in terms of the number concentration. As noted above, this number varies during coalescence, and it is therefore convenient to use the number concentration before coalescence takes place (i.e., N_0). Thus, the length scale is $N_0^{-1/3}$. Hence, we introduce nondimensional volumes by means of $v' = N_0 v$, so that

$$K(v', w') = \pi U_0 \left(\frac{3}{4\pi N_0}\right)^{2/3}\left(v'^{1/3} + w'^{1/3}\right)^2 E(v', w')S(v', w')|u_v' - u_w'| \tag{10.5.25}$$

It should be noted that, while the variables appearing here are nondimensional, $K(v', w')$ continues to have the dimensions of $L^3 T^{-1}$.

For the coalescence equations, we also need nondimensional number concentrations and time. For the number concentration of particles of volume v, we put $v_v = N_v/N_0^2$; and, for the time scale, we chose the time taken by the fluid to move a distance equal to $N_0^{-1/3}$. Thus, we put $t' = U_0 N_0^{1/3} t$. Hence, the rate equation for the total number of particles can then be expressed as

$$\frac{d\overline{N}}{dt'} = -\frac{1}{2}\frac{N_0^{2/3}}{U_0}\int_0^\infty\int_0^\infty v_v v_w K(v', w')dv'\, dw' \qquad (10.5.26)$$

Use of (10.5.25) gives the explicit result

$$N_0^{-2/3}\frac{d\overline{N}}{dt'} = -\frac{\pi}{2}\left(\frac{3}{4\pi}\right)^{2/3}\int_0^\infty\int_0^\infty v_v v_w \left(v'^{1/3} + w'^{1/3}\right)^2$$
$$\times E(v', w')S(v', w')\left|u_v' - u_w'\right|dv'\, dw' \qquad (10.5.27)$$

These equations, as well as the corresponding equations for dv_v/dt', apply to all cases where the particles move in a straight line, far from one another, as a result of imposed forces or fluid velocity fields. Thus, in addition to the collision and sticking factors, we see that $K(v, w)$, and, therefore, $\overline{N}$, depends on the specific nature of the motions through the relative velocity, which must be determined at every instant. This is not possible for arbitrary motions. But, for linear motions (e.g., acoustic), the relative velocity can be easily obtained from the expressions derived in Chapter 4. The simplest of such expressions is that obtained when Stokes' law is applicable, and as we know, this is the case for aerosols. As it turns out, acoustical coalescence is most efficient in aerosols.

Acoustic Agglomeration of Aerosols

We consider a plane, monochromatic sound wave propagating in a polydisperse aerosol having a volume concentration ϕ_v. If the frequency of the wave is ω and its velocity amplitude of the wave is U_0, the velocity amplitude of a spherical particle is

$$u_{p0}/U_0 = [1 + (\omega\tau_d)^2]^{-1/2}$$

where τ_d is the dynamic relaxation time corresponding to that particle. As discussed in Chapter 4, this simple result shows that, for a given frequency, particles having different sizes will move with different velocities. Although the above equation cannot be used to determine whether a given pair will collide, it shows that the frequency of the wave plays an important role in the process. This is seen as follows. First, the average distance between uniformly distributed particles is relatively large, even for moderate volume concentrations. Thus, no collisions occur at very low or very high frequencies because all particles move the same amount in the first case and are at rest in the second. On the other hand, intermediate frequencies can produce coalescence, and one of the questions that arises is whether a frequency exists that results in more rapid coalescence. We also wish to determine the changes in both number concentration and size distribution, if any, that are induced by the sound wave.

To proceed, we need the relative velocity, the collision efficiency, and the sticking factor. The first of these can be easily obtained by using complex notation [see equation (4.4.5)] to express both velocities and then taken the real part. Thus,

$$u'_v - u'_w = \frac{\omega(\tau_{d,v} - \tau_{d,w})}{\sqrt{1 + (\omega\tau_{d,v})^2}\sqrt{1 + (\omega\tau_{d,w})^2}} \sin(\omega t - \varphi_v - \varphi_w) \quad (10.5.28)$$

Here, $\tau_{d,v}$ and $\tau_{d,w}$ are the dynamic relaxation times for particles of volumes v and w, respectively, $\tan\varphi_v = \omega\tau_{d,v}$, and $\tan\varphi_w = \omega\tau_{d,w}$. It is advantageous to express the relaxation times in terms of a common value and the nondimensional volume of the corresponding particle. We thus write, for example,

$$\omega\tau_{d,v} = \omega\tau_{d,0}\, v'^{2/3} \quad (10.5.29)$$

where

$$\tau_{d,0} = \left(\frac{3}{4\pi}\right)^{2/3} \frac{2N_0^{-2/3}}{v_f \delta} \quad (10.5.30)$$

Reference to (4.4.3) shows that this is equal to the dynamic relaxation time for a particle of radius $1/N_0^{1/3}$. With these definitions, we can write (10.5.28) as

$$u'_v - u'_w = \omega\tau_{d,0} \frac{\left(v'^{2/3} - w'^{2/3}\right)}{\sqrt{1 + (\omega\tau_{d,0})^2 v'^{4/3}}\sqrt{1 + (\omega\tau_{d,0})^2 w'^{4/3}}} \sin(\omega t - \varphi_v - \varphi_w)$$

$$(10.5.31)$$

The collision and sticking factors are not known and for simplicity we take them to be equal to 1. Assuming that $E(v, w) = 1$ overestimates the collisions between some particle pairs, but underestimates the number of collisions between particles of nearly equal size, for which E is known to be larger than unity. This underestimate is probably more significant because collisions between particles of nearly equal size lead to the fastest particle growth. The assumption that $S(v, w)$ is equal to 1 generally overestimates the number of collisions that result in a union of the particles if the collisions are very energetic (i.e., if the approach velocity is large). But, for acoustic motions, where the collisions take place with slow relative velocities, the assumption is adequate for droplets. For rigid particles, the assumption is less well founded because colliding particles can bounce more easily. Furthermore, if agglomeration takes place, the resulting particle will not be spherical as assumed.

Thus, using these assumptions and (10.5.31) for the relative velocity, we obtain

$$K(v', w')$$
$$= \pi U_0 \left(\frac{3}{4\pi N_0}\right)^{2/3} \omega\tau_{d,0} \frac{|v'^{2/3} - w'^{2/3}|(v'^{1/3} + w'^{1/3})}{\sqrt{1 + (\omega\tau_{d,0})^2 v'^{4/3}}\sqrt{1 + (\omega\tau_{d,0})^2 w'^{4/3}}} |\sin(\omega t - \varphi_v - \varphi_w)|$$

$$(10.5.32)$$

This result applies at every instant, as implied by the appearance of the circular function. But experimental observations show that the coalescence process takes place on a time scale that is much larger than the period of the acoustic wave. Typically, the acoustic period is of the order of $1/1,000$ sec, whereas the time scale for coalescence is of the order of seconds. This means that, to study the coalescence process, it is sufficient to consider it over the longer time scale and ignore whatever effects take place during any one single period. To do this, we may average the coalescence equations over one acoustic period, during which the values of ν_v, ν_v, $\overline{N}$, and their time derivatives may be taken as constants. The result of this averaging is given below. However, since it is the collision kernel that determines whether collisions occur or not, it useful to first consider its time average. Thus, using $\langle|\sin(\omega t - \varphi_v - \varphi_w|\rangle = 2/\pi$, we obtain

$$\langle K(v', w')\rangle = 2U_0 \left(\frac{3}{4\pi N_0}\right)^{2/3} \frac{\omega\tau_{d,0}|v'^{2/3} - w'^{2/3}|(v'^{1/3} + w'^{1/3})}{\sqrt{1 + (\omega\tau_{d,0})^2 v'^{4/3}}\sqrt{1 + (\omega\tau_{d,0})^2 w'^{4/3}}} \tag{10.5.33}$$

As expected, this shows that the rate at which collisions take place increases with the imposed gas velocity. More importantly, we see that, for a given value of U_0', that rate depends only on the frequency, ω, of the waves and on the initial number concentration, or equivalently, on the interparticle distance. Finally, it should be noted that the collision kernel vanishes when the particles are of the same size. This occurs because the velocities adopted by the model do not include interactive effects.

Now, for a given aerosol, the initial number concentration is fixed, and we may study the dependence on the frequency through the product $\omega\tau_{d,0}$. This dependence is nonlinear because $\langle K(v', w')\rangle$ increases with $\omega\tau_{d,0}$ when $\omega\tau_{d,0} \ll 1$, and decreases when $\omega\tau_{d,0} \gg 1$. Hence, $\langle K(v', w')\rangle$ must have a maximum for some value of $\omega\tau_{d,0}$, thus defining an optimum frequency for acoustic agglomeration. Taking the derivative of (10.5.33) with respect to $\omega\tau_0$ and equating to zero, we find that a maximum exists for $(\omega\tau_{d,0})^4(vw)^{4/3} = 1$. This has only one real and positive root, namely

$$(\omega\tau_{d,0})_{opt} = (vw)^{-1/3} \tag{10.5.34}$$

While succinct, the result is not too useful from the practical standpoint because it simply states that, for each pair of values of v and w, (i.e., for each particle pair), there is a different optimum frequency. Therefore, since in a continuous distribution there are infinitely many such pairs, there is no unique frequency that optimizes the agglomeration process for the aerosol as a whole. On the other hand, real distributions have particles whose diameters fall between some extreme values. Thus, the ranges of values that v and w can have are finite. Furthermore, for distributions that are not too wide, a representative size can be used instead of v and w to define an optimum frequency. This can be done as follows. First, express the nondimensional values of v and w in terms of the corresponding particle radii, a_v and a_w, respectively. Thus, $v' = 4/3\pi N_0 a_v^3$. Next use the value of $\tau_{d,0}$ given by (10.5.30). These give

$$\omega_{opt} = \frac{1}{(2a_v a_w/9\nu_f \delta)} \tag{10.5.35}$$

Now, if the distribution is not wide, we may use for these variable radii the values corresponding to some representative mean (e.g., one of those introduced in Section 10.2). For simplicity, we select the geometric mean radius of the distribution, a_g. Another possible choice would be the radius corresponding to the peak of the distribution function, a^*, but the geometric mean is more appropriate, at least for typical aerosol distributions. Also, if the distributions are not wide, the two average values are nearly equal. Thus, putting $a_v = a_w = a_g$ in the above equation, we see that the denominator on the right-hand side is equal to the dynamic relaxation time τ_g for a particle having that radius. Hence, we find that the optimum frequency is

$$\omega_{opt} \approx 1/\tau_{d,g} \tag{10.5.36}$$

Although approximate, this gives an indication of the frequencies required to optimize acoustic coalescence. For example, for an aerosol having $a_g = 1\ \mu m$, the optimum frequency, as defined above, would be about 11 kHz. However, as time increases, the aerosol distribution evolves, with larger and larger particles being created, which means that $\tau_{d,g}$ increases with time, so that ω_{opt} decreases accordingly. Hence, an optimum frequency selected on the basis of an initial mean diameter ceases to be as effective at latter times.

Let us now return to (10.5.27) and average it over one acoustic period, remembering that over this time v_v, v_v, $\overline{N}$, and their time derivatives may be taken as constants. Thus,

$$\frac{d\overline{N}}{dt'} = -\frac{1}{2}\frac{N_0^{2/3}}{U_0} \int_0^\infty \int_0^\infty v_v \langle K(v', w') \rangle v_w \, dv' \, dw' \tag{10.5.37}$$

To describe the coalescence that takes place in a given aerosol, we need to integrate this equation, and this requires the values of v_v at all times. We will later present the results of some numerical calculations that show the evolution of an initially known size distribution in an aerosol. Before we do that, we consider (10.5.37) at $t = 0$ (i.e., before the coalescence process begins). Here, the distribution function is known and can be used to obtain the initial rate of change of $\overline{N}$, thus enabling us to see the effects of the frequency. Thus, if $v_v^{(0)}$ represents the nondimensional particle volume distribution function, we have

$$\left(\frac{d\overline{N}}{dt'}\right)_0 = -\left(\frac{3}{4\pi}\right)^{2/3} \omega\tau_{d,0} \int_0^\infty \int_0^\infty$$

$$\times \frac{\left(v'^{1/3} + w'^{1/3}\right)\left|v'^{2/3} - w'^{2/3}\right| v_v^{(0)} v_w^{(0)}}{\sqrt{1 + (\omega\tau_{d,0})^2 v'^{4/3}}\sqrt{1 + (\omega\tau_{d,0})^2 w'^{4/3}}} dv' \, dw' \tag{10.5.38}$$

To proceed, we need an explicit particle volume distribution function, and for this purpose we choose a log-normal distribution. This type has been found to describe some aerosols. Mathematically, the log-normal distribution is a normal (i.e., Gaussian), distribution for the logarithm of the sizes or volumes. For the diameters, that distribution

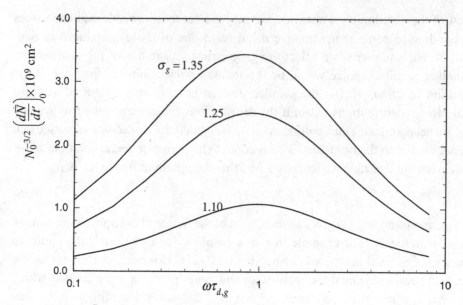

Figure 10.5.6. $(d\bar{N}/dt')$ at $t' = 0$ for three log-normal size distributions.

is given by

$$n_D dD = \frac{1}{\sqrt{2\pi}\,\ln\sigma_{D_g}} \exp\left[-\frac{1}{2}\left[\frac{\ln D/D_g}{\ln\sigma_{D_g}}\right]^2\right] d(\ln D) \qquad (10.5.39)$$

where D_g is the geometric mean particle diameter, and σ_{D_g} is the geometric standard deviation. The log-normal volume distribution is given by a similar expression:

$$n_v dv = \frac{1}{\sqrt{2\pi}\,\ln\sigma_{v_g}} \exp\left[-\frac{1}{2}\left[\frac{\ln v/v_g}{\ln\sigma_{v_g}}\right]^2\right] d(\ln v) \qquad (10.5.40)$$

with $\ln\sigma_{v_g} = 3\ln\sigma_{D_g}$. This can be used in (10.5.38) to determine the initial rate at which $\bar{N}$ decays. The integrals can be evaluated numerically, and Fig. 10.5.6 shows the results for three lognormal distributions having a geometric mean diameter, D_g, equal to 1 μm, but having different standard deviations. As we see in the figure, the maximum value of the scaled derivative occurs at $\omega\tau_{d,g} \approx 1$ for $\sigma_{D_g} = 1.10$, in agreement with the heuristic approximation given above. For larger standard deviations (i.e., for wider distributions), the maximum occurs, as expected, at a lower value of $\omega\tau_{d,g}$. Since this maximum occurs for $\omega\tau_{d,g} < 1$, and since $\tau_{d,g}$ is the same for both curves, we see that the optimum frequency required for the wider distribution is lower than that for the narrow one. Furthermore, since all distributions widen as coalescence takes place, we also see that lower frequencies are required if optimum results are desired.

Also implied by the results shown in the figure is the fact that the scaled value of $(d\bar{N}/dt')_0$ is independent of the initial number concentration, N_0 and of the imposed velocity in the sound wave. But, in terms of the original variables, we have

$$\left(\frac{dN}{dt}\right)_0 = -U_0 N_0^2 f(\omega\tau_{d,g}, \sigma_{D_g}) \tag{10.5.41}$$

where $f(\omega\tau_{d,g}, \sigma_{D_g})$ is function having a maximum at some value of $\omega\tau_{d,g}$, as shown in the figure. We note from the above equation that the initial rate of decrease is proportional to the square of the initial number concentration, as was the case for Brownian coalescence.

When the elapsed time is small, (10.5.41) may be used to estimate the initial decrease in the number of particles. This is about $\Delta N \approx U_0 N_0^2 f(\omega\tau_{d,g}, \sigma_g)\Delta t$. Thus, if N_0 is 10^5 cm^{-3}, and U_0 is 500 cm/sec, we see that in 1 sec the total number of particles decreases by about 5% if $\omega\tau_{d,g} = 1$, $D_g = 1\,\mu$m and $\sigma_{D_g} = 1.1$. Two factors account for the significant decrease. First, at the selected frequency, 1 sec corresponds to about 10^5 acoustic periods. Hence, even if the number of collisions is insignificant, the cumulative effect is large. Second, the chosen value of U_0 corresponds to a fairly intense acoustic wave. While it is clear that large sound amplitudes can cause more rapid coalescence, such amplitudes are larger than the monochromatic theory requires. We will return to these matters later, after we discuss some numerical results, applicable at larger times.

Numerical Results for $t \geq 0$. Let us now consider the evolution of the size distribution in an aerosol for $t > 0$ (i.e., the variations with time of the total number of particles, N, and the number of particles having a diameter in a given size range, $N_D dD$). These are prescribed by the coalescence equations. Numerical solutions of these may be obtained by first dividing the range of integration into L sections, each having a given diameter, or volume. In nondimensional form the equations for the sectional numbers are

$$\frac{dv_m}{dt'} = -v_m \sum_{i=1}^{L} K_{m-1,i} v_i + \frac{1}{2} \sum_{i=1}^{m-1} v_{m-1} K_{m-i,i} v_i, \quad m = 1, 2, \cdots, L \tag{10.5.42}$$

The rate equation for the total number may be evaluated from

$$\frac{d\overline{N}}{dt'} = -\sum_{m=1}^{L} \frac{dv_m}{dt'} \tag{10.5.43}$$

Thus, the system consists of $L + 1$ ordinary differential equations, which can be solved numerically, once the coalescence kernel is specified. Because, as pointed out previously, particles having large sizes are created by the process, it follows that a large number of sections is necessary to account for the particle distribution at large times. If this number is not sufficiently large, some particles will not be accounted for. Thus, in a numerical integration of the equations, it is necessary to verify that no particles are lost. One way of ensuring that all particles are accounted for, is to verify at each instant that the aggregate volume occupied by the particles remains constant.

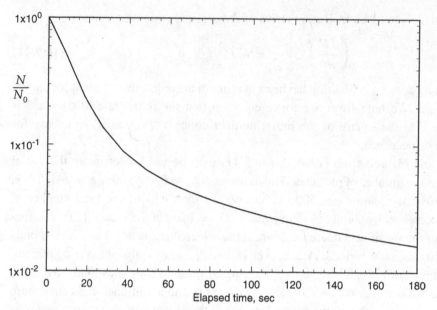

Figure 10.5.7. Variations of $\overline{N}$ with frequency.

Since the total volume occupied by the particles is initially equal to V_0, and, at all times equal to $V = \sum_{m=1}^{L} v_m$, we must have $V(t)/V_0 = 1$.

Several numerical procedures exist that have been developed to integrate the discretized coalescence equations in a variety of conditions, and the reader is referred to the works cited in the bibliography for information on them. The results shown in the figures that follow are intended to illustrate only some of the main features of the coagulation induced by sound waves. They refer to the evolution of a slightly modified log-normal distribution having a geometric mean diameter, D_g, equal to 1 μm, and a standard deviation equal to 1.1. The modification consisted in chopping the distribution below 0.35 μm and above 6 μm. In between those two values, the distribution was divided into $L = 16^3$ sections. This number allows study of the evolution of the distribution for about 3 minutes, with only a small particle loss.

In the calculations, the number concentrations in each of the sections is computed first. These numbers are then added to obtain the total number concentration at each instant. Figure 10.5.7 shows the variations of $\overline{N} = N/N_0$ with time. We see that $\overline{N}$ initially decreases rapidly, with half the number of particles disappearing in about 10 sec. As time increases, the decrease rate is slowed considerably because of the significantly lower number of particles and the driving frequency becoming less suitable for inducing collisions. Nevertheless, we see that, in 3 minutes, the number of particles has decreased by about 98%.

The variations of $\overline{N}$ displayed in Fig. 10.5.7 show that the acoustic coalescence process can roughly be divided into three distinct stages. The first, marked by a rapid decrease of the number of particles that is proportional to the elapsed time, is followed by a transition region where the decrease rate slows down continuously. Eventually, a

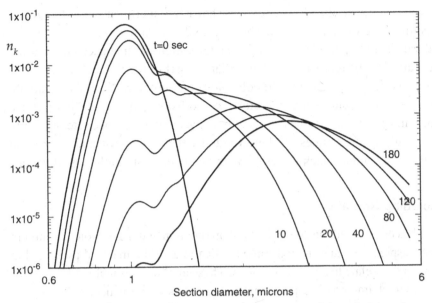

Figure 10.5.8. Variations of the size distribution function at several times during coalescence.

stage is reached where $\overline{N}$ decreases as $t^{-2/3}$. This is marked by the nearly straightline trend observed at the largest times shown in the figure. Although the demarcation of the three regions is not sharp, we anticipate that each corresponds to markedly different size distributions. In Fig. 10.5.8, we show the size distribution at various times. To see the effects of the sound wave more clearly, the nondimensional section numbers, ν_k, are shown on a logarithmic scale. The figure also includes the initial distribution that shows the absence of large particles at $t = 0$. But, almost immediately, we see the appearance of larger particles not initially present. As time proceeds, the size distribution continues to be shifted toward larger values, and we see that a secondary maximum appears at some larger diameter. The distribution function is then said to be *bimodal*. At the same time, the initial distribution begins to disappear. These features occur during the second, or transition stage. At later times yet, the size distribution approaches a log-normal shape, like the initial one, but much wider.

One final word about the calculations. Figure 10.5.8 shows that, for the largest times shown, the distribution function is cut at about 6 μm. This is due to the limited number of sections used in the calculations. While the number of particles larger than that diameter is small, it is evident that they are being produced by collisions between smaller particles and that their absence in the calculations imply a faster decrease of the total number.

Large-Amplitude Waves

The above results refer to sound waves whose amplitudes are sufficiently small. But the amplitudes used in practice are far from being small. In fact, their amplitudes are so large that they develop shock fronts every wavelength. Such waves produce very rapid coalescence, even at small frequencies, as shown by experiments reported by this author some time ago (Temkin, 1970). The reasons for such high rates are not well understood, but include particle drift, wake capture, high-particle acceleration, and turbulence. Particle drift refers to the nonoscillatory part of the motion of a particle. This component effectively increases the collection volume of a particle. Wake capture can also increase that volume and therefore the collision efficiency. High-particle acceleration can accentuate the relative velocity between particles, because their drag is affected strongly by unsteadiness. It should be added that this type of coalescence can also be produced by single shock waves of weak to moderate strength (e.g., thunder).

10.6 Concluding Remarks

This chapter has considered the coefficients of attenuation and dispersion of sound in polydisperse suspensions, and treated some applications of the material presented in earlier chapters that relate to acoustic motions. These applications included the use of acoustic waves to characterize suspensions, and to induce coalescence.

APPENDIX A

Material and Transport Properties of Some Substances at 1 atm and 20°C

	c_s (ms^{-1})	ρ (gm cm^{-3})	c_p [J/(gm K)]	γ	β (K^{-1})	k [J/(gm s K)]	ν (cm^2/s)
Water	1484	0.9982	4.182	1.003	2.10×10^{-4}	5.90×10^{-3}	1.00×10^{-2}
Air	344	0.00121	1.012	1.401	3.48×10^{-3}	2.54×10^{-4}	1.50×10^{-1}
Helium	1000	0.00018	4.990	1.667	3.47×10^{-3}	2.50×10^{-4}	5.00×10^{-2}
Toluene	1360	0.87	1.652	1.350	1.04×10^{-3}	1.60×10^{-3}	5.86×10^{-3}
Polysterene	2380	1.05	1.190	1.005	2.50×10^{-5}	1.15×10^{-2}	–
Silica	4100	2.20	0.836	1.001	3.15×10^{-5}	1.02×10^{-2}	–
Alumina	6200	3.97	0.227	1.000	2.60×10^{-8}	1.70×10^{-1}	–

APPENDIX B

Useful Formulas from Vector Analysis

In the following formulas, $\mathbf{A}, \mathbf{B}, \mathbf{C}$, and $\mathbf{D}$ represent vector functions; ϕ and ψ represent scalar functions.

General Relations

$$\mathbf{A \cdot B} \times \mathbf{C} = \mathbf{B \cdot C} \times \mathbf{A} - \mathbf{C \cdot A} \times \mathbf{B}$$

$$\mathbf{A} \times \mathbf{B} \times \mathbf{C} = (\mathbf{A \cdot C})\mathbf{B} - (\mathbf{A \cdot B})\mathbf{C}$$

$$(\mathbf{A} \times \mathbf{B}) \cdot (\mathbf{C} \times \mathbf{D}) = (\mathbf{A \cdot C})(\mathbf{B \cdot D}) - (\mathbf{A \cdot D})(\mathbf{B \cdot C})$$

$$(\mathbf{A} \times \mathbf{B}) \times (\mathbf{C} \times \mathbf{D}) = (\mathbf{A} \times \mathbf{B \cdot D})\mathbf{C} - (\mathbf{A} \times \mathbf{B \cdot C})\mathbf{D}$$

$$\nabla \times \nabla\phi = 0$$

$$\nabla \cdot \nabla \times \mathbf{B} = 0$$

$$\nabla(\phi + \psi) = \nabla\phi + \nabla\psi$$

$$\nabla(\phi\psi) = \psi\nabla\phi + \phi\nabla\psi$$

$$\nabla \cdot (\mathbf{A} + \mathbf{B}) = \nabla \cdot \mathbf{A} + \nabla \cdot \mathbf{B}$$

$$\nabla \times (\mathbf{A} + \mathbf{B}) = \nabla \times \mathbf{A} + \nabla \times \mathbf{B}$$

$$\nabla \cdot (\phi\mathbf{A}) = \mathbf{A} \cdot \nabla\phi + \phi\nabla \cdot \mathbf{A}$$

$$\nabla \times (\phi\mathbf{A}) = \nabla\phi \times \mathbf{A} + \phi\nabla \times \mathbf{A}$$

$$\nabla(\mathbf{A \cdot B}) = (\mathbf{A} \cdot \nabla)\mathbf{B} + (\mathbf{B} \cdot \nabla)\mathbf{A} + \mathbf{A} \times (\nabla \times \mathbf{B}) + \mathbf{B} \times (\nabla \times \mathbf{A})$$

$$\nabla \cdot (\mathbf{A} \times \mathbf{B}) = \mathbf{B} \cdot \nabla \times \mathbf{A} - \mathbf{A} \cdot \nabla \times \mathbf{B}$$

$$\nabla \times (\mathbf{A} \times \mathbf{B}) = \mathbf{A}\nabla \cdot \mathbf{B} - \mathbf{B}\nabla \cdot \mathbf{A} + (\mathbf{B} \cdot \nabla)\mathbf{A} - (\mathbf{A} \cdot \nabla)\mathbf{B}$$

$$\nabla \times (\nabla \times \mathbf{A}) = \nabla\nabla \cdot \mathbf{A} - \nabla^2\mathbf{A}$$

Special Relations

If $\mathbf{x} = \mathbf{i}x + \mathbf{j}y + \mathbf{k}z$ in the position vector of a point (x, y, z) and $r = |\mathbf{x}|$, then:

$$\nabla \cdot \mathbf{x} = 3$$
$$\nabla \times \mathbf{x} = 0$$
$$\nabla r = \mathbf{x}/r$$
$$\nabla(1/r) = -\mathbf{x}/r^3$$
$$(\mathbf{A} \cdot \nabla)\mathbf{x} = \mathbf{A}$$

Integral Relations

In the following formulas, V is a volume bounded by a closed surface S, having a unit normal vector $\mathbf{n}$ directed outward from V:

$$\int_S \phi \mathbf{n} \, dS = \int_V \nabla\phi \, dV$$

$$\int_S \mathbf{A} \cdot \mathbf{n} \, dS = \int_V \nabla \cdot \mathbf{A} \, dV$$

$$\int_S \mathbf{n} \times \mathbf{A} \, dS = \int_V \nabla \times \mathbf{A} \, dV$$

Let S be an open surface with unit normal $\mathbf{n}$, bounded by the closed contour C having a line element $d\mathbf{I}$. Then,

$$\oint_C \phi \, d\mathbf{l} = \int_S \mathbf{n} \times \nabla\phi \, dS$$

$$\oint_C \mathbf{A} \cdot d\mathbf{l} = \int_S \mathbf{n}(\nabla \times \mathbf{A}) \, dS$$

APPENDIX C

Explicit Expressions for Some Quantities in Spherical Polar Coordinates

Coordinate symbols	r	θ	φ
Line elements	dr	$r\,d\theta$	$r\sin\theta\,d\varphi$
Orthonormal vectors along coordinate axes	$\mathbf{e}_r$	$\mathbf{e}_\theta$	$\mathbf{e}_\varphi$
Velocity components	u_r	u_θ	u_φ
Components of $\nabla\phi$	$\dfrac{\partial\phi}{\partial r}$	$\dfrac{1}{r}\dfrac{\partial\phi}{\partial\theta}$	$\dfrac{1}{r\sin\theta}\dfrac{\partial\phi}{\partial\psi}$
Components of $\mathbf{B}$	B_r	B_θ	B_φ
Components of $\nabla\times\mathbf{B}$	$\dfrac{1}{r\sin\theta}\left[\dfrac{\partial(B_\varphi\sin\theta)}{\partial\theta}-\dfrac{\partial B_\theta}{\partial\varphi}\right]$	$\dfrac{1}{r}\left[\dfrac{1}{\sin\theta}\dfrac{\partial B_r}{\partial\varphi}-\dfrac{\partial(r B_\varphi)}{\partial r}\right]$	$\dfrac{1}{r}\left[\dfrac{\partial(r B_\theta)}{\partial r}-\dfrac{\partial B_r}{\partial\theta}\right]$

Components of $\nabla^2\mathbf{B}$

$$\mathbf{e}_r\cdot\nabla^2\mathbf{B}=\nabla^2 B_r-\frac{2B_r}{r^2}-\frac{2}{r^2\sin\theta}\frac{\partial(B_\theta\sin\theta)}{\partial\theta}-\frac{2}{r^2\sin\theta}\frac{\partial B_\varphi}{\partial\varphi}$$

$$\mathbf{e}_\theta\cdot\nabla^2\mathbf{B}=\nabla^2 B_\theta+\frac{2}{r^2}\frac{\partial B_r}{\partial\theta}-\frac{B_\theta}{r^2\sin^2\theta}-\frac{2\cos\theta}{r^2\sin^2\theta}\frac{\partial B_\varphi}{\partial\varphi}$$

$$\mathbf{e}_\varphi\cdot\nabla^2\mathbf{B}=\nabla^2 B_\varphi+\frac{2}{r^2\sin\theta}\frac{\partial B_r}{\partial\varphi}+\frac{2\cos\theta}{r^2\sin^2\theta}\frac{\partial B_\theta}{\partial\varphi}-\frac{B_\varphi}{r^2\sin^2\theta}$$

Special operators

$$\nabla \cdot \mathbf{B} = \frac{1}{r^2}\frac{\partial(r^2 B_r)}{\partial r} + \frac{1}{r\sin\theta}\frac{\partial(\sin\theta\, B_\theta)}{\partial\theta} + \frac{1}{r\sin\theta}\frac{\partial B_\varphi}{\partial\varphi}$$

$$\nabla^2\phi = \frac{1}{r^2}\frac{\partial}{\partial r}\left(r^2\frac{\partial\phi}{\partial r}\right) + \frac{1}{r^2\sin\theta}\frac{\partial}{\partial\theta}\left(\sin\theta\frac{\partial\phi}{\partial\theta}\right) + \frac{1}{r^2\sin^2\theta}\frac{\partial^2\phi}{\partial\varphi^2}$$

Components of rate-of-strain tensor

$$e_{rr} = \frac{\partial u_r}{\partial r}, \quad e_{\theta\theta} = \frac{1}{r}\frac{\partial u_\theta}{\partial\theta} + \frac{u_r}{r}, \quad e_{\varphi\varphi} = \frac{1}{r\sin\theta}\frac{\partial u_\varphi}{\partial\varphi} + \frac{u_r}{r} + \frac{u_\theta \cot\theta}{r}$$

$$e_{\theta\varphi} = e_{\varphi\theta} = \frac{\sin\theta}{2r}\frac{\partial}{\partial\theta}\left(\frac{u_\varphi}{\sin\theta}\right) + \frac{1}{2r\sin\theta}\frac{\partial u_\theta}{\partial\varphi}$$

$$e_{\varphi r} = e_{r\varphi} = \frac{1}{2r\sin\theta}\frac{\partial u_r}{\partial\varphi} + \frac{r}{2}\frac{\partial}{\partial r}\left(\frac{u_\varphi}{r}\right)$$

$$e_{r\theta} = e_{\theta r} = \frac{r}{2}\frac{\partial}{\partial r}\left(\frac{u_\theta}{r}\right) + \frac{1}{2r}\frac{\partial u_r}{\partial\theta}$$

APPENDIX D

Some Properties of the Spherical Bessel Functions

For n equal to an integer, the solution to

$$f''(z) + \frac{2}{z} f'(z) + \left[1 - \frac{n(n+1)}{z} \right] f(z) = 0$$

may be written in terms of the *spherical Bessel functions of the first kind*:

$$j_n(z) = \sqrt{\pi/2z}\, J_{n+1/2}(z)$$

the spherical Bessel functions of the second kind:

$$y_n(z) = \sqrt{\pi/2z}\, Y_{n+1/2}(z)$$

and the spherical Bessel functions of the third kind:

$$h_n^{(1)}(z) = j_n(z) + i y_n(z) = \sqrt{\pi/2z}\, H_{n+1/2}^{(1)}(z)$$

$$h_n^{(2)}(z) = j_n(z) - i y_n(z) = \sqrt{\pi/2z}\, H_{n+1/2}^{(2)}(z)$$

The pairs j_n, y_n and $h_n^{(1)}$, $h_n^{(2)}$ are linearly independent.

Explicit Expressions

When the order n is small, the spherical Bessel functions may be expressed in terms of elementary functions:

$n = 0$

$$j_0(z) = \frac{\sin z}{z}$$

$$y_0(z) = -\frac{\cos z}{z}$$

$$h_0^{(1)}(z) = -\frac{i}{z} e^{iz}, \quad h_0^{(2)}(z) = -\frac{i}{z} e^{-iz}$$

$n = 1$

$$j_1(z) = \frac{\sin z}{z^2} - \frac{\cos z}{z}$$

$$y_1(z) = -\frac{\cos z}{z^2} - \frac{\sin z}{z^2}$$

$$h_1^{(1)}(z) = -\frac{1}{z}\left(1 + \frac{i}{z}\right)e^{iz}, \quad h_1^{(2)}(z) = -\frac{1}{z}\left(1 - \frac{i}{z}\right)e^{-iz}$$

Recurrence Relations

For $n > 2$, the spherical Bessel functions, and their derivatives, may be obtained from the following relations, applicable to j_n, y_n, $h_n^{(1)}$, and $h_n^{(2)}$:

$$f_{n+1}(z) = \frac{2n+1}{z}f_n(z) - f_{n-1}(z)$$

$$f_n'(z) = f_{n-1}(z) - \frac{n+1}{(z)}f_n(z)$$

Wronskians

$$W[j_n(z), y_n(z)] = j_{n+1}(z)y_n(z) - j_n(z)y_{(n+1)}(z) = 1/z^2$$

$$W[h_n^{(1)}(z), h_n^{(2)}(z)] = h_{n+1}^{(1)}(z)h_n^{(2)}(z) - h_n^{(1)}(z)h_{(n+1)}^{(2)}(z) = -2i/z^2$$

Expressions for Small Arguments

When $z \to 0$, the leading terms in the series expansion of $j_n(z)$ and $y_z(z)$ are

$$j_n(z) \simeq \frac{z^n}{1 \cdot 3 \cdot 5 \cdots (2n+1)}\left[1 - \frac{z^2/2}{1!(2n+3)} + \cdots\right]$$

$$y_n(z) \simeq \frac{1 \cdot 3 \cdot 5 \cdots (2n-1)}{z^{n+1}}\left[1 - \frac{z^2/2}{1!(1-2n)} - \cdots\right]$$

For the cases $n = 0$ and $n = 1$, these give

$$j_0(z) \simeq 1 - z^2/6, \qquad j_1(z) \simeq \frac{1}{3}z - z^3/30$$

$$y_0(z) \simeq -1/z + z/2, \qquad y_1(z) \simeq -1/z^2 - \frac{1}{2}$$

Legendre Polynomials

The relevant differential equation here is (5.3.17), or

$$(1 - x^2)\frac{d^2\Theta}{dx^2} - 2x\frac{d\Theta}{dx} + n(n+1)\Theta = 0$$

The solutions to this equation are the Legendre polynomials of the first and second kinds $P_n(x)$ and $Q_n(x)$, respectively. Explicit forms for low values of n are:

$n = 0$

$$P_0(x) = 1, \quad Q_0(x) = \frac{1}{2}\ln\left(\frac{1+x}{1-x}\right)$$

$n = 1$

$$P_1(x) = x, \quad Q_1(x) = \frac{x}{2}\ln\left(\frac{1+x}{1-x}\right) - 1$$

$n = 2$

$$P_2(x) = \frac{1}{2}(3x^2 - 1), \quad Q_2(x) = \frac{3x^2 - 1}{4}\ln\left(\frac{1+x}{1-x}\right) - \frac{3x}{2}$$

Recurrence Relations for $P_n(x)$

$$(n+1)P_{n+1}(x) = (2n+1)x\,P_n(x) - n\,P_{n-1}(x)$$

$$(x^2 - 1)\frac{d\,P_n(x)}{dx} = nx\,P_n(x) - n\,P_{n-1}(x)$$

These recurrence relations are also satisfied by Q_n.

Special Relations and Values

$$P_{-n-1}(x) = P_n(x)$$

$$P_n(-x) = P_n(x)$$

Bibliography

BOOKS

A. BASIC PHYSICS AND MATHEMATICS

Fluid Mechanics

BATCHELOR, G. K. 1967 *An Introduction to Fluid Dynamics*, Cambridge.
LAMB, H. 1932 *Hydrodynamics*, Dover.
LANDAU, L. D. AND LIFSHITZ, E. M. 1959 *Fluid Mechanics*, Pergamon.
LIEPMANN, H. W. AND ROSHKO, A. 1957 *Elements of Gasdynamics*, Wiley.
VINCENTI, W. G. AND KRUGER, C. H., JR., 1963 *Introduction to Physical Gas Dynamics*, Wiley.

Heat and Thermodynamics

EPSTEIN, P. S. 1937 *Textbook of Thermodynamics*, Wiley.
CALLEN, H. 1960 *Thermodynamics*, Wiley.
CARSLAW, H. S. AND JAEGER, J. C. 1959 *Conduction of Heat in Solids*, Oxford.
LANDAU, L. D. AND E. M. LIFSHITZ, E. M. 1958 *Statistical Physics*, Pergamon.
PIPPARD, A. B. 1961 *The Elements of Classical Thermodynamics*, Cambridge.
PRIGOGINE, I. 1967 *Thermodynamics of Irreversible Processes*, 3rd Ed., Interscience.
WOODS, L. C. 1975 *The Thermodynamics of Fluid Systems*, Oxford.

Mathematics and Mathematical Tables

ABRAMOWITZ, M. AND STEGUN, I. 1964 *Handbook of Mathematical Functions*, Dover.
ARIS, R. 1989 *Vectors, Tensors, and the Basic Equations of Fluid Mechanics*, Dover.
GRADSHTEYN, I. S. AND RYZHIK, I. M. 1965 *Tables of Integrals, Series and Products*, Academic.
JEFFREYS, H. AND JEFFREYS, B. S. 1956 *Methods of Mathematical Physics*, Cambridge.
MORSE, P. M. AND FESHBACH, H. 1953 *Methods of Theoretical Physics*, Vols. I and II, McGraw-Hill.
SMITH, M. G. 1966 *Laplace Transform Theory*, van Nostrand.
TRICOMI, F. G. 1985 *Integral Equations*, Dover.

Mechanics

GOLDSTEIN, H. 1959 *Classical Mechanics*, Addison-Wesley.
LANDAU, L. D. AND E. M. LIFSHITZ, E. M. 1960 *Mechanics*, Pergamon.

PIPPARD, A. B. 1978 *The Physics of Vibration*, Cambridge.
SYNGE, J. L. AND GRIFFITH, B. A. 1959 *Principles of Mechanics*, McGraw-Hill.

Sound

BHATIA, A. B. 1967 *Ultrasonic Absorption*, Dover.
HERZFELD, K. F. AND LITOVITZ, T. A. 1959 *Absorption and Dispersion of Ultrasonic Waves*, Academic.
MORSE, P. M. AND INGARD, K. U. 1972 *Theoretical Acoustics*, McGraw-Hill.
PRICE, A. D. 1989 *Acoustics – An Antroduction to Its Physical Principles and Applications*, Acoustical Society of America.
SKUDRZYK, E. 1971 *The Foundations of Acoustics*, Springer.
STRUTT, J. W. 1896 *The Theory of Sound*, Vol. II, 2nd Ed., Dover.
TEMKIN, S. 2001a *Elements of Acoustics*, Acoustical Society of America.

B. PARTICLES AND SUSPENSIONS

CLIFT, R., GRACE, J. R., AND WEBER, M. E. 1978 *Bubbles, Drops and Particles*, Academic.
CROW, C., SUMMERFELD, M., AND TSUJI, Y. 1998 *Multiphase Flows with Droplets and Particles*, CRC.
DAVIES, C. N. 1968 *Aerosol Science*, Academic.
DENNIS, S., Ed. 1976 *Handbook on Aerosols* National Technical Information Service, Document No. TID 26608.
DRAKE, R. L. 1974 *A General Mathematical Survey of the Coagulation Equation* (Vol. 3 of *International Reviews in Aerosol Physics and Chemistry*, Hidy, G. M. and Brock, J. R., Eds.), Pergamon.
DUFOUR, L. AND DUFAY, R. 1963 *Thermodynamics of Clouds*, Academic.
EINSTEIN, A. 1926 *Investigations on the Theory of the Brownian Movement*, R. Fürth, Ed., Dover.
FRIEDLANDER, S. K. 1977 *Smoke, Dust and Haze: Fundamentals of Aerosol Behavior*, Wiley.
FROHN, A. AND ROTH, N. 2000 *Dynamics of Droplets*, Springer.
FUCHS, N. A. 1959 *Evaporation and Droplet Growth in Gaseous Media*, Pergamon.
FUCHS, N. A. 1964 *The Mechanics of Aerosols*, Dover.
HAPPEL, J. AND BRENNER, H. 1965 *Low Reynolds Number Hydrodynamics*, Prentice Hall.
HUNTER, R. J. 1987 *Foundations of Colloid Science*, Vols. I and II, Oxford.
ISHII, M. 1975 *Thermo-fluid Dynamic Theory of Two-phase Flow*, Eyrolles.
JOSEPH, D. D. AND SCHAEFFER, D. G. Eds., 1990 *Two Phase Flows and Waves*, Springer.
KIM, S. AND KARRILA, S. J. 1990 *Microhydrodynamics*, Butterworth-Heinemann.
KIRKWOOD, J. G. 1967 *Macromolecules*, P. L. Auer, Ed., Gordon and Breach.
LEIGHTON, T. G. 1994 *The Acoustic Bubble*, Academic.
MASON, B. J. 1971 *The Physics of Clouds*, Oxford.
MEDNIKOV, E. 1965 *Acoustic Coagulation and Precipitation of Aerosols*, Plenum.
MILLIKAN, R. A. 1917 *The Electron*, Chicago.
NAKORYAKOV, V. E., POKUSAEV, B. G., AND SHREIBER, I. R. 2000 *Wave Propagation in Gas-Liquid Media*, CRC.
OSEEN, C. W. 1927 *Neuere Methoden in der Hydrodynamik*, Leipzig.
PRUPPACHER, H. R. AND KLETT, J. D. 1978 *Microphysics of Clouds and Precipitation*, D. Reidel.
RUSSEL, W. B. 1987 *The Dynamics of Colloidal Systems*, Wisconsin.
RUSSEL, W. B., SAVILLE, D. A., AND SCHOWALTER, W. R. 1990 *Colloidal Dispersions*, Cambridge.
SOO, S. L. 1967 *Fluid dynamics of Multiphase Systems*, Blaisdell.
SOO, S. L. 1989 *Particulate and Continuum – Multiphase Fluid Mechanics*, Hemisphere.
UNGARISH, M. 1993 *Hydrodynamics of Suspensions*, Springer.

VAN DE HULST, H. C. 1957 *Light Scattering by Small Particles*, Wiley.
YOUNG, F. R. 1989 *Cavitation*, McGraw-Hill.

REVIEW ARTICLES

BATCHELOR, G. K. 1974 Transport properties of two phase materials with random structure, *Ann. Rev. Fluid Mech.* **6**, 227–255.

BATCHELOR, G. K. 1976 Developments in microhydrodynamics, in *Theoretical and Applied Mechanics*, Proceedings of the 14th IUTAM Congress, Delft, The Netherlands, Koiter, W. T., Ed., North-Holland.

BRADY, J. F. 1988 Stokesian dynamics, *Ann. Rev. Fluid Mech.* **20**, 111–157.

BRENNER, H. 1970 Rheology of two-phase systems, *Ann. Rev. Fluid Mech.* **2**, 137–176.

CLIFT, R. AND GAUVIN, W. H. 1971 Motion of entrained particles in gas streams, *Can. J. Chem. Eng.* **19**, 439–448.

DAVIS, R. H. AND ACRIVOS, A. 1985 Sedimentation of non-colloidal particles at low Reynolds numbers, *Ann. Rev. Fluid. Mech.* **17**, 91–118.

DREW, D. A. 1983 Mathematical modeling of two-phase flow, *Ann. Rev. Fluid Mech.* **15**, 261–291.

HERCZYNSKI, R. AND PIENKOWSKA, I. 1980 Toward a statistical theory of suspension, *Ann. Rev. Fluid Mech.* **12**, 237–269.

HUGHES, R. R. AND GILLILAND, E. R. 1952 The mechanics of drops, *Chem. Eng. Prog.*, **48**, 497–504.

IGRA, O. AND BEN-DOR, O. 1988 Dusty shock waves, *Appl. Mech. Revs.* **41**, 379–437.

LANE, W. R. AND GREEN, H. L. 1956 The mechanics of drops and bubbles, in *Surveys in Mechanics, The G. I. Taylor 70th Anniversary Volume*, Batchelor, G. K. and Davies, R. M., Eds., Cambridge.

LAVENDA, B. H. 1985 Brownian motion, *Sci. Amer.* **252**, 70–85.

LEAL, L. G. 1980 Particle motions in a viscous fluid, *Ann. Rev. Fluid Mech.* **12**, 435–476.

LOHSE, D. 2003 Bubble puzzles, *Phys. Today* **56**, 36–41.

MARBLE, F. E. 1970 Dusty gases, *Ann. Rev. Fluid Mech.* **2**, 397–448.

PLESSET, M. S. AND PROSPERETTI, A. 1977 Bubble dynamics and cavitation, *Ann. Rev. Fluid Mech.* **9**, 145–185.

RUDINGER, G. 1973 Wave propagation in suspensions of solid particles in gas flow, *Appl. Mech. Revs.* **26**, 273–279.

RUSSEL, W. B. 1981 Brownian motion of small particles suspended in liquids, *Ann. Rev. Fluid Mech.* **13**, 425–455.

SPIELMAN, L. A. 1977 Particle capture from low-speed laminar flows, *Ann. Rev. Fluid Mech.* **9**, 297–319.

TOROBIN, L. B. AND GAUVIN, W. H. 1959 Fundamentals of solid-gas flow. III. Accelerated motion of a particle in a fluid, *Can. J. Chem. Eng.* **38**, 224–236.

WEINBAUM, S., GANATOS, P., AND YAN, Z. Y. 1990 Numerical multipole and boundary integral equation techniques in Stokes flow, *Ann. Rev. Fluid Mech.* **22**, 275–316.

WIJNGAARDEN, L. VAN. 1972 One dimensional flow of liquids containing small gas bubbles, *Ann. Rev. Fluid Mech.* **4**, 369–396.

RESEARCH ARTICLES

Acoustics of Suspensions

General suspensions

ALLEGRA, J. R. AND HAWLEY, S. A. 1972 Attenuation of sound in suspensions and emulsions: Theory and experiments, *J. Acoust. Soc. Amer.* **51**, 1545–1564.

CHAMBRE, P. L. 1954 Speed of a plane wave in a gross mixture, *J. Acoust. Soc. Amer.* **26**, 329–331.

CHOW, J. C. F. 1964 Attenuation of acoustic waves in dilute emulsions and suspensions, *J. Acoust. Soc. Am.* **36**, 2395–2401.

EPSTEIN, P. S. 1941 On the absorption of sound by suspensions and emulsions, in *Contributions to Applied Mechanics, Theodore von Karman Anniversary Volume*, California Institute of Technology, pp. 162–188.

EPSTEIN, P. S. AND CARHART, R. R. 1953 The absorption of sound in suspensions and emulsions, I. Water fog in air, *J. Acoust. Soc. Amer.* **25**, 553–565.

HAY, A. E. AND SCHAAFSAMA, A. S. 1989 Resonance scattering in suspensions, *J. Acoust. Soc. Amer.* **85**, 1124–1138.

MORFEY, C. L. 1968 Sound attenuation by small particles in a fluid, *J. Sound Vib.* **8**, 156–170.

TEMKIN, S. 1992 Sound speeds in suspensions in thermodynamic equilibrium, *Phys. Fluids A* **4**, 2399–2409.

TEMKIN, S. 1996 Viscous attenuation of sound in dilute suspensions of rigid particles, *J. Acoust. Soc. Amer.* **100**, 825–831.

TEMKIN, S. 1998 Sound propagation in dilute suspensions of rigid particles, *J. Acoust. Soc. Amer.* **103**, 838–849.

TEMKIN, S. 2000 Attenuation and dispersion of sound in dilute suspensions of spherical particles, *J. Acoust. Soc. Amer.* **108**, 126–146.

TEMKIN, S. 2002 Erratum: Attenuation and dispersion of sound in dilute suspensions of spherical particles, *J. Acoust. Soc. Amer.* 111, 1126–1128.

Aerosols

CHU, B. T. 1960 Thermodynamics of a dusty gas and its applications to some aspects of wave propagation in the gas, Brown University, Division of Engineering Report No. DA-4761/1.

COLE, J., III AND DOBBINS, R. A. 1970 Propagation of sound through atmospheric fog, *J. Atmos. Sci.* **27**, 426–434.

COLE, J. E., III AND DOBBINS, R. A. 1971 Measurements of the attenuation of sound in a warm air fog, *J. Atmos. Sci.* **28**, 202–209.

DAVIDSON, G. A. AND SCOTT, D. S. 1973 Finite-amplitude acoustics of aerosols, *J. Acoust. Soc. Amer.* **53**, 1717–1729.

DAVIDSON, G. A. AND SCOTT, D. S. 1973 Finite-amplitude acoustic phenomena in aerosols from a single governing equation, *J. Acoust. Soc. Amer.* **54**, 1331–1342.

FONER, S. N. AND NALL, B. H. 1975 Attenuation of sound by rigid spheres: Measurements of the viscous and thermal components of attenuation and comparison with theory, *J. Acoust. Soc. Amer.* **57**, 59–66.

GUMEROV, N. A., IVANADEV. A. I., AND NIGMATULIN, R. I. 1988 Sound waves in monodisperse gas-particle or vapour droplet mixtures, *J. Fluid Mech.* **193**, 53–74.

HENLEY, D. C. AND HOIDALE, G. B. 1973 Attenuation and dispersion of acoustic energy by atmospheric dust, *J. Acoust. Soc. Amer.* **54**, 437–445.

ISHII, R. AND MATSUHISA, H. 1983 Steady reflection, absorption and transmission of small disturbances by a screen of dusty gas, *J. Fluid Mech.* **130**, 259–277.

KNUDSEN, V. O., WILSON, J. V., AND ANDERSON, N. S. 1948 The attenuation of audible sound in fogs and smokes, *J. Acoust. Soc. Amer.* **20**, 849–857.

MARBLE, F. E. AND WOOTEN, D. C. 1970 Sound attenuation in a condensing vapor, *Phys. Fluids.* **13**, 2657–2664.

MARGULIES, T. S. AND SCHWARZ, W. H. 1994 A multiphase continuum theory for sound wave propagation through dilute suspensions of particles, *J. Acoust. Soc. Amer.* **96**, 319–331.

SEWELL, C. J. T. 1910 On the extinction of sound in a viscous atmosphere by small obstacles of cylindrical and spherical form, *Phil. Trans. R. Soc. Lond.* **A210**, 239–270.

TEMKIN, S. 1993 Particle force and heat transfer in a dusty gas sustaining an acoustic wave, *Phys. Fluids* **5**, 1296–1304.

TEMKIN, S. AND DOBBINS, R. A. 1966a Attenuation and dispersion of sound by particulate relaxation processes, *J. Acoust. Soc. Amer.* **40**, 317–324.

TEMKIN, S. AND DOBBINS, R. A. 1966b Measurement of the attenuation and dispersion of sound by an aerosol, *J. Acoust. Soc. Amer.* **40**, 1016–1024.

ZINK, J. W. AND DELSASSO, L. P. 1958 Attenuation and dispersion of sound by solid particles suspended in a gas, *J. Acoust. Soc. Amer.* **30**, 765–771.

Bubbly liquids

BATCHELOR, G. K. 1967 Compression waves in a suspension of gas bubbles in liquid, *Fluid Dynamics Transactions* **4**, 425–445. Institute of Fundamental Technical Research, Polish Academy of Science, Warsaw.

CAFLISCH, R. E., MIKSIS, M. J., PAPANICOLAOU, G. C., AND TING, L. 1985 Wave propagation in liquids at finite volume fraction, *J. Fluid Mech.* **160**, 1–14.

CARSTENSEN, E. L. AND FOLDY, L. L. 1947 Propagation of sound through a liquid containing bubbles, *J. Acoust. Soc. Amer.* **19**, 481–501.

CHEYNE, S. A., STEBBINGS, C. T., AND ROY, R. A. 1995 Phase velocity measurements in bubbly liquids using a fiber optic laser interferometer, *J. Acoust. Soc. Amer.* **97**, 1621–1624.

COMMANDER, K. W. AND PROSPERETTI, A. 1989 Linear pressure waves in bubbly liquids: Comparison between theory and experiments, *J. Acoust. Soc. Amer.* **85**, 732–746.

CRESPO, A. 1969 Sound and shock waves in liquids containing bubbles, *Phys. Fluids* **12**, 2274–2282.

DRUMHELLER, D. S., KIPP, M. E., AND BEDFORD, A. 1982 Transient wave propagation in bubbly liquids, *J. Fluid Mech.* **119**, 347–365.

ELLER, A. I. 1970 Damping constants of pulsating bubbles, *J. Acoust. Soc. Amer.* **47**, 1469–1470.

FLOYD, E. R. 1981 Thermodynamic corrections to the velocity of propagation in a bubbly medium, *J. Acoust. Soc. Amer.* **70**, 1748–1751.

FOX, F. E., CURLEY, S. R., AND LARSON, G. S. 1955 Phase velocity and absorption measurements in water containing air bubbles, *J. Acoust. Soc. Amer.* **27**, 534–539.

GAUNAURD, G. C. AND H. ÜBERALL, H. 1981 Resonance theory of bubbly liquids, *J. Acoust. Soc. Amer.* **69**, 362–370.

HSIEH, D. Y. AND PLESSET, M. S. 1961 On the propagation of sound in liquids containing gas bubbles, *Phys. Fluids* **4**, 970–975.

KARPLUS, H. B. 1961 The velocity of sound in liquids containing gas bubbles, Research and Development/Report ARF-4132-12, Atomic Energy Commission.

KARPOV, S., PROPSERETTI, A., AND OSTROVSKY, L. 2003 Nonlinear wave interaction in bubbly layers, *J. Acoust. Soc. Amer.* **113**, 1304–1316.

KIEFFER, S. 1977 Sound speed in liquid-gas mixtures: Water-air and water-steam, *J. Geophys. Res.* **82**, 2895–2904.

MIKSIS, M. J. AND TING, L. 1989 Effects of bubbly layers on wave propagation, *J. Acoust. Soc. Amer.* **86**, 2349–2358.

NICHOLAS, M., ROY, R. A., CRUM, L. A., OGUZ, H., AND PROSPERETTI, A. 1994 Sound emission by a laboratory bubble cloud, *J. Acoust. Soc. Amer.* **95**, 3171–3181.

NIGMATULIN, R. I., KHABEEV, N. S., AND HAI, Z. N. 1988 Waves in liquids with vapour bubbles, *J. Fluid Mech.* **186**, 85–117.

NOORDZIJ, L. AND WIJNGAARDEN, L. VAN. 1974 Relaxation effects, caused by relative motion, on shock waves in gas-bubble/liquid mixtures, *J. Fluid Mech.* **66**, 115–143.

SANGANI, A. S. 1991 A pairwise interaction theory for determining the linear acoustic properties of dilute bubbly liquids, *J. Fluid Mech.* **232**, 221–284.

SILBERMAN, E. 1957 Sound velocity and attenuation in bubbly mixtures measured in standing wave tubes, *J. Acoust. Soc. Amer.* **29**, 925–933.

TEMKIN, S. 1990 Attenuation and dispersion of sound in bubbly liquids via the Kramers-Kronig relations, *J. Fluid Mech.* **211**, 61–72.

Emulsions

EVANS, J. M. AND ATTENBOROUGH, K. 1997 Coupled phase theory for sound propagation in emulsions, *J. Acoust. Soc. Amer.* **102**, 278–282.

FUKUMOTO, Y. AND T. IZUYAMA, T. 1992 Thermal attenuation and dispersion of sound in a periodic emulsion, *Phys. Rev. A* **46**, 4905–4921.

HEMAR, Y., HOCQUART, R., AND PALIERNE, J. F. 1998 Frequency-dependent compressibility in emulsions: Probing interfaces using Isakovitch sound absorption, *Europhys. Lett.* **42**, 253–258.

ISAKOVITCH, M. A. 1948 On the propagation of sound in emulsions, *Zh. Exper. i Teor. Fiz.* **18**, 907–912.

McCLEMENTS, D. J. AND POVEY, M. J. W. 1989 Scattering of ultrasound by emulsions, *J. Phys. D: Appl. Phys.* **22**, 38–47.

ONUKI, A. 1991 Sound propagation in phase-separating fluids, *Phys. Rev. A* **43**, 6740–6755.

Hydrosols

ATKINSON, C. M. AND KYTÖMAA, H. K. 1992 Acoustic wave speed and attenuation in suspensions, *Int. J. Multiphase Flow* **18**, 577–592.

GIBSON, R. L. AND TOKSÖZ. M. N. 1989 Viscous attenuation of acoustic waves in suspensions, *J. Acoust. Soc. Amer.* **85**, 1925–1934.

HARKER, A. H. AND TEMPLE, J. A. G. 1988 Velocity and attenuation of ultrasound in suspensions of particles in fluids, *J. Phys. D* **21**, 1576–1588.

HAY, A. E. AND MERCER, D. G. 1985 On the theory of sound scattering and viscous absorption in aqueous suspensions at medium and short wavelengths, *J. Acoust. Soc. Amer.* **78**, 1761–1771.

HERZFELD, K. F. 1930 Propagation of sound in suspensions, *Phil. Mag.* **9** (7th Series), 752–768.

HOLMES, A. K. AND CHALLIS, R. E. 1993 A wide bandwidth study of ultrasound velocity and attenuation in suspensions: Comparison of theory and experimental measurements, *J. Colloid. Interface Sci.* **156**, 261–268.

HOLMES, A. K., CHALLIS, R. E., AND WEDLOCK, D. J. 1994 A wide bandwidth ultrasonic study of suspensions: The variation of velocity and attenuation with particle size, *J. Colloid. Interface Sci.* **168**, 339–348.

URICK, R. J. 1947 A sound velocity method for determining the compressibility of finely divided substances, *J. Appl. Phys.* **18**, 983–987.

URICK, R. J. 1948 The absorption of sound of irregular particles, *J. Acoust. Soc. Amer.* **20**, 283–289.

URICK, R. J. AND AMENT, W. S. 1948 The propagation of sound in composite media, *J. Acoust. Soc. Amer.* **21**, 115–119.

Breakup, Collision, and Coalescence

ANILKUMAR, A.V., LEE, C. P., AND WANG, T. G. 1993 Stability of an acoustically levitated and flattened drop: An experimental study, *Phys. Fluids* **5**, 2763–2774.

ASHGRIZ, N. AND POO, J. Y. 1990 Coalescence and separation in binary collisions of liquid drops, *J. Fluid Mech.* **221**, 183–204.

BEAL, S. K. 1972 Turbulent agglomeration of suspensions, *J. Aerosol Sci.,* **3**, 113–125.

BLANCHARD, D. C. 1954 Bursting of bubbles at an air-water interface, *Nature* **173**, 1048.

BRADLEY, S. G. AND STOW, C. D. 1978 Collision between liquid drops, *Phil. Trans. R. Soc. Lond. A* **287**, 635–678.

BRENN, G. AND FROHN, A. 1989 Collision and merging of two equal spheres of propanol, *Exp. Fluids* **7**, 441–446.

CROWE, C. T. AND WILLOUGHBY, P. G. 1966 A mechanism for particle growth in a rocket nozzle, *AIAA J.* **4**, 1677–1678.

DANILOV, S. D. AND MIRNOV, M. A. 1992 Breakup of a droplet in a high-intensity sound field, *J. Acoust. Soc. Amer.* **92**, 2747–2755.

FALKOVICH, G., FOUXON, A., AND STEPANOV, M. G. 2002 Acceleration of rain initiation by cloud turbulence, *Nature* **419**, 151–154.

GAST, L. 1991 Capillary instability of a liquid ring. Ph.D. thesis, Rutgers University.

GELBARD, F., MONDY, L. A., AND OHRT, S. E. 1991 A new method for determining hydrodynamic effects on the collision of two spheres, *J. Statist. Phys.* **62**, 945–960, 1991.

GELBARD, F. AND SEINFELD, J. 1980 Simulation of multicomponent aerosol dynamics, *J. Colloid Interface Sci.* **78**, 485–501.

HOFFMANN, T. L. AND KOOPMANN, G. H. 1997 Visualization of acoustic particle interaction and agglomeration: Theory and experiments, *J. Acoust. Soc. Amer.* **101**, 3421–3429.

KLETT, J. D. AND DAVIS, M. H. 1973 Theoretical collision efficiencies of cloud droplets at small Reynolds numbers, *J. Atmos. Sci.* **30**, 107–117.

MARBLE, F. E. 1964 Mechanism of particle collision in the one-dimensional dynamics of gas-particle mixtures, *Phys. Fluids* **7**, 1270–1282.

MARBLE, F. E. 1967 Droplet agglomeration in rocket nozzles caused by particle slip and collision, *Astronautica Acta*, **13**, 159–166.

MEGARIDIS, C. M. AND DOBBINS, R. A. 1990 A bimodal integral solution of the dynamic equation for an aerosol undergoing simultaneous particle inception and coagulation, *Aerosol Sci. Technol.* **12**, 240–255.

MEHTA, H. 1980 Droplet drag and breakup in shock-wave induced unsteady flows. Ph.D. thesis, Rutgers University.

PINSKY, M., KHAIN, A., AND SHAPIRO, M. 2001 Collision efficiency of drops in a wide range of Reynolds numbers, *J. Atmos. Sci.* **58**, 742–766.

REICHMAN, J. 1973 A study of the motion, deformation and breakup of accelerating water droplets. Ph.D. thesis, Rutgers University.

REICHMAN, J. M. AND TEMKIN, S. 1974 A study of the deformation and breakup of accelerating water droplets, *Proceedings of the International Colloquim on Drops and Bubbles*, California Institute of Technololgy, *Jet Propul. Lab.* **2**, 446–464.

SAFFMAN, P. G. AND TURNER, J. S. 1956 On the collision of drops in turbulent clouds, *J. Fluid Mech.* **1**, 16–30, 1956. (Corrigendum, *J. Fluid Mech.* **196**, 599, 1988.)

SCOTT, D. S. 1975 A new approach to the acoustic conditioning of industrial aerosol emissions, *J. Sound Vib.* **43**, 607–619.

SHAW, D. T. AND TU, K. W. 1979 Acoustic particle agglomeration due to hydrodynamic interaction between monodisperse aerosols, *J. Aerosol Sci.* **10**, 317–328.

SHUSTER, K., FICHMAN, M., GOLDSHTEIN, A., AND GUTFINGER, C. 2002 Agglomeration of submicrometer particle in weak periodic shock waves, *Phys. Fluids* **14**, 1802–1805.

SZUMOWSKI, A. P. AND FALKOWSKI, K. 1995 Photographic study of the shock-induced dispersion of microscopic gas bubbles, *Phys. Fluids* **7**, 2529–2531.

TELFORD, J. W., THORNDIKE, N. S., AND BOWEN, E. G. 1955 The coalescence of small water drops, *Q. J. R. Met. Soc.* **81**, 241–250.

TEMKIN, S. 1969 Cloud droplet collision induced by thunder, *J. Atmos. Sci.* **26**, 776.

TEMKIN, S. 1970 Droplet agglomeration induced by weak shock waves, *Phys. Fluids* **13**, 1639–1641.

TEMKIN, S. 1994 Gasdynamic agglomeration of aerosols. I. Acoustic waves, *Phys. Fluids* **6**, 2294–2303.

WAGNER, P. E. AND KERKER, M. 1977 Brownian coagulation of aerosols in rarefied gases, *J. Chem. Phys.* **66**, 638–646.

WARREN, D. AND SEINFELD, J. H. 1985 Simulations of aerosol size distribution evolution in systems with simultaneous nucleation, condensation and coagulation, *Aerosol Sci. Technol.* **4**, 31–43.

WHELPDALE, D. M. AND LIST, R. 1971 The coalescence process in raindrop growth, *J. Geophys. Res.* **76**, 2386–2856.

ZHANG, X. AND DAVIS, R. H. 1991 The rate of collisions due to Brownian or gravitational motion of small drops, *J. Fluid Mech.* **230**, 479–504.

Bubble and Droplet Formation

ADAM, J. R., CATANEO, R., AND SEMONIN, R. G. 1971 The production of equal and unequal size droplet pairs, *Rev. Sci. Instr.* **42**, 1847–1849.

ATCHLEY, A. AND PROSPERETTI, A. 1989 The crevice model of bubble nucleation, *J. Acoust. Soc. Amer.* **86**, 1065–1084.

BASSETT, J. D. AND BRIGHT, A. W. 1976 Observations concerning the mechanism of atomization in an ultrasonic fountain, *J. Aerosol Sci.* **7**, 47–51.

BRENN, G. AND LACKERMEIER, U. 1997 Drop formation from a vibrating orifice generator driven by modulated electrical signals, *Phys. Fluids* **9**, 3658–3669.

CHAUDHARY, K. C. AND MAXWORTHY, T. 1980 The nonlinear instability of a liquid jet. Part 3. Experiments on satellite drop formation and control, *J. Fluid Mech.* **96**, 287–297.

CLARK, C. J. AND DOMBROWSKI, N. 1971 On the formation of drops from the rims of fan spray sheets, *J. Aerosol Sci.* **3**, 173–183.

DEANE, G. B. AND STOKES, M. D. 2002 Scale dependence of bubble creation mechanisms in breaking waves, *Nature* **418**, 839–844.

DIJKSMAN, J. F. 1984 Hydrodynamics of small tubular pumps, *J. Fluid Mech.* **139**, 173–191.

DOMBROWSKI, N. AND NEALE, N. D. 1974 Formation of streams of uniform drops from fan spray pressure nozzles, *J. Aerosol Sci.* **5**, 551–555.

DONNELLY, R. J. AND GLABERSON, W. 1966 Experiments on the capillary instability of a liquid jet, *Proc. R. Soc. A* **290**, 547–556.

HIBLING, J. H. AND HEISTER, S. D. 1996 Droplet size control in liquid jet breakup, *Phys. Fluids* **8**, 1574–1581.

LONGUET-HIGGINS, M. S., KERMAN, B. R., AND LUNDE, K. 1991 The release of air bubbles from an underwater nozzle, *J. Fluid Mech.* **230**, 365–390.

ORME, M. 1991 On the genesis of droplet microspeed dispersions, *Phys. Fluids A* **3**, 2936–2947.

PANDIT, A. B. AND DAVIDSON, A. B. 1990 Hydrodynamics of the rupture of thin liquid films, *J. Fluid Mech.* **212**, 11–24.

PHILIPSON, K. 1973 On the production of monodisperse particles with a spinning disc, *Aerosol Sci.* **4**, 51–57.

THORPE, S. A. 1987 Internal waves and whitecaps, *Nature* **330** 740–742.

THORPE, S. A. AND HUMPHRIES, P. N. 1980 Bubbles and breaking waves, *Nature* **283**, 463–465.

TOPP, M. N. 1973 Ultrasonic atomization – A photographic study of the mechanism of disintegration, *Aerosol Sci.* **4**, 17–25.

Cavitation and Sonoluminescence

BARBER, B. P. AND PUTTERMAN, S. J. 1991 Observation of synchronous picosecond sonoluminescence, *Nature* **352**, 318–320.

CRUM, L. A. 1994 Sonoluminescence, sonochemistry, and sonophysics, *J. Acoust. Soc. Amer.* **95**, 559–562.

DIDENKO, Y. T., MCNAMARA, W. B., III, AND SUSLIK, K. S. 2000 Molecular emission from single-bubble sonoluminescence, *Nature* **407**, 877–879.

DIDENKO, Y. T. AND SUSLIK, K. S. 2002 The energy efficiency of formation of photons, radicals and ions during single-bubble cavitation, *Nature* **418**, 394–397.

FLYNN, H. G. 1975 Cavitation dynamics. I. A mathematical formulation, *J. Acoust. Soc. Amer.* **57**, 1379–1396.

FLYNN, H. G. 1975 Cavitation dynamics. II. Free pulsations and models for cavitation bubbles, *J. Acoust. Soc. Amer.* **58**, 1160–1170.

FLYNN, H. G. AND CHURCH, C. C. 1988 Transient pulsations of small gas bubbles in water, *J. Acoust. Soc. Amer.* **84**, 985–998. [Erratum, *J. Acoust. Soc. Amer.* **84**, 1863–1876.]

LÖFSTEDT, R., BARBER, B. P., AND PUTTERMAN, S. J. 1991 Toward a hydrodynamic theory of sonoluminescence, *Phys. Fluids* **A 5**, 2911–2927.

NEPPIRAS, E. A. 1980 Acoustic cavitation, *Phys. Rep.* (Review Section of Physics Letters) **61**, 159–251.

PROSPERETTI, A. 1986 Physics of acoustic cavitation in *Frontiers of Physical Acoustics*, Sette, D., Ed., North-Holland, pp. 145–188.

PUTTERMAN, S. J. 1995 Sonoluminescence: Sound into light. *Sci. Amer.*, **272**, 46–51.

SUSLICK, K. S. 1989 The chemical effects of ultrasound, *Sci. Amer.*, **260**, 80–86.

Evaporation and Condensation

DUKOWICZ, J. K. 1984 Drag of evaporating or condensing droplets in low Reynolds number flow, *Phys. Fluids* **27**, 1351–1358.

GOOSENS, H. W. J., CLEIJNE, J. W., SMOLDERS, H. J., AND van DONGEN, M. E. 1988 Shock wave induced evaporation of water droplets in a gas-droplet mixture, *Exp. Fluids* **6**, 561–568.

HILL, P. E. 1966 Condensation of water vapour during supersonic expansion in nozzles, *J. Fluid Mech.* **25**, 593–620.

MARBLE, F. E. 1969 Some gasdynamics problems in the flow of condensing vapors, *Astronaut. Acta* **14**, 585–614.

MARSTON, P. L. 1979 Evaporation-condensation resonance frequency of oscillating vapor bubbles, *J. Acoust. Soc. Amer.* **66**, 1516–1521.

MOUZURKEWICH, M. 1986 Aerosol growth and the condensation coefficient for water: A review, *Aerosol Sci. Technol.* **5**, 223–236.

OKUYAMA, M. AND ZUNG, J. T. 1967 Evaporation-condensation coefficient for small droplets, *J. Chem. Phys.* **46**, 1580–1585.

Heat Transfer

FRANKIEL, N. A. AND ACRIVOS, A. 1968 Heat and mass transfer from small spheres and cylinders freely suspended in shear flow, *Phys. Fluids* **11**, 1913–1918.

MICHAELIDES, E. E. AND FENG, Z. 1994 Heat transfer from a rigid sphere in a nonuniform flow and temperature field, *Int. J. Heat Mass Transfer* **37**, 2069–2076.

PARKS, J. M., ABLOW, C. M., AND WISE, H. 1966 Temperature distribution within a particle during evaporation, *AIAA J.* **4**, 1032–1035.

Miscellaneous

BORMAN, S. AND CURTIUS, J. 2002 Lasing on a cloudy afternoon, *Nature* **418**, 826–827.

CHING, B., GOLAY, M. W., AND JOHNSON, T. J. 1984 Droplet impacts upon liquid surfaces, *Science* **226**, 535–537.

DIEBOLD, G. J. AND WESTERVELT, P. J. 1988 The photoacoustic effect generated by a spherical droplet in a fluid, *J. Acoust. Soc. Amer.* **84**, 2245–2251.

DREYFUSS, D. AND TEMKIN, S. 1983 Charge separation during rupture of small water drops in transient flows: Shock tube measurements and application to lightning, *J. Geophys. Res.* **88**, 10,993–10,998.

FRANZ, G. J. 1959 Splashes as sources of sound, *J. Acoust. Soc. Amer.* **31**, 1080–1096.

GREENSPAN, M. 1956 Propagation of sound in five monatomic gases, *J. Acoust. Soc. Amer.* **28**, 644–648.

LIKHTEROV, L. 1998 High-frequency acoustic noise emitted by initial impact of solid sphere falling onto a liquid surface, *Phys. Fluids* **10**, 321–323.

LONGUET-HIGGINS, M. S. 1990 An analytical model of sound production by raindrops, *J. Fluid Mech.* **214**, 395–410.

O'BRIEN, R. W. 1988 Electro-acoustic effects in a dilute suspension of spherical particles, *J. Fluid Mech.* **190**, 71–86.

PENNER, S. S. AND LI, T. 1967 Thermodynamic considerations of droplets and bubbles, *AIAA J.* **5**, 1528–1529.

PUMPHREY, H. C., CRUM, L., AND BJØRNØ, L. 1989 Underwater sound produced by individual drop impacts and rainfall, *J. Acoust. Soc. Amer.* **85**, 1518–1526.

RICHARD, D., CLANET, C., AND QUÉRE, D. 2002 Contact time of a bouncing drop, *Nature* **417**, 811.

ROY, R., DAVIDSON, J. F., AND TUPONOGOV, V. G. 1990 The velocity of sound in fluidized beds, *Chem. Eng. Sci.* **45**, 3233–3245.

SHAFER, N. E. AND ZARE, R. N. 1991 Through a beer glass darkly, *Phys. Today* **44**, 48–52.

SHIRLEY, B. M., SHELDON, J. W., AND KRANC, S. C. 1972 Particle charging behind shock waves in suspensions, *AIAA J.* **10**, 1110–1111.

Model Equations

BIESHEUVEL, A. AND VAN WIJNGAARDEN L. 1984 Two-phase equations for a dilute dispersion of gas bubbles in a liquid, *J. Fluid Mech.* **148**, 301–318.

BILICKI, Z. AND KESTIN, J. 1990 Physical aspects of the relaxation model in two-phase flows, *Proc. R. Soc. Lond. A* **428**, 379–397.

CAFLISCH, R. E., MIKSIS, M. J., PAPANICOLAOU, G. C., AND TING, L. 1985 Effective equations for wave propagation in bubbly liquids, *J. Fluid Mech.* **153**, 259–273.

DRUMHELLER, D. S. AND BEDFORD, A. 1979 A theory of bubbly liquids, *J. Acoust. Soc. Amer.* **66**, 197–208.

DRUMHELLER, D. S. AND BEDFORD, A. 1980 A theory of liquids with vapor bubbles, *J. Acoust. Soc. Amer.* **67**, 186–200.

KRAIKO, A. N. AND STERNIN, L. E. 1965 Theory of flows of a two velocity continuous medium containing solid or liquid particles, *Prikladnaia Matematika i Mekhanica* **29**, 418–429.

MARBLE, F. E. 1963 Dynamics of a gas containing small solid particles, in *Proceedings of the 5th AGARD Colloquim on Combustion and Propulsion*, Braunschweig, Pergamon, pp. 175–215.

MIKSIS, M. J. AND TING, L. 1992 Effective equations of multiphase flows – Waves in a bubbly liquid, *Adv. Appl. Mech.* **28**, 141–261.

NIGMATULIN, R. I. 1982 Mathematical modeling of bubbly liquid motion and hydrodynamical effects in wave propagation phenomenon, *Appl. Sci. Res.* **38**, 267–289.

O'BRIEN, R. W. 1990 The electroacoustic equations for a colloidal suspension, *J. Fluid Mech.* **212**, 81–93.

PANTON, R. 1968 Flow properties for the continuum viewpoint of a non-equilibrium gas-particle mixture, *J. Fluid Mech.* **31**, 273–303.

WIJNGAARDEN, L. VAN. 1968 On the equations of motion for mixtures of liquid and gas bubbles, *J. Fluid Mech.* **33**, 465–474.

WIJNGAARDEN, L. VAN. 1976 Some problems in the formulation of the equations for gas/liquid flows, in *Theoretical and Applied Mechanics*, Proceedings of the 14th IUTAM Congress, Delft, The Netherlands, Koiter, W. T., Ed., North-Holland., pp. 249–260.

YURKOVETSKY, Y. AND BRADY, J. F. 1996 Statistical mechanics of bubbly liquids, *Phys. Fluids* **8**, 881–895.

ZHANG, D. Z. AND PROSPERETTI, A. 1994 Ensemble phase-averaged equations for bubbly flows, *Phys. Fluids* **6**, 2956–2970.

Numerical Methods

CHANG, E. R. AND MARTIN, M. R. 1994 Unsteady flow about a sphere at low to moderate Reynolds number. Part 1. Oscillatory motion, *J. Fluid Mech.* **277**, 347–379.

CROWE, C. T. 1982 Numerical models for dilute gas-particle flows, *J. Fluids Eng.* **104**, 297–303.

ISHII, R., UMEDA, Y., AND YUHI, M. 1989 Numerical analysis of gas-particle two-phase flows, *J. Fluid Mech.* **203**, 475–515.

KAMATH, V. AND PROSPERETTI, A. 1989 Numerical integration methods in gas-bubble dynamics, *J. Acoust. Soc. Amer.* **85**, 1538–1548.

Particle Interactions

BATCHELOR, G. K. AND GREEN, J. T. 1972 The hydrodynamic interaction of two small freely-moving spheres in a linear flow field, *J. Fluid Mech.* **56**, 375–400.

BRADY, J. F., PHILLIPS, R. J., LESTER, J. C, AND BOSSIS, G. 1988 Dynamic simulation of hydro-dynamically interacting suspensions, *J. Fluid Mech.* **195**, 257–280.

ECKER, G. 1985 Interactions of droplet pairs in weak shock-wave induced unsteady flows. Ph.D. thesis, Rutgers University.

GOREN, S. 1983 Resistance and stability of a line of particles moving near a wall, *J. Fluid Mech.* **132**, 185–196.

HABER, S. G., HETSRONI, G., AND SOLAN, A. 1973 On the low Reynolds number motion of two droplets, *Int. J. Multiphase Flow* **1**, 57–71.

KANEDA, Y. AND ISHII, K. 1982 The hydrodynamic interaction of two spheres moving in an unbounded fluid at small but finite Reynolds number, *J. Fluid Mech.* **124**, 209–217.

MAZUR, P. AND SAARLOOS, W. VAN. 1982 Many-sphere interactions and mobilities in a suspension, *Phys. A* **115**, 21–57.

MAZUR, P. AND SAARLOOS, W. VAN. 1983 Many-sphere hydrodynamic interactions. II. Mobilities at finite frequency, *Phys. A* **120** 77–102.

STEINBERGER, E. H., PRUPPACHER, H. R., AND NEIBURGER, M. 1968 On the hydrodynamics of pairs of spheres falling along their line of centers in a viscous medium, *J. Fluid Mech.* **34**, 809–819.

STIMSON, M. AND JEFFREY, G. B. 1926 The motion of two spheres in a viscous fluid, *Proc. R. Soc. Lond. A* **111**, 110–116.

TAM, C. K. W. 1969 The drag on a cloud of spherical particles in low Reynolds number flow, *J. Fluid Mech.* **38**, 537–546.

TEMKIN, S. AND G. ECKER, G. Z. 1989 Droplet interactions in a shock-wave flow field, *J. Fluid Mech.* **202**, 467–497.

WIJNGAARDEN, L. VAN. 1976 Hydrodynamic interaction between gas bubbles in liquid (with an appendix by D. J. Jeffrey), *J. Fluid Mech.* **77**, 27–44, 1976.

Sedimentation

AUZERAIS, F. M., JACKSON, R., RUSSEL, W. B., AND MURPHY, W. F. 1990 The transient settling of stable and flocculated dispersions, *J. Fluid Mech.* **221**, 613–639.

BATCHELOR, G. K. 1972 Sedimentation in a dilute dispersion of spheres, *J. Fluid Mech.* **52**, 245–268.

BRADY, J. F. AND DURLOFSKY, L. J. 1988 The sedimentation rate of disordered suspensions, *Phys. Fluids* **31**, 717–727.

HABER, S. AND HETSRONI, G. 1981 Sedimentation in a dilute dispersion of small drops of various sizes, *J. Colloid Interface Sci.* **79**, 56–75.

HANRATTY, T. J. AND BANDUKWALA, A. 1957 Fluidization and sedimentation of spherical parti-cles, *AIChE J.* **3**, 293–296.

Single-Particle Motions

Translational

ALEXEEV, A. AND GUTFINGER, C. 2003 Particle drift in a resonance tube – A numerical study, *J. Acoust. Soc. Amer.* **114**, 1357–1365.

BOYADZHIEV, L. 1973 On the movement of a spherical particle in a vertically oscillating liquid, *J. Fluid Mech.* **57**, 545–548.

BURROWS, F. M. 1983 Calculation of the primary trajectories of seeds and other particles in strong winds, *Proc. R. Soc. Lond. A* **389**, 15–66.

CHESTER, W. AND BREACH, D. R. (with an appendix by I. Proudman) 1969 On the flow past a sphere at low Reynolds number. *J. Fluid Mech.* **37**, 751–760.

CORRSIN, S. AND LUMLEY, J. 1956 On the equation of motion for a particle in turbulent fluid, *Appl. Sci. Res. Sec. A*, **6**, 114–116.

DANILOV, S. D. AND MIRNOV, M. A. 2000 Mean force on a small sphere in a sound field in a viscous fluid, *J. Acoust. Soc. Amer.* **107**, 143–153.

GALINDO, V. AND GERBETH, G. 1993 A note on the force on an accelerating spherical drop at low Reynolds number, *Phys. Fluids A* **5**, 3290–3292.

GOLDSTEIN, S. 1929 The steady flow of viscous fluid past a fixed spherical obstacle at small Reynolds numbers, *Proc. R. Soc. A* **123**, 216–235.

GUCKER, F. T. AND DOYLE, G. J. 1956 The amplitude of aerosol droplets in a sonic field, *J. Phys. Chem.* **60**, 989–996, 1956.

HOENIG, S. A. 1957 Acceleration of dust particles by shock waves, *J. Appl. Phys.* **28**, 1218–1219.

INGEBO, R. D. 1956 Drag coefficients for droplets and solid spheres in clouds accelerating in air streams, *NACA* Technical Note TN 3762.

ISHII, R. 1984 Motion of small particles in a gas flow, *Phys. Fluids* **27**, 33–41.

KANG, I. S. AND LEAL, L. G. 1988 The drag coefficient for a spherical bubble in a uniform streaming flow, *Phys. Fluids* **31**, 233–241.

KANWAL, R. P. 1964 Drag on an axially symmetric body vibrating slowly along its axis in a viscous fluid, *J. Fluid Mech.* **19**, 631–636.

KARANFILIAN, S. K. AND KOTAS, T. J. 1978 Drag on a sphere in unsteady motion in a liquid at rest, *J. Fluid Mech.* **87**, 85–96.

KIM, S. S. 1977 An experimental study of droplet reponse to weak shock waves. Ph.D. thesis, Rutgers University.

LEGENDRE, D. AND MAGNAUDET, J. 1997 A note on the lift force on a spherical bubble or drop in a low Reynolds number shear flow, *Phys. Fluids* **9**, 3572–3574.

LIU, D. Y., ANDERS, K., AND FROHN, A. 1988 Drag coefficients of single droplets moving in an infinite droplet chain on the axis of a tube, *Int. J. Multiphase Flow* **14**, 217–232.

LOVALENTI, P. M. AND BRADY, J. 1993 The force on a bubble, drop, or particle in arbitrary time-dependent motion at small Reynolds number, *Phys. Fluids A* **5**, 2104–2116.

MAGNAUDET, J., RIVERO, M., AND FABRE, J. 1995 Accelerated flows past a rigid sphere of a spherical bubble. Part 1. Steady straining flow, *J. Fluid Mech.* **284**, 97–135.

MAXEY, M. R. AND RILEY, J. J. 1983 Equations of motion for a small rigid sphere in nonuniform flow, *Phys. Fluids* **26**, 883–889.

MAXWORTHY, T. 1965 Accurate measurements of sphere drag at low Reynolds numbers, *J. Fluid Mech.* **23**, 369–372.

MEI, R., KLAUSNER, J. F., AND LAWRENCE, C. J. 1994 A note on the history force on spherical bubble at finite Reynolds number, *Phys. Fluids* **6**, 418–420.

MEI, R., LAWRENCE, C. J., AND ADRIAN, R. J. 1991 Unsteady drag on a sphere at finite Reynolds number with small fluctuations in the free-stream velocity, *J. Fluid Mech.* **233**, 613–631.

MORSI, S. A. AND ALEXANDER, A. J. 1972 An investigation of particle trajectories in two-phase flow systems, *J. Fluid Mech.* **55**, 193–208.

OCKENDON, J. R. 1968 The unsteady motion of a small sphere in a viscous fluid, *J. Fluid Mech.* **34**, 229–239.

ODAR, F. 1966 Verification of the proposed equation for calculation of the forces on a sphere accelerating in a viscous fluid, *J. Fluid Mech.* **25**, 591–592.

ODAR, F. AND HAMILTON, W. S. 1964 Forces on a sphere accelerating in a viscous fluid, *J. Fluid Mech.* **18**, 302–314.

OGDEN, T. L. AND JAYAWEERA, K. O. L. F. 1971 Drag coefficients of water droplets decelerating in air, *Q. R. Met. Soc.* **97**, 571–574.

PEARCEY, T. AND HILL, G. W. 1956 The accelerated motion of droplets and bubbles, *Austr. J. Phys.* **9**, 19–30.

PROUDMAN, I. AND PEARSON, R. R. A. 1957 Expansions at small Reynolds numbers for the flow past a sphere and circular cylinder, *J. Fluid Mech.* **2**, 237–262.

RALLISON, J. M. Note on the Faxen relations for a particle in Stokes flow, *J. Fluid Mech.* **88**, 529–533, 1978.

RUDINGER, G. 1974 Penetration of particles into a constant cross flow, *AIAA J.* **12**, 1138–1140.

SANO, T. 1981 Unsteady flow past a sphere at low Reynolds number, *J. Fluid Mech.* **112**, 433–441.

SELBERG, B. P. AND NICHOLLS, J. A. 1968 Drag coefficient of small spehrical particles, *AIAA J.* **6**, 401–408.

TAYLOR, K. J. 1976 Absolute measurement of acoustic particle velocity, *J. Acoust. Soc. Amer.* **59**, 691–694.

TEMKIN, S. 1972 On the response of a sphere to an acoustic pulse, *J. Fluid Mech.* **54**, 339–349.

TEMKIN, S. 1993 Particle force and heat transfer in a dusty gas sustaining an acoustic wave, *Phys. Fluids* **5**, 1296–1304.

TEMKIN, S. AND KIM, S. S. 1980 Droplet motion induced by weak shock waves, *J. Fluid Mech.* **96**, 133–157.

TEMKIN, S. AND LEUNG, C.-M. 1976 On the velocity of a rigid sphere in a sound wave, *J. Sound Vib.*, **49**, 75–92.

TEMKIN, S. AND MEHTA, H. K. 1982 Droplet drag in an accelerating and decelerating flow, *J. Fluid Mech.* **116**, 297–313.

THOMAS, P. J. 1992 On the influence of the Basset history force on the motion of a particle through a fluid, *Phys. Fluids A* **4**, 2090–2093.

VAINSHTEIN, P., FICHMAN, M., AND PNUELI, D. 1992 On the drift of aerosol particles in sonic fields, *J. Aerosol Sci.* **23**, 631–637.

Pulsational

CRUM, L. A. 1983 The polytropic exponent of gas contained within air bubbles pulsating in a liquid, *J. Acoust. Soc. Amer.* **73**, 116–120.

CRUM, L. A. AND PROSPERETTI, A. 1983 Nonlinear oscillations of gas bubbles in liquids: An interpretation of some experimental results, *J. Acoust. Soc. Amer.* **73**, 121–127.

DEVIN, C., Jr., 1959 Survey of thermal, radiation, and viscous damping of pulsating air-bubbles in water, *J. Acoust. Soc. Amer.* **31**, 1654–1667.

EXNER, M. L. AND HAMPE, W. 1953 Experimental determination of the damping of pulsating air bubbles in water, *Acustica* **3**, 67–72.

KENNARD, E. H. 1943 Radial motion of water surrounding a sphere of gas in relation to pressure waves, in *Underwater Explosion Research*, Vol. II. The Gas Globe, Office of Naval Research.

LAUTERBORN, W. 1976 Numerical investigation of nonlinear oscillations of gas bubbles in liquids, *J. Acoust. Soc. Amer.* **59**, 283–293.

MEYER, VON E. AND SKUDRZYK, E. 1953 Über die akustischen eigenschaften von gasblasenschleiern in wasser, *Acustica* **3**, 435–440.

MINNAERT, M. 1933 On musical air bubbles and the sounds of running water, *Phil. Mag.* **XVI** (7th Series), 235–248.

PROSPERETTI, A. 1977 Thermal effects and damping mechanisms in the forced radial oscillations of gas bubbles in liquids, *J. Acoust. Soc. Amer.* **61**, 17–27.

PROSPERETTI, A. 1987 The equation of bubble dynamics in a compressible liquid, *Phys. Fluids* **30**, 3626–3628.

PROSPERETTI, A. 1988 Nonlinear bubble dynamics, *J. Acoust. Soc. Amer.* **83**, 502–514.

PROSPERETTI, A. 1991 The thermal behaviour of oscillating gas bubbles, *J. Fluid Mech.* **222**, 587–616.

PROSPERETTI, A. AND LEZZI A. 1986 Bubble dynamics in a compressible liquid. Part 1. First-order theory, *J. Fluid Mech.* **168**, 457–478.

PROSPERETTI, A. AND LEZZI, A. 1987 Bubble dynamics in a compressible liquid. Part 2. Second-order theory, *J. Fluid Mech.* **185**, 289–321.

SCHIFFERS, W. P., EMMONY, D. C., AND SHAW, S. J. 1997 Visualization of the fluid flow around a laser generated oscillating bubble, *Phys. Fluids* **9**, 3201–3208.

STRASBERG, M. 1956 Gas bubbles as sources of sound in liquids, *J. Acoust. Soc. Amer.* **28**, 20–26.

TEMKIN, S. 1999 Radial pulsations of a fluid particle in a sound wave, *J. Fluid. Mech.* **380**, 1–38.

TEMKIN, S. 2001b Corrigendum: Radial pulsations of a fluid sphere in a sound wave, *J. Fluid. Mech.* **430**, 407–410.

Rotational

LEE, C. P., LYELL, M. J., AND WANG, T. G 1985 Viscous damping of the oscillations of a rotating simple drop, *Phys. Fluids* **28**, 3187–3188.

RUBINOW, S. I. AND KELLER, J. B. 1961 The transverse force on a spinning sphere in a viscous fluid, *J. Fluid Mech.* **11**, 447–459.

Shape deformations

CHIU, H. H. 1970 Dynamics of deformation of liquid drops, *Astronautica Acta* **15**, 199–213.

COX, R. G. 1969 The deformation of a drop in a general time-dependent fluid flow, *J. Fluid Mech.* **37**, 601–623.

LUNDGREN, T. S. AND MANSOUR, N. N. 1988 Oscillations of drops in zero gravity with weak viscous effects, *J. Fluid Mech.* **194**, 479–510.

MARSTON, P. L. 1980 Shape oscillations and static deformation of drops and bubbles driven by modulated radiation stresses – Theory, *J. Acoust. Soc. Amer.* **67**, 15–26.

Coupled-mode motions

FAXEN, H. 1921 Einwirkung der Gefäßwände auf den Widerstend gegen die Bewegung einer kleinen Kugel in einer Zähen Flussigkeit. Diss. Upsala. (See Oseen, 1927, Section 9.)

FFOWCS WILLIAMS, J. E. AND GUO, Y. P. 1991 On resonant non-linear bubble oscillations, *J. Fluid Mech.* **224**, 507–529.

LONGUET-HIGGINS, M. S. 1989 Monopole emission of sound by asymmetric bubble oscillations. Part 1. Normal modes, *J. Fluid Mech.* **201**, 525–541.

LONGUET-HIGGINS, M. S. 1989 Monopole emission of sound by asymmetric bubble oscillations. Part 2. An initial-value problem, *J. Fluid Mech.* **201**, 543–565.

LONGUET-HIGGINS, M. S. 1991 Resonance in nonlinear bubble oscillations, *J. Fluid Mech.* **224**, 531–549.

MAGNAUDET, J. AND LEGENDRE, D. 1998 The viscous drag force on a spherical bubble with a time-dependent radius, *Phys. Fluids* **10**, 550–554.

WATANABE, T. AND KUKITA, Y. 1993 Translational and radial motions of a bubble in an acoustic standing wave field, *Phys. Fluids A* **5**, 2682–2688.

Size Distributions

BRAZIER, R., SPARKS, R. S. J., CAREY, S. N., SIGURDSSON, H., AND WESTGATE, J. A. 1983 Bi-modal grain size distribution and secondary thickening in air-fall ash layers, *Nature* **301**, 115–119.

GELBARD, F., TAMBOUR, Y., AND SEINFELD, J. H. Sectional representation for simulating aerosol dynamics, *J. Colloid Interface Sci.* **76**, 541–556, 1980.

R. JAENICKE AND DAVIES, C. N. 1976 The mathematical expression of the size distribution of atmospheric aerosols, *J. Aerosol Sci.* **7**, 255–259.

LEE, R. E., Jr., 1972 The size of suspended particulate matter in air, *Science* **178**, 567–575.

MUGELE, R. A. AND EVANS, H. D. 1951 Droplet size distribution in sprays, *Indust. Eng. Chem.*, **43**, 1317–1324.

SMITH, J. E. AND JORDAN, M. L. 1964 Mathematical and graphical interpretation of the log-normal law for particle size distribution analysis, *J. Colloid Sci.*, **19**, 549–559, 1964.

WEICKMANN, H. K. AND AUFM KAMPE, H. J. 1953 Physical properties of cumulus clouds, *J. Met.* **10**, 204. [See Mason, B. K., *The Physics of Clouds*, pp. 98–103.]

Size Measurement Techniques

DAVIS, M. C. 1978 Coal slurry diagnostics by ultrasound transmission, *J. Acoust. Soc. Amer.* **64**, 406–410.

DOBBINS, R. A. 1963 Measurement of mean particle size in a gas-particle flow, *AIAA J.* **1**, 1940–1942.

DOBBINS, R. A., CROCCO, L., AND GLASSMAN, I. 1963 Measurement of mean particle sizes in sprays from diffractively scattered light, *AIAA J.* **1**, 1882–1886.

DOBBINS, R. A. AND TEMKIN, S. 1967 Acoustical measurements of aerosol particle size and concentration, *J. Colloid Interface Sci.* **25**, 329–333.

DUKHIN, A. S. AND GOETZ, P. 2001 Acoustic and electroacoustic spectroscopy for characterizing concentrated dispersions and emulsions, *Adv. Colloid Interface Sci.* **92**, 73–132.

DURAISWAMI, R., PRABHUKUMAR, S., AND CHAHINE, G. L. 1998 Bubble counting using an inverse acoustic scattering method, *J. Acoust. Soc. Amer.* **104**, 2699–2717.

KÖNIG, G. ANDERS, K. AND FROHN, A. 1986 A new light-scattering technique to measure the diameter of periodically generated moving droplets, *J. Aerosol Sci.* **17**, 157–167.

MEDWIN, H. 1970 In-situ acoustic measurements of bubble populations in coastal waters, *J. Geophys. Res.* **75**, 599–611.

MEDWIN, H. 1977 Acoustical determinations of bubble-size spectra, *J. Acoust. Soc. Amer.* **62**, 1041–1044.

NEWHOUSE, V. L. AND SHANKAR, P. M. 1984 Bubble size measurements using the nonlinear mixing of two frequencies, *J. Acoust. Soc. Amer.* **75**, 1473–1477.

TEMKIN, S. AND REICHMAN, J. M 1972 A new technique to photograph small particles in motion, *Rev. Sci. Instr.* **43**, 1456–1459.

Stokeslets and Other Singularities

JEFFREY, D. J. 1992 The calculation of the low Reynolds number resistance functions for two unequal spheres, *Phys. Fluids A* **4**, 16–29.

KIM, S. AND MIFFLIN, R. T. 1985 The resistance and mobility functions of two equal spheres in low-Reynolds-number flow, *Phys. Fluids* **28**, 2033–2045.

MAUL, C. AND KIM, S. 1994 Image systems for a Stokeslet inside a rigid spherical container, *Phys. Fluids* **6**, 2221–2223.

POZRIKIDIS, C. 1989 A singularity method for unsteady linearized flow, *Phys. Fluids A* **1**, 1508–1520.

Suspension Motions (Nonacoustic)

Aerosols (dusty gases)

BURESTI, G. AND CASAROSA, C. 1989 One dimensional adiabatic flow of equilibrium gas-particle mixtures in long vertical ducts with friction, *J. Fluid Mech.* **203**, 251–272.

CARRIER, G. F. 1958 Shock waves in a dusty gas, *J. Fluid Mech.* **4**, 376–382.

HAMAD, H. AND FROHN, A. 1980 Structure of fully dispersed waves in dusty gases, *J. Appl. Math. Phys. (ZAMP)* **31**, 66–82.

HEALY, J. V. AND YANG, H. T. 1972 The Stokes problems for a suspension of particles, *Astronautica Acta* **17**, 851–856.

ISHII, R., HATTA, N., UMEDA, Y., AND YUHI, M. 1990 Supersonic gas-particle two-phase flow around a sphere, *J. Fluid Mech.* **221**, 453–483, 1990.

ISHII, R. AND UMEDA, Y. 1987 Nozzle flows of gas-particle mixtures, *Phys. Fluids* **30**, 752–760, 1987.

LIU, J. T. C. 1965 On the hydrodynamic stability of parallel dusty gas flow, *Phys. Fluids* **8**, 1939–1945.

LIU, J. T. C. 1967 Flow induced by the impulsive motion of an infinite flat plate in a dusty gas, *Astronautica Acta,* **13**, 369–377.

MIURA, H. AND GLASS, I. I. 1982 On a dusty-gas shock tube, *Proc. R. Soc. Lond. A* **382**, 373–388.

MIURA, H. AND GLASS, I. I. 1983 On the passage of a shock wave through a dusty-gas layer, *Proc. R. Soc. Lond. A* **385**, 85–105.

MURRAY, J. D. 1967 Some basic aspects of one-dimensional incompressible particle-fluid two-phase flows, *Astronautica Acta* **13**, 417–430.

NEILSON, J. H. AND GILCHRIST, A. 1968 An analytical and experimental investigation of the velocities of particles entrained by the gas flow in nozzles, *J. Fluid Mech.* **33**, 131–149.

RUDINGER, G. 1964 Some properties of shock relaxation in gas flows carrying small particles, *Phys. Fluids* **7**, 658–663.

RUDINGER, G. 1965 Some effects of finite particle volume on the dynamics of gas-particle mixtures, *AIAA J.* **3**, 1217–1222.

RUDINGER, G. 1969 Relaxation in gas-particle flows, in *Nonequilibrium Flows*, P. Wegner, Ed., Marcel Dekker pp. 119–161.

RUDINGER, G. 1970 Gas-particle flow in convergent nozzles at high loading ratio, *AIAA J.* **8**, 1288–1294.

RUDINGER, G. AND CHANG, A. 1964 Analysis of nonsteady two-phase flow, *Phys. Fluids* **7**, 1747–1754.

SAFFMAN, P. G. 1962 On the stability of laminar flow of a dusty gas, *J. Fluid Mech.* **13**, 120–128.

SAUERWEIN, H. AND FENDELL, F. E. 1965 Method of characteristics in two-phase flow, *Phys. Fluids* **8**, 1564–1565.

SCHMITT-VON SCHUBERT, B. 1969 Existence and uniqueness of normal shock waves in gas-particle mixtures, *J. Fluid Mech.* **38**, 633–655.

TOMITA, Y., TASHIO, H., DEGUCHI, K., AND JOTAKI, T. 1980 Sudden expansion of gas-solid two-phase flow in a pipe, *Phys. Fluids* **23**, 663–666.

Bubbly liquids

HSIEH, D.-Y. Some aspects of dynamics of bubbly liquids, *Appl. Sci. Res.* **38**, 305–312, 1982.

ISHII, R., UMEDA, Y., MURATA, S., AND SHISHIDO, N. 1993 Bubbly flows through a converging-diverging nozzle, *Phys. Fluids A* **5**, 1630–1643.

KAMEDA, M. AND MATSUMOTO, Y. 1996 Shock waves in a liquid containing small gas bubbles, *Phys. Fluids* **8**, 322–335.

KENNARD, E. H. 1941 *Report on Underwater Explosions*, David W. Taylor Model Basin, pp. 1–51.

OMTA, R. 1987 Oscillations of a cloud of bubbles of small and not so small amplitude, *J. Acoust. Soc. Amer.* **82**, 1018–1033.

WIJNGAARDEN, L. VAN. On the structure of shock waves in liquid-bubble mixtures, *Appl. Sci. Res.* **22**, 366–381.

WIJNGAARDEN, L. VAN AND KAPTEYN, C. 1990 Concentration waves in dilute bubble/liquid mixtures, *J. Fluid Mech.* **212**, 111–137.

Transport Properties

BATCHELOR, G. K. AND GREEN, J. T. 1972 The determination of the bulk stress in a suspension of spherical particles to order c^2, *J. Fluid Mech.* **56**, 401–427.

BEENAKKER, C. W. J. 1984 The effective viscosity of a concentrated suspension of spheres (and its relation to diffusion), *Phys. A* **128**, 48–81.

BIESHEUVEL, A. AND SPOELSTRA, S. 1989 The added mass coefficient of a dispersion of spherical gas bubbles in a liquid, *Int. J. Multiphase Flow* **15**, 911–924.

O'BRIEN, R. W. 1979 A method for the calculation of the effective transport properties of suspensions of interacting particles, *J. Fluid Mech.* **91**, 17–39.

TAYLOR, G. I. 1954 The two coefficients of viscosity for a liquid containing air bubbles, *Proc. R. Soc. Lond. A* **226**, 34–39.

TAYLOR, G. I. 1954 The two coefficients of viscosity for a liquid containing air bubbles, *Proc. R. Soc. Lond. A* **226**, 34–39.

Author Index

Subject Index

Symbol Index

Latin Letters

A_n	Acceleration number
$b = k_0 a$	Nondimensional external wavenumber
b_i	Nondimensional internal wavenumber
$b_{i,0}$	Value of b_i at radial resonance frequency
C_D	Drag coefficient
$C_D^{(S)}$	Drag coefficient based on Stokes' law
$C_D^{(O)}$	Drag coefficient based on Oseen's approximation
C_{DS}	Drag coefficient for steady motions
C_R	Coefficient of reflection
C_T	Coefficient of transmission
c_p	Suspension's constant-pressure specific heat
c_{pf}, c_{pp}	Constant-pressure specific heats for fluid and particle materials
$c_s(0)$	Equilibrium isentropic sound speed in suspensions
$c_s(\omega)$	Nonequilibrium isentropic sound speed in suspensions
c_{sf}, c_{sp}	Isentropic sound speeds for fluid and particle materials
$c_s(\infty)$	Frozen-equilibrium sound speed in dilute suspensions
$c_T(0)$	Equilibrium isothermal sound speed in suspensions
$c_T(\omega)$	Nonequilibrium isothermal sound speed in suspensions
c_{Tf}, c_{Tp}	Isothermal sound speeds for fluid and particle materials
$c_{\kappa f}$	Phase velocity of thermal waves in fluid
c_{vf}	Phase velocity of viscous waves in fluid
D_{nm}^{n-m}	n-m power of generalized average diameter
D_∞	Largest diameter present in particle distribution
D_f/Dt	Time derivative following fluid phase
D_p/Dt	Time derivative following particulate phase
d	Distance between particles centers
$\hat{d}_\mu$	Nondimensional viscous damping coefficient
$\hat{d}$	Nondimensional total damping coefficient
$\hat{d}_{ac}$	Nondimensional acoustic damping coefficient
$\hat{d}_T$	Nondimensional isothermal damping coefficient
$\hat{d}_{th}$	Nondimensional thermal damping coefficient

E_{kin}	Kinetic energy
E_p	Internal energy of particles
E_{pul}	Pulsational energy
E_{rot}	Rotational energy of particle
$E(a_i,a_j)$	Collision efficiency
E_0	Reference oscillatory energy
e	Internal energy of suspension per unit mass
e_f, e_p	Internal energies per unit mass
$\langle \dot{e} \rangle_{loss}$	Energy dissipation rate, per unit mass
$\dot{e}_{th}$	Instantaneous thermal energy dissipation rate
$\langle \dot{e}_{th} \rangle_f$	Time-averaged thermal dissipation rate in fluid
$\langle \dot{e}_{th} \rangle_p$	Time-averaged thermal dissipation rate in particle
$\mathbf{F}$	Total force on particle
$F(q,q_i)$	Function defined by (3.5.15)
$\mathbf{F}_D$	Drag force
$\mathbf{F}_L$	Lift force
$\mathbf{F}_p$	Fluid force on particle
$\mathbf{F}_r$	Radial force on pulsating surface
$\mathbf{f}_p(\mathbf{x},t)$	Particulate phase force on fluid phase, per unit suspension volume
$\mathbf{f}_{Sp}$	Surface force on a particle
$G(q_i)$	Function defined by (3.5.20)
$g(C_D)$	Empirical drag decrease due to interactions, (4.12.6)
$g(\delta)$	Function of the fluid-to-density ratio, (4.9.5)
h	Heat capacity ratio, (3.2.22)
$h_n^{(1)}(z)$	Spherical Bessel function of the third kind, denoted in text by $h_n(z)$
I	Sphere's moment of inertia
$\Im\{f\}$	Imaginary part of complex function f
$j_n(z)$	Spherical Bessel function of the first kind
Kn	Knudsen number
$K_s(0)$	Suspension's equilibrium isentropic compressibility
$K_s(\omega)$	Nonequilibrium isothermal compressibility of suspension
K_{sf}, K_{sp}	Isentropic compressibilities of fluid and particle materials
$K_T(0)$	Suspension's equilibrium isothermal compressibility
K_{Tf}, K_{Tp}	Isothermal compressibilities of fluid and particle materials
$K(v_j, v_k)$	Coagulation kernel
k	Complex wavenumber
k_f, k_f	Thermal conductivities
k_0	Equilibrium wavenumber
M_0	Added mass for pulsating particle
M_1	Approach Mach number of in flow of a gas across normal shock
M_{pul}	Pulsational mass, equal to M_0
M_n	nth moment of particle size distribution function
$\overline{M}_1$	Approach Mach number of inflow of a dusty gas across normal shock
m_p	Mass of one particle
$\dot{m}$	Mass flow rate
$\dot{m}_f, \dot{m}_p$	Mass flow rates of fluid and particulate phases

$N_D \Delta D$	Number of particles of size in range ΔD
N_k	Number of particles in kth section
$\mathbf{N}_p$	Torque on particle
N_s	Ratio of fluid-to-particle materials isentropic compressibilities
N_T	Ratio of fluid-to-particle materials isothermal compressibilities
N_0	Total number of particles in a unit suspension volume
n_D	Nondimensional particle size distribution function
n'_D	Truncated particle size distribution function
$n_k = N_k/N_0$	Nondimensional sectional numbers
$O_{ij}(\mathbf{x} - \mathbf{x}')$	Oseen tensor
$P_n(x)$	Legendre polynomial of order n
P'_f	Pressure fluctuation in fluid without particles
p_f	Fluid pressure
p'_f	Fluctuation of the pressure in the fluid
p_{fs}	Pressure on fluid's side of particle-fluid interface
$p_{f\infty}$	Fluid's pressure far from pulsating particle
$\overline{p}_p$	Volume-averaged particle pressure
p'_p	Fluctuation of the pressure in particle
p_{ps}	Pressure on particle's side of particle-fluid interface
p_s	Pressure field produced by a stokeslet
$\dot{Q}_p$	Heat transfer rate, by conduction, to particle
$\dot{\mathbf{q}}_k$	Heat flux
$q = (1 + i)z_f$	Value of Ka in fluid
$q_i = (1 + i)z_p$	Value of Ka in particle
$\dot{q}_p(\mathbf{x},t)$	Particulate phase heat transfer rate to fluid phase
$\mathfrak{R}$	Universal gas constant for dusty gas
$\mathfrak{R}\{f\}$	Real part of complex function f
$\dot{R}$	Radial velocity of particle surface
$S(v, w)$	Sticking factor in coalescence efficiency
$\dot{S}_{total}$	Total rate of entropy increase
s	Entropy of suspension, per unit mass
s_f	Fluid's entropy per unit mass
s_p	Particles' entropy per unit mass
Re	Reynolds number
T_0	Ambient temperature
$\overline{T}_p$	Volume average of particle temperature
$T = \overline{T}'_p/\Theta'_f$	Complex ratio of average particle temperature to fluid temperature
T_f, T_p	Fluid and particle temperatures
$T_s = T'_s/\Theta'_f$	Complex ratio of surface temperature to fluid temperature
T_0	Ambient temperature
$\mathbf{U}_s$	Radial velocity of pulsating surface
U_r	Relative translational velocity
$\dot{U}_r$	Relative translational acceleration
$\mathbf{u}_S$	Velocity field produced by a stokeslet
$\mathbf{u}_f, \mathbf{u}_p$	Fluid and particle velocities; also of fluid and particulate phases
$u_{p\infty}$	Terminal velocity of a particle

$V = u_p/U_f$	Complex ratio of particle-to-fluid oscillatory velocities
$V^{(C)}$	Complex ratio of particle-to-fluid oscillatory velocities in an ideal fluid
$V^{(S)}$	Complex ratio of particle-to-fluid oscillatory velocities as given by Stokes' law
$V^{(V,I)}$	Complex ratio of particle-to-fluid oscillatory velocities as given by the unsteady Stokes' result
v_p	Particle volume
We	Weber number
$\dot{W}_{in}$	Rate at which fluid does work on particle
$\dot{W}_{out}$	Rate at which particle does work on fluid
$X(\omega)$	Real part of $(k/k_0)^2$
$\mathbf{x}_s$	Position vector of point on particle's surface
$Y(\omega)$	Imaginary part of $(k/k_0)^2$
y	Ratio of particle radius to viscous penetration depth into fluid
z_f	Ratio of particle radius to thermal penetration depth in external fluid
z_p	Ratio of particle radius to thermal penetration depth into particle

Greek Letters

α	Attenuation coefficient, cm^{-1}
$\hat{\alpha} = \alpha c_{sf}/\omega$	Nondimensional attenuation based on c_{sf}
$\overline{\alpha} = \alpha c_s(0)/\omega$	Nondimensional attenuation based on $c_s(0)$
$\overline{\alpha}_0$	Attenuation estimate for the K-K method
$\hat{\alpha}_{ac}$	Attenuation due to acoustic radiation
$\hat{\alpha}_{th}$	Thermal attenuation
$\hat{\alpha}_{mol}$	Molecular attenuation in fluid
$\hat{\alpha}_{tr}$	Translational attenuation
$\hat{\alpha}_{pul}$	Pulsational attenuation
$\widehat{\alpha}$	Attenuation for polydisperse suspensions
β	Coefficient of thermal expansion; also damping coefficient in Chapter 6
$\hat{\beta}$	Dispersion coefficient
$\hat{\beta}_{ac}$	Acoustic contribution to the dispersion coefficient
$\hat{\beta}_{th}$	Thermal contribution to the dispersion coefficient
$\hat{\beta}_{tr}$	Translational contribution to the dispersion coefficient
$\widehat{\beta}$	Dispersion coefficient for polydisperse suspensions
β_f, β_p	Coefficients of thermal expansion for fluid and particle materials
β_{ac}	Acoustic radiation damping coefficient
β_μ	Viscous damping coefficient for pulsations
γ	Suspension's specific heat ratio
$\overline{\gamma}$	Suspension specific heat ratio for aerosols and bubbly liquids
γ_f, γ_p	Fluid and particle materials specific heat ratios
δV	Element of suspension volume
$\delta V_f, \delta V_p$	Volumes occupied by fluid and particles in a suspension element
δM	Mass of suspension element
$\delta M_f, \delta M_p$	Masses of fluid and particles in a suspension element
$\delta = \rho_f/\rho_p$	Fluid-to-particle material density ratio

$\delta\tau$	Material suspension volume
δ_{ij}	Kronecker delta
δ_{vf}	Viscous penetration depth into fluid
$\delta_{\kappa f}$	Thermal penetration depth into particle
$\delta_{\kappa p}$	Thermal penetration depth into particle
δE	Internal energy of suspension element
δE_f	Internal energy of fluid in suspension element
δE_p	Internal energy of particles in suspension element
δS	Entropy of suspension element
δS_f	Entropy of fluid in suspension element
δS_p	Entropy of particles in suspension element
ε	Radial displacement of particle surface
ε_f	Amplitude of fluid displacement in a plane sound wave
$\zeta'(\omega)$	Real part of response function in the K-K method
$\zeta'(\infty)$	High-frequency limit of the real part of response function in the K-K method
η_m	Mass loading
Θ'_f	Temperature fluctuation in fluid without particles
θ	Polar angle
θ'_f	Temperature fluctuation in fluid
κ	Polytropic index
κ_r	Real part of the polytropic index
κ_i	Imaginary part of the polytropic index
κ_f, κ_p	Thermal diffusivities of fluid and particle materials
$\Delta = \nabla \cdot \mathbf{u}$	Rate of expansion
Δ_c	Capillary constant
Δ'_f	Fluid's density fluctuation in a fluid without particles
$(\Delta N_k)_{gain}$	Number increase due to coalescence
$(\Delta N_k)_{loss}$	Number decrease due to coalescence
λ_p	Interparicle separation
μ_f	Coefficient of viscosity for fluid
ν_f	Kinematic viscosity of fluid
$\nu_v = N_v/N_0^2$	Nondimensional number concentration
ν_k	Nondimensional sectional number concentration
ξ	Thermal property ratio, (6.6.10)
$\Pi = p'_p/P'_f$	Complex ratio of pressure fluctuations
ρ	Suspension density
ρ_0	Equilibrium density of suspension
ρ_f, ρ_p	Materials densities
ρ'_f	Fluctuation of fluid density
$\overline{\rho}_p$	Volume-averaged particle density
σ	Surface tension; density of two-phase medium
σ_0	Equilibrium density of two-phase medium, $= \rho_0$
σ_a	Absorption scattering cross-section
σ_s	Scattering cross-section
$\sigma_{a,ac}$	Acoustic absorption scattering cross-section
$\sigma_{a,th}$	Thermal absorption scattering cross-section

$\sigma_{a,\mu}$	Viscous absorption scattering cross-section
$\sigma_{r\theta}$	(r, θ) component of the stress tensor
σ_f, σ_p	Densities of the fluid and particulate phases
σ_{ij}	(i, j) component of the stress tensor
σ_{v_g}	Standard deviation for log-normal particle volume distribution
σ_r	Radial, or (r, r) component of the stress tensor
τ_d	Dynamic relaxation time for translational motions
τ_t	Thermal relaxation time
$\tau_f = \theta_f' / \Theta_f'$	Nondimensional temperature fluctuation in fluid
τ_v	Viscous relaxation time in fluid
τ_κ	Thermal relaxation time in fluid
Φ_f	Viscous dissipation function
ϕ	Velocity potential
ϕ_m	Mass concentration of particles in a suspension
ϕ_v	Volume concentration of particles in a suspension
χ	Response function for radial pulsations
χ_T	Isothermal response function for radial pulsations
ψ	Stream function
Ω_p	Angular velocity of particle
ω	Circular frequency
ω_0	Natural frequency for radial pulsations
ω_n	nth natural frequency for surface vibrations of a droplet
ω_{opt}	Optimum frequency for acoustic coagulation
ω_{T0}	Isothermal natural frequency for radial pulsations